Ball Lightning

An Unsolved Problem in Atmospheric Physics

Ball Lightning
An Unsolved Problem in Atmospheric Physics

Mark Stenhoff

B.Sc., M.Phil. (Lond.), C.Phys., M.Inst.P., F.R.A.S., F.R.Met.S.

Springer Science+Business Media, LLC

Library of Congress Cataloging-in-Publication Data

Stenhoff, Mark.
Ball lightning : an unsolved problem in atmospheric physics / Mark Stenhoff.
p. cm.
Includes bibliographical references and index.

DOI 10.1007/978-0-306-47092-9
1. Ball lightning. I. Title.

QC966.7.B3 S83 1999
551.56'34--dc21

99-047977

Originally published by Kluwer Academic/Plenum Publishers, New York in 1999
MyCopy version of the original edition 1999

10 9 8 7 6 5 4 3 2 1

Foreword: Ball Lightning

Looking back at twentieth-century science, one may discern three broad frontiers of research—the very large, the very small, and the very complex. Thus astronomy and cosmology continue to spring surprises in the form of puzzling new objects in the depths of the universe. In the microscopic realm, particle physicists have struggled to make sense of a plethora of subatomic fragments. Finally, the burgeoning subject of complexity theory hints at deep linkages between physics and biology, and touches on topics as diverse and bewildering as chaos, turbulence, and even ecology.

We are so used to dramatic developments being concentrated in these three frontier categories that it comes as a bit of a surprise to find an unexplained phenomenon that falls outside of any of them. Ball lightning isn't very large, very small, or seemingly very complex, yet it has baffled some of the best investigators in the business. Here, it seems, under our very noses, is the sort of mystery that should have been solved a century ago, but wasn't.

I have never witnessed ball lightning personally, but I have read many reports of it and interviewed a number of witnesses. Although it can take a variety of forms, a typical lightning ball is a glowing sphere a few centimeters across that floats across ground for a few seconds before either fading away or exploding. Most alarming of all are those cases where a ball is seen inside a room or even in an aircraft.

The bizarre nature of the phenomenon has led to a certain amount of neglect in scientific circles. Some years ago it was common for scientist to pooh-pooh the whole subject. Attempts were made to explain ball lightning reports away as optical illusion, will o' the wisps, or hoaxes. Today, however, the weight of evidence is compelling that a genuine unexplained physical phenomenon underlies the majority of sightings.

Hardest to understand is how such a large amount of energy can be confined in a spherical form and remain stable for so long. Although estimates are rather unreliable, some lightning balls appear to pack a lot of punch, and can cause serious injury and damage. There have been many attempts to model ball lightning as some sort of energetic glowing plasma trapped by suitably configured electric and magnetic fields. Ironically, physicists have for years tried to confine hot plasmas electromagnetically, in order to create a controlled nuclear fusion reaction that might serve as a potential power supply. After 40 years of effort, their attempts are still plagued by plasma instabilities. Yet if the reports of ball lightning are to be believed, nature has found a way to create stable plasma balls without any sophisticated equipment.

The phenomenon becomes really puzzling when the peculiar properties of ball lightning are examined. How is a ball of plasma able to pass through a window pane without disruption? Or glide down the aisle of an aircraft? Why should such a ball shoot up a chimney, roll along electric power lines, or bounce along the ground? What produces the curious hissing or sizzling sound reported in some sightings?

Mark Stenhoff is a physicist who has a long professional association with the subject of ball lightning and related phenomena. He has personally studied a great many cases and thoroughly researched the literature. In this book he provides us with a welcome review of the phenomenon and the different theories put forward to explain it. He relates some of the more startling incidents in a sober and methodical fashion, and makes some useful suggestions for how our understanding of the mystery can be advanced.

The subject of ball lightning falls on the edge of what we might call respectable science. Because ball lightning has so far proved impossible to recreate in the laboratory (aside from some very short-lived small balls), most of the information available has been gleaned from eyewitness reports. Since members of the public, when presented with an unexpected and alarming phenomenon, are notoriously unreliable, it is hard to know how much credence to give the details of the reports. Might the witnesses have exaggerated the size, duration, or brightness of the ball? Could the reported damage have been caused by something else, like an associated conventional lightning strike?

It must be remembered that untrained observers also report flying saucers, ghosts, poltergeists, and alien beings. Are scientists supposed to take these seriously too? Indeed, Stenhoff points out that some reports of unidentified flying objects (UFOs) bear many of the hallmarks of ball lightning, so if the puzzle of ball lightning is solved, then at least some UFOs may become IFOs (identified flying objects). In spite of its slightly wacky overtones, ball lightning deserves to be taken seriously. There can be little doubt that some interesting and as yet ill-understood physical phenomenon is taking place, and we may learn some very interesting physics by studying the reports more carefully.

Ball lightning researchers are fond of citing the case of meteorite falls, long derided by scientists as stories by crackpots, but eventually accepted as a real astronomical phenomenon. The comparison is pertinent. Both meteorites and ball lightning are unpredictable, alarming, and transient, and both emanate from the sky, the source of so many mysteries. A vast amount of important scientific information has been gleaned from the study of meteorites. A similar cornucopia may await us in the phenomenon of ball lightning.

Paul Davies
Adelaide, South Australia

*Paul Davies was formerly Professor of Natural Philosophy at The University of Adelaide. He took early retirement from the university to devote himself full-time to his popular book writing and media activities.

Preface

Down comes a deluge of sonorous hail,
Or prone-descending rain. Wide-rent, the clouds
Pour a whole flood, and yet, its flame unquenched,
Th'unconquerable lightning struggles through.
Ragged and fierce, or in red whirling balls,
And fires the mountains with redoubled rage.
Black from the stroke, above, the smould'ring pine
Stands a sad shattered trunk; and, stretched below,
A lifeless group the blasted cattle lie.

James Thompson, "The Seasons" (1727)

I have been investigating ball lightning for more than two decades. I published a ball lightning report in *Nature* in 1976 that received worldwide publicity and consequently many people wrote to me with accounts of their own experiences. Within a very short time, I had accumulated about 200 firsthand accounts, and the file has continued to grow steadily since then. Several things impressed me. Few of those who wrote to me had any detailed foreknowledge of ball lightning at the time of their observation. Nonetheless, once reports of other phenomena such as St. Elmo's fire had been eliminated, the remaining descriptions were remarkably consistent. Furthermore, nearly all who contacted me were keen to have an explanation of what they had seen and seemed entirely sincere.

Most of my ball lightning research has been at the grassroots level of analyzing eyewitness reports and conducting field visits. For several years, I was director of the Ball Lightning Division of the Tornado and Storm Research Organisation (TORRO). TORRO continues to do invaluable work in documenting various interesting meteorological phenomena, and its publication, the *Journal of Meteorology*, has become one of the most popular journals in which to publish ball lightning reports. It increasingly carries papers on theoretical and experimental aspects of the subject.

Over the past couple of decades, I have become aware that the reliability of ball lightning reports is very variable. Some investigators, however, have accepted as ball lightning any phenomenon so reported by the observer. Others have taken the line that all ball lightning reports may be explained in terms of more familiar phenomena. My position is intermediate—perhaps that of an agnostic. I am conscious that my approach will not satisfy investigators at either extreme of polarization. Elsewhere, I refer to ball lightning research as an immature field of study and discuss this in the context of science seen as a social activity. Perhaps it is because ball lightning research has yet to evolve as a fully systematic scientific activity that discussions are often emotive, and differing points of view are defended with a fervor more usually associated with political or religious extremism. Although I will not be surprised by some emotive responses to this book, I hope that its conclusions will point the way toward more streamlined and effective research.

Acknowledgments

I am especially indebted to a number of people who have helped considerably with the production of this book. I am grateful to my colleagues at TORRO, especially Adrian James and Mike Rowe, who provided many ball lightning reports. Adrian collected the majority of reports in Chapter 5.

James Allen and Marie Barnes gave extensive and painstaking assistance with bibliographical research. Dr. James Dale Barry provided assistance, advice, and permission to incorporate the substantial bibliography from his excellent book *Ball Lightning and Bead Lightning* (1980a) within the updated bibliography presented here.

Professor Paul Davies kindly consented to write the foreword. Dr. Eric Wooding provided advice, encouragement, and his extensive files of ball lightning materials. Jonathan Kearley provided practical help, and Martin Willson checked my translations of Arago's work.* Professor Martin Uman, Professor Neil Charman, and Dr. Earle Williams gave advice on a number of matters. Professor Eric Witalis, Dr. Geoffrey Endean, Dr. Peter Handel, Dr. Geert Dijkhuis, Dr. Pierre Laroche, and others have kindly checked sections of the chapters on theories, but I take full responsibility for any errors!

I am grateful to Dr. Richard Orville and the American Meteorological Society for kind permission to quote extensively from their forthcoming *Glossary of Meteorology* (Glickman 1999). A number of publishers and photographers have kindly given permission for reproduction of photographs, and these are credited in the text. I am also indebted to all those who have sent me their own ball lightning reports and tolerated lengthy questionnaires, interviews, and house visits.

Mark Stenhoff†

*Dimitar Netchev also assisted with translation.

†The author may be contacted c/o Royal Meteorological Society, 104 Oxford Road, Reading, Berkshire RG1 7LL, UK.

Contents

Chapter 1

The Study of Ball Lightning

Ball lightning has been a matter of controversy among scientists for more than 160 years. Since 1838, several thousand reports of ball lightning have been documented and about two thousand scientific papers about ball lightning have been published. It may thus seem surprising that after all these years of scientific study and discussion, the problem of explaining ball lightning reports remains intractable. There is no consensus that any model of ball lightning fully predicts its reported characteristics, or that any experiment fully replicates its reported behavior. Ball lightning, if it exists, is a phenomenon that is both short lived and unpredictable. Most ball lightning reports are from casual observers.

The difficulty of evolving a plausible theory of ball lightning or carrying out experiments, together with the method by which empirical data about natural ball lightning are acquired, has led some scientists to the conclusion that all reports of ball lightning may be explained in terms of well-understood phenomena. They would argue that there is no problem to be solved here. Other scientists share the view of Arago, expressed more than 140 years ago (Arago, 1854), that ball lightning represents "one of the most inexplicable problems of physics today."

There have been many claims that ball lightning represents a significant hazard and that it has caused death, injury, and severe damage. These factors, together with the persistent difficulties encountered by theoreticians and experimentalists in this field, justify a careful scrutiny of the evidence for ball lightning. This chapter provides an overview of this subject.

1.1 Definitions

Ball lightning is reported to have the following characteristics: it is associated with thunderstorms. It is luminous and roughly spherical, with a modal diameter of 20–50 cm and a lifetime of several seconds. It moves independently through the air, often in a horizontal direction. A *ball lightning report* is a description of an alleged phenomenon that is identified by the observer as ball lightning, or that possesses some or all of the above characteristics. A *ball lightning event* is defined here as the stimulus for a report of ball lightning which, after evaluation, cannot be explained in terms of an already understood phenomenon such as St. Elmo's fire or an optical illusion.

1.2 Explaining Ball Lightning Reports

Clearly, not all ball lightning reports describe ball lightning events. Many other physical phenomena share some of the same characteristics. Luminous sources which may be misidentified as ball lightning include corona discharge (St. Elmo's fire), ignis fatuus (will-o'-the-wisp), other meteorological phenomena, combustion of materials initiated by a lightning flash to the ground, astronomical sources (including meteors—especially bolides and "fireballs"), automobile headlights, soap bubbles, streetlights, airplane landing lights, airborne debris, satellite reentries, birds, insect swarms, pyrotechnics (fireworks and flares), and blimps and weather balloons. In addition, there is the possibility of optical or perceptual distortion, for example, an afterimage generated by a lightning flash. Such possible explanations for ball lightning reports are discussed in Chapter 3. The question of whether all ball lightning reports can be explained in terms of such familiar phenomena is discussed in Chapter 10.

1.3 History of the Scientific Study of Ball Lightning

There are many historical accounts that may refer to ball lightning, but early sources usually fail to distinguish among conventional linear lightning, ball lightning, St. Elmo's fire, and meteors. The term "fireball" is used to describe both bright meteors and ball lightning. Until the late eighteenth century, meteors were thought to have a meteorological rather than an extraterrestrial origin, and this doubtless served to increase the confusion with ball lightning. The term "thunderbolt" is also ambiguous, suggesting something materially tangible, and has been used to refer to both ball and linear lightning. Even in modern eyewitness reports of ball lightning, one sometimes finds confusion among these various phenomena.

In the course of this historical overview, many models developed to explain ball lightning are briefly described, with no attempt at a critical review. Many of the earliest theories may be considered archaic and in a modern context, unworthy of further development. Others are discussed in more detail in Chapters 11, 12, and 13.

The interpretation of linear lightning as an electrical discharge was first suggested by the experiments of Benjamin Franklin (1751, 1774) in the mideighteenth century. Until this time, there had been no attempt to provide a physical explanation of ball lightning. In an attempt to reproduce one of Franklin's experiments, Professor G. W. Richmann, according to one eyewitness, was struck and killed by a ball of blue fire the size of a fist (see Chapter 5).

Although the electrical nature of linear lightning was soon accepted, Pieter von Muschenbroek (who had pursued research in electrostatics and is one of those credited with the invention of the Leyden jar, an early form of electrical capacitor) interpreted ball lightning not as an electrical phenomenon but as an "agglomeration of combustible materials" descending from the upper atmosphere. These materials were thought to have originated from within the Earth and vaporized, rising to high altitudes, where they would condense and thus descend. This descent was thought to be accompanied by an increase in temperature that eventually caused ignition or explosion (Muschenbroek 1769).

In 1838, astronomer and physicist Dominique François Jean Arago (1786–1853) published a survey of twenty reports of ball lightning (Arago 1838). Arago made many distinguished contributions to physics. At the age of only 23, he was elected a member of the Académie de Sciences, of which he subsequently became permanent secretary. He was appointed to the Bureau of Longitudes. He was professor of analytical geometry at the Ecole Polytechnique from 1809 to 1830 and director of the Paris Observatory. He carried out studies of magnetism and established the relationship between the aurora borealis and variations in terrestrial magnetism. His later studies were in astronomy, and he worked with Fresnel on the polarization of light.

In the same year as his first publication on ball lightning, he suggested a crucial experiment to decide between the particle and wave theories of light. Arago's theory of light predicted that the velocity of light should decrease as it passes into a denser medium. In 1838, he proposed an experiment to compare the velocity of light in different media. However, experimental difficulties meant that Arago was not in a position to attempt his experiment until 1850. By then his sight had deteriorated, so it was left to others to carry out a refined version of the experiment. Fizeau and Foucault obtained successful results before Arago died.

In *Le Tonnerre*, Arago considered the question of the existence of ball lightning. He wrote:

> Have these globes of fire or fireballs . . . really existed? Was not the spherical form attributed to them an optical illusion? Would not a flash of [conventional linear] lightning, if we suppose it to be cylindrical, if its direction were exactly towards an observer, appear to him to be circular or at least globular?
>
> This objection would have weight if the spheroidal form had only ever been seen by those who, being situated *exactly* in the path of the lightning, should have been struck by it. But an observer placed outside the path of the lightning, viewing it transversely, and seeing

> it strike a neighboring or distant house, could not attribute to it the form of a globe, unless it really were globular. These positional circumstances were almost always applicable in the following examples. The objection does not therefore deserve further attention. (Arago 1854, pp. 38–39)

Elsewhere, he wrote:

> The *éclairs en boule* [ball lightning—plural] of which I have cited so many examples, and which are so remarkable, first for the slowness and uncertainty of their movements, and next for the extent of the damage which they cause in exploding, appear to me to be one of the most inexplicable problems of physics today.
>
> These balls or globes of fire seem to be agglomerations of ponderable substances, strongly impregnated with the matter of lightning. How are such agglomerations formed? In what regions do they originate? Whence are derived the substances of which they are composed? What is their nature? Why do they sometimes pause for some time in their course, and afterwards rush off with great rapidity? . . . In the face of all these questions, science remains silent. (Arago 1854, p. 219)

If ball lightning exists, it is quite remarkable that nearly 150 years later, questions of its nature remain unresolved. Following the publication of Arago's monograph, the earliest hypotheses concerning the nature of ball lightning were proposed. Besnou (1852) suggested that the explosive decay of ball lightning could be explained by the formation of nitrogen triiodide in electrical discharges. Michael Faraday (1833, 1839) remarked that ball lightning bore little resemblance to the electrical discharge phenomena with which he was familiar. He therefore concluded that ball lightning was not an electrical discharge. He did not, however, deny its existence.

In contrast, Sir W. Snow Harris (1843) argued in his monograph *On the Nature of Thunderstorms* that ball lightning was simply a brush or glow discharge like St. Elmo's fire. Poey (1855) was the first to suggest a model for ball lightning based on electrostatic charging of droplets and dust.

De Tessan (1859a,b) published a model, widely accepted at the time, based on the structure of the Leyden jar. Positive and negative charges were seen as being separated by an insulating layer of dry air. Ozone and the observed luminosity of the ball were produced by the recombination of the charges across the leaky, dielectric layer. The sphere was thought to be in stable equilibrium. This was by virtue of the balance of radial forces: the coulomb force between the opposite charges, the increased pressure of the dielectric wall caused by electrostatic attraction, the reduced internal pressure, and the coulomb repulsion of the like charges in the inner part of the wall. Penetration of the ball by a conductor would cause its explosive decay.

In the same year, models were proposed in which the interaction of two oppositely directed lightning channels produced a vortex (Coulvier-Gravier 1859,

Moigno 1859). Similar models have been proposed much more recently (see Chapter 12).

Camille Flammarion (1842–1925) was a French astronomer who entered the Paris Observatory in 1858 and founded the observatory of Juvisy in 1883. He was a great popularizer of science who wrote books on astronomy, ballooning, and psychic research. In 1874, he published a collection of ball lightning reports, fifty in number, in his book *L'Atmosphère* (Flammarion 1874), which is journalistic and flamboyant.

Pfeil (1886a,b) suggested that cosmic dust became impregnated with combustible gas as it passed through protuberances from the sun (Singer 1971). The impregnated dust mixed with ice crystals and descended through the upper atmosphere, forming localized electrified clouds. These would start to burn as they neared the Earth's surface, gradually diminishing in size. Either quiet burning or violent explosive decay could take place—the former because of unreactive components in the mixture that inhibit combustion, or the former by the generation of an electric spark when the ball is in the vicinity of a good conductor.

The period from 1870 to 1910 was one in which much discussion in the scientific literature concerned direct current electrical discharges. There were numerous experimental studies at that time of such discharges and attempts to relate these to ball lightning (e.g., Hesehus 1876a,b, 1899, 1900a–d, 1901, 1903; du Moncel 1853, 1854a–d, 1857, 1881, 1882a–c; Planté 1860, 1868, 1872, 1873, 1875a–f, 1876a–j, 1877a–d, 1878a–c, 1879a,b, 1884a–i, 1885a,b, 1887, 1888, 1890, 1891, 1901; Righi 1891a–f, 1892,1895, 1896a–d; Toepler 1897, 1898a–c, 1899, 1900a–d, 1901a–e, 1902, 1903, 1916, 1917a,b, 1926a,b, 1929, 1944, 1954, 1959, 1960). The year 1910 saw the publication of Townsend's important monograph concerning his theories of electrical discharges (Townsend 1910).

Thornton (1911), on the other hand, used various observational properties of ball lightning to argue that "the principal but perhaps not the only constituent of lightning balls is an aggregation of ozone and partially dissociated oxygen thrown off from negatively charged clouds by an electric surge, after a heavy lightning discharge."

In the period from 1910 to 1955, there was review and discussion of all the existing models of ball lightning. The decade from 1924 to 1934 saw the publication of a series of papers discussing so-called "fulminating matter" as a model to explain the explosive decay of some ball lightning. Fulminating matter was the term then used to describe the substance of the lightning channel in which atmospheric constituents were heated to exceptionally high temperatures. These theories developed from Arago's hypothesis published 70 years earlier (Arago, 1854) and were the forerunners of modern plasma theories. The term "plasma," from a Greek word meaning "something molded or fabricated," was defined in 1928 by Irving Langmuir as the state of matter found in electrical discharges.

In 1923, Walther Brand published a monograph entitled *Der Kugelblitz* (Brand 1923), which was the most extensive survey of ball lightning at the time. This volume was part of a series entitled "Probleme der Kosmischen Physik." Brand collected about 600 ball lightning reports from the university libraries of Berlin, Marburg, and Gottingen, and the Naval Observatory in Hamburg, Germany. Of these, he selected the best-established 215 reports in the 100 years before publication of his book and presented the full original text for 108 of these. From the 215 reports, he compiled a statistical summary of the frequency of different properties. He then reviewed the theories proposed to explain the phenomenon from the time of Muschenbroek through to that of Toepler. Brand stated that there were no known photographs of ball lightning at the time.

In 1928, Bottlinger suggested that the high electric field intensities in thunderstorms could impart sufficient kinetic energies to charged particles so that they would initiate thermonuclear reactions. Ball lightning could be a manifestation of such reactions (Bottlinger 1928). Similar models continued to receive occasional discussion in the literature until recent times (e.g., Altschuler, House, and Hildner 1970).

Two widely publicized observations of ball lightning were published in the 1930s. Professor J. C. Jensen (1933a,b) of the University of Nebraska published four photographs he had taken during a thunderstorm on August 30, 1930, of a "shapeless mass of lavender color which seemed to float downwards," and which he considered to be ball lightning. These appeared in the wake of a flash of conventional linear lightning. Jensen had been engaged in a study of lightning during the period 1929–1931 and these studies entailed photographing a large number of night-time thunderstorms. These are among the most commonly published photographs of alleged ball lightning. The pyrotechnic appearance of the phenomena, as evidenced by the photographs and Jensen's description, has led some investigators to suspect that they were fireworks and that Jensen was perhaps hoaxed by some of his students (Singer 1971, Berger 1973, Barry 1980). (see Chapter 9).

Another much-reported observation was the so-called "tub-of-water" event, which was not first reported through a scientific journal, but as a letter to the editor of the English newspaper, *The Daily Mail.*

This event, one of the most oft-cited in literature on this subject, led to an estimate of ball lightning energy in excess of 1 MJ (Goodlet 1937). It thus had a very significant influence in shaping discussion of possible models for ball lightning. It is discussed in more detail in Chapter 11.

During the same decade, Humphreys (1936) published a survey of ball lightning reports and drew some skeptical conclusions. To judge from his earlier writings on the subject, Humphreys (1929) had previously been of the opinion that the number and high quality of ball lightning reports indicated that it could not be an optical illusion. Having secured 280 reports, he wrote in 1936 that "not one of these

A Thunderstorm Mystery

—Explained by the Astronomer Royal

To the Editor of "The Daily Mail"

SIR,—During a thunderstorm I saw a large, red hot ball come down from the sky. It struck our house, cut the telephone wires, burnt the window frame, and then buried itself in a tub of water which was underneath.

The water boiled for some minutes afterwards, but when it was cool enough for me to search I could find nothing in it.

Dorstone, Hereford. W. MORRIS.

Ball Lightning

Here is the Astronomer Royal's explanation of the above:

It would seem that your correspondent saw a very rare phenomenon, known as globular or ball lightning. This is much the rarest of all forms of lightning and has never satisfactorily been explained.

The lightning in this form appears as a luminous ball of fire, anything from a few inches to two or three feet in diameter. It usually remains visible for several seconds, but exceptionally it can be seen for one or two minutes; it generally moves slowly and ultimately bursts with a loud report like a bomb.

It is not, of course, a "thunderbolt," and your correspondent naturally found no material object in the tub.

H. SPENCER JONES

Figure 1.1. Letter appearing in *The Daily Mail* (London) November 5, 1936. (Reproduced with the kind permission of *The Daily Mail*.)

many accounts is unmistakably that of what ball lightning is supposed to be, that is, a leisurely moving and approximately spherical body of gas, electrons, or what not, luminous by virtue of its electrical state or condition." He dismissed all reports "except, perhaps, two or three" as due to a few causes: "brilliant flash at the point struck; persistence of vision; broken discharge path, giving separate flashes; meteorites; will-o'-the-wisp; falling molten metal; lightning seen end-on; brush discharge." Such explanations for ball lightning reports are found in Chapter 3, and skeptical attitudes of scientists and others are discussed in Chapter 10.

Neugebauer (1937) published a novel ball lightning model based on quantum mechanics. He carried out a quantitative analysis of a macroscopically neutral sphere composed of free electrons and positive gaseous ions, the latter being formed from atmospheric molecules. The mass density was equal to that of the atmosphere. Using a quantum mechanical argument, he showed that such a mass of ionized gas could exist in metastable equilibrium, being held together by the quantum mechanical exchange energy of the electron gas, a weak attractive force between electrons of opposite spin.

In 1955, the Russian physicist and later Nobel prize winner in physics, Pyotr Kapitza (1894–1984), developed a suggestion made earlier by others (de Jans 1912, Marchant 1930, Cerrillo 1943) that ball lightning might be caused by an electromagnetic standing wave. Ionized regions generated by the pattern of nodes and antinodes caused the ball (Kapitza 1955, 1958). Kapitza's paper generated a good deal of interest, and in the period from 1955 to 1970 all contemporary ball lightning models were subjected to much critical appraisal and underwent further development and refinement.

Arabadji proposed a nuclear theory in which cosmic rays were focused into small regions by thunderstorm electric fields. Ball lightning would be formed in these regions, he suggested, by self-sustaining nuclear reactions. (Arabadji 1956). Dauvillier (1957) proposed an alternative in which ball lightning or "lightning remnants" are made up of radioactive carbon-14 created from atmospheric nitrogen by the action of thermal neutrons liberated in lightning.

The greatest emphasis now, though, was on models based on plasma and ionized gases. In 1957, Shafranov quantitatively analyzed the equilibrium magnetohydrodynamic (MHD) conditions for containment of plasma by internal and external magnetic fields (Shafranov 1957). From his work a wide range of plasma machines were developed and these experimental studies continue today in the quest to achieve controlled thermonuclear fusion. His paper also raised interest in the possibility of stable plasma configurations in the forms of toroids and spheres. These were called *plasmoids*. Experimental confirmation that plasmoids could be generated was subsequently forthcoming (Bostick 1956, 1957; Wells 1964, 1966).

Thus, the models of ball lightning based on ionized gas and plasma were now divided into two groups—those that involved a self-contained energy source (plasmoids) and those that were fed with external energy (discharge models of various kinds).

Hill (1960) was of the opinion that ball lightning was not a plasma phenomenon, but rather a region of strongly inhomogeneous distribution of space charge in the form of a highly ionized gas in molecular form.

Further surveys and reviews of ball lightning were published in the 1960s (McNally 1961, Dewan 1964, 1966; Rayle 1966; Barry 1966). The results of these and more recent surveys are discussed in Section 1.4.

The vortex plasmoid models were developed in more detail by Bruce (1963a,b) and Wooding (1963). Both researchers suggested formation mechanisms for such plasmoids. Bruce proposed the escape of plasma from the lightning channel, whereas Wooding suggested the formation of a plasma vortex ring by the striking or penetration of a solid surface by a lightning flash. Wooding (1972) subsequently used the analog of the ablation of a solid surface by a high-powered laser pulse as support for his theory.

The dc discharge models also underwent further development with the publication of a paper suggesting that ball lightning was a nonlinear dc phenomenon produced by the introduction of a dielectric inhomogeneity into the dc uniform electric field (Finkelstein and Rubenstein 1964). A mathematical theory of ball lightning based on this model and taking into account the nonlinear electrical and thermal conductivities of heated air was published (Uman and Helstrom 1966). Later, Uman acknowledged that such models could not explain the formation of ball lightning within enclosed metal structures such as aircraft (Uman 1968a). A discussion of the cooling process for heated spheres of air indicated that this process alone was insufficient to explain such properties as the steady luminosity of ball lightning (Lowke, Uman, and Liebermann 1969).

The Finkelstein–Rubenstein model was further developed in papers by Powell and Finkelstein (1969, 1970), who provided experimental laboratory evidence for the model and proposed a structure based on three regions: (1) a central region in which Townsend multiplication occurs, (2) an intermediate region with a radial increase in temperature, and (3) the surrounding air. (Townsend's collision theory of ionization proposed that collisions by electrons accelerated in an electric field produce secondary ions, thus carrying a current through a gas.)

Singer's book, *The Nature of Ball Lightning*, appeared in 1971. This was a concise yet comprehensive and balanced discussion of the subject that reviewed all the important models proposed until that date (Singer 1971). In the same year, Argyle (1971) revived the suggestion that ball lightning was an afterimage phenomenon. This view was seriously questioned in ensuing correspondence from a physicist who had seen ball lightning at close proximity in an aircraft (Jennison 1971), from an optical physicist (Charman 1971b), and from an astronomer (Davies 1971).

Also in the same year, Ashby and Whitehead (1971) published a new kind of model based on the suggestion that micrometeorites of antimatter might be directed toward the ground by the electric field beneath a thundercloud. The annihilation of the antimatter then provided the energy for the formation of ball lightning. Ashby and Whitehead carried out experimental observations over a 12-month period and detected four possible events of this kind. Crawford (1971) suggested that these events could be explained by cosmic rays.

Berger (1973), an experienced lightning observer, said that in many years of observing conventional linear lightning he had never seen ball lightning. He expressed skepticism about its existence. In the decade following this expression of doubt, however, some of the best-authenticated ball lightning observations in the history of the subject were reported in the literature. This had a significant effect in changing the climate of scientific opinion about the existence of ball lightning.

In 1975, photographs were published taken by the U.S. Prairie Meteorite Network, several of which seemed to have recorded ball lightning (Tompkins and Rodney 1977, 1980; Tompkins, Rodney, and Gooding 1975, 1976). A videotape sequence of what may have been the development of ball lightning from a multiple-stroke lightning flash was recorded in 1976 in South Africa in the course of a research program by the National Electrical and Engineering Research Institute (Eriksson 1977a,b). Photographs were published that seemingly showed luminous, spherical objects formed in the vicinity of the decaying plasma channel of lightning triggered by rockets with attached, grounded wires fired into thunderclouds (Fieux, Gary, and Hubert 1975, Hubert 1975a). These events are described in more detail in Chapter 9.

In 1976, a report was published in which a woman in Smethwick in the Midlands region of England was hit by ball lightning that damaged her clothing and burned her (Stenhoff 1976). From the damage and her description of the event, Wooding (1976) and later Barry (1980) estimated the energy and energy density of the ball lightning and found both to be significantly lower than the estimates obtained from the Morris (1936) event. The Smethwick event is discussed in more detail in Chapter 5.

The increased respectability of ball lightning studies was confirmed in 1977 by the inclusion of a chapter on the subject in a substantial, two-volume review of the physics of lightning and lightning protection (Golde 1977). A substantial review of ball lightning was published in 1979 (Charman), and a further book reviewing ball lightning appeared in 1980 (Barry). This included a bibliography of nearly 2000 references and a very comprehensive collection of published photographs.

Professor Sir Brian Pippard, FRS, published an account of ball lightning reported by several independent witnesses at the University of Cambridge's Cavendish Laboratory in August 1982 (Pippard 1982). This case is more thoroughly described in Chapter 10.

Until 1988, scientific papers about ball lightning had been limited to an occasional, single paper presented at a conference in the contexts of electrostatics or lightning physics. On January 25, 1952, a geophysics discussion at the Royal Astronomical Society in London was chiefly devoted to lightning phenomena, including ball lightning (Gold 1952). In 1988, the First International Symposium on Ball Lightning took place in Tokyo, and the proceedings were published (Ohtsuki 1989). Since then, subsequent International Symposia on Ball Lightning have taken place in Budapest, Hungary in 1990 (Dijkhuis 1991b), in Los Angeles, California in 1992 (Dijkhuis 1993b), in Canterbury, England in 1995 (Dijkhuis 1996), and in Tsugawa Town, Japan in 1997 (Ohtsuki and Ofuruton 1997).

The Tornado and Storm Research Organisation (TORRO) (Bradford-on-Avon, England) held a conference on ball lightning in Oxford, England, in 1992 (Stenhoff 1992). An interdisciplinary congress on ball lightning entitled "Vizotum" was held in Salzburg, Austria, in 1993 (Keul 1993). A review of ball lightning that drew heavily on Charman's review, but included some subsequent developments, was published in 1995 (Kikuchi 1995).

1.4 Reported Characteristics

It is the apparently exotic and paradoxical properties of ball lightning and the very dramatic nature of many eyewitness descriptions that have fascinated scientists and made the subject so controversial. The several statistical surveys of ball lightning (Table 1.1) have mostly adopted what Rayle (1966) called the "lexicographical" approach. By this, Rayle meant that he "accept[ed] as ball lightning any phenomenon that an observer has so identified." However, surveys based on the study and evaluation of reports have revealed that a large proportion of such phenomena may be explained without the need to invoke ball lightning (Humphreys 1936, Stenhoff 1988b).

Statistical surveys based on the "lexicographical" approach thus describe the properties of ball lightning *reports*, not of ball lightning *events*. However, these findings have, incorrectly, been used to describe the characteristics of ball lightning modeled in a number of theories. Campbell (1993) writes of statistical surveys, "If the observations are not of ball lightning (or only partly of ball lightning), then the data are useless." Jennison (1997) points out that "many of the quoted parameters are taken from the raw statistics gleaned from the accounts of many untrained observers. The small minority of really good reports are far more important than the statistical results of a thousand 'typical' reports. Indeed, the important parameters tend to be suppressed in the statistics because the reliable observations are relatively few in number." The present author strongly supports this view.

The signal-to-noise ratio in a statistical survey of this kind is very difficult to establish. It varies from one report to another, and even across the parameters

Table 1.1. Some Statistical Surveys of Ball Lightning Reports

Survey	Year	Number of reports	Notes
Brand	1923	215	Selected from a collection of 600 ball lightning reports in the university libraries of Berlin, Marburg, and Gottingen (Brand 1923).
Humphreys	1936	280	Humphreys 1936
McNally	1961, 1966	498	Based on a survey of 15,923 Union Carbide Nuclear Company personnel in Oak Ridge, Tennessee (McNally 1961, 1966).
Barry	1967	400	Barry 1967a
Rayle	1966	112	Based on a survey of personnel at National Aeronautics and Space Administration (NASA) Lewis Research Center, Cleveland, Ohio (Rayle 1966).
Charman	1979	68	Unpublished survey; summary of some results given in Charman (1979).
Egely	1987	380	Analysis of Hungarian ball lightning reports (Egely 1989).
Stakhanov, Bychkov, and Keul	1991	2500	Russian–Austrian data bank (Bychkov, Smirnov, and Stridjev 1993).
Hubert	1996	253	Hubert 1996

described by each report because some properties are perceived and recalled more accurately than others. In statistical surveys, individual reports are usually given equal weighting irrespective of such factors as the time between an alleged event and a report, which, as Chapter 3 explains, may be of critical importance. Jennison (1992) and Amirov and Bychkov (1997) discuss the use of statistics weighted by reliability.

1.4.1 Environmental Conditions and Incidence

Ball lightning has been reported in the open air, indoors, and within aircraft.

Rayle (1966) argued that ball lightning was not a particularly rare phenomenon. His survey suggested that ball lightning reports are comparable in frequency to that of direct observations of cloud-to-ground (CG) lightning. Almost 95% of 97 ball lightning reports stated that the events occurred during thunderstorms, and of these, 97% of the respondents stated that the storm was of at least average violence. Almost 43% of 89 reports described heavy rainfall just before the observation, and nearly 70% said the rainfall was medium to heavy.

In two different surveys, nearly 85% of 444 observers (McNally 1966) and 73% of 95 observers (Rayle 1966) reported that ball lightning were seen following a lightning flash, and, of these, nearly 90% said that the flash was to the ground.

1.4.2 Form

Ball lightning is described as spheroidal in 88–89% of reports (Rayle 1966). Other shapes described have included ellipsoids, rings, rods, and irregular forms (Rayle 1966, Amirov, Bychkov, and Strizhev 1995).

1.4.3 Diameter

Statistical surveys of ball lightning reports have given fairly consistent data concerning reported diameter, which has a modal value between 20 and 50 cm (see Figs. 1.2 and 1.3). Rayle (1966) found that estimates of ball lightning diameter followed a log-normal distribution (see Section 1.4.14), with a median value of 36 cm for his survey and 25 cm for McNally's (1960, 1966). This log-normal relationship was supported by Dijkhuis (1992a,b). Amirov and Bychkov (1997c) found that an increase in humidity, or, to a lesser extent, an increase of electric field strength decreased both ball lightning diameter and the lifetime of the ball.

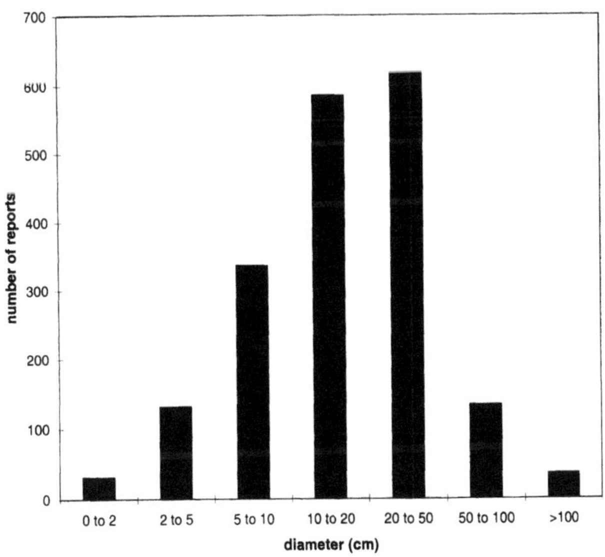

Figure 1.2. Frequency distribution for diameter in 1869 reports of ball lightning (Bychkov et al. 1993).

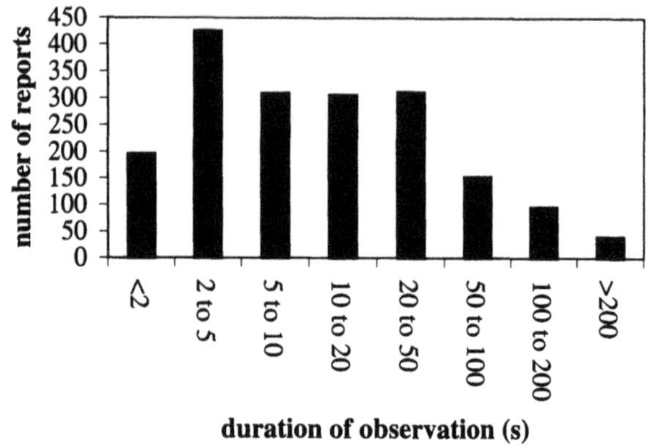

Figure 1.3. Frequency distribution for diameter in 446 reports of ball lightning (McNally 1966).

1.4.4 Duration of Observations

There is a similar high level of consistency in statistical data concerning reported durations of observations of ball lightning. Modal values lie between 2 and 5 s (see Figs. 1.4 and 1.5). Rayle (1966) remarks that durations refer to the time the ball was in sight, not the lifetime of the ball because the majority of observers did not see the formation and/or the decay of the ball. Dijkhuis (1992a,b) reported a log-normal distribution for ball lightning lifetime (see Section 1.4.14). Amirov and Bychkov (1997c) found that reported lifetimes of ball lightning were greater for larger diameters and, as already stated, that an increase in humidity, or, to a

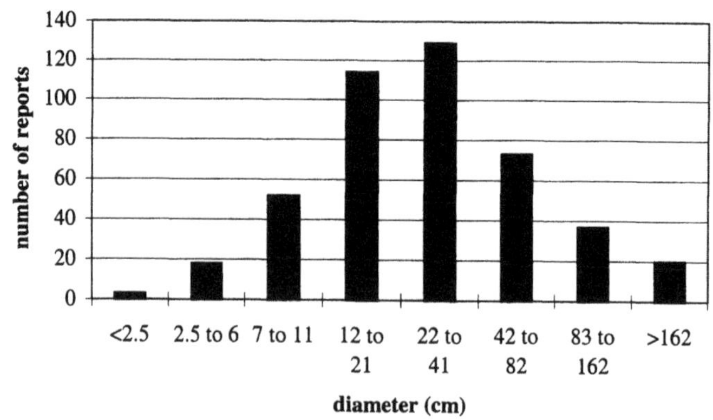

Figure 1.4. Frequency distribution for duration of observation in 1836 reports of ball lightning (Bychkov et al. 1993).

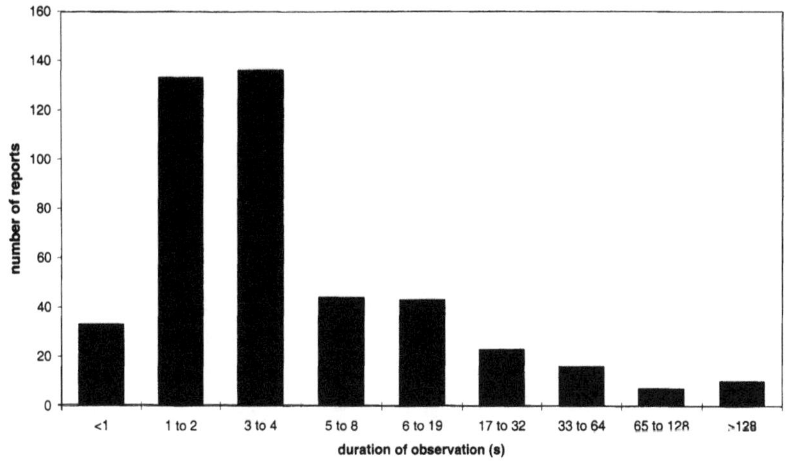

Figure 1.5. Frequency distribution for duration of observation in 445 reports of ball lightning (McNally 1966).

lesser extent, an increase in electric field strength, decreased both the observation time and the diameter of ball lightning.

1.4.5 Luminosity

Most reports describe ball lightning as "bright enough to be clearly visible in daylight." Rayle (1966) found that this was so in 60% of 110 reports. He found that 76% of 90 reports stated that the ball was uniformly illuminated across its surface. In only 11% was there an indication of limb-darkening.

Dijkhuis (1992a,b) reported a log-normal distribution for ball lightning luminosity (see Section 1.4.14). Amirov and Bychkov (1997c) found that the reported diameter, lifetime, and radiation power of ball lightning depended on whether it was seen indoors or outdoors.

1.4.6 Color

A poor level of correlation is found in different groups of statistical data about reported color (see Fig. 1.6). A Spearman rank correlation test applied to four major surveys yields correlation coefficients between −0.21 and +0.79 (see Table 1.2). Perceptual uncertainties concerning color are discussed in Chapter 3.

1.4.7 Constancy of Appearance

Rayle (1966) found strong evidence of constancy of reported size (87% of 103 reports), luminosity (87% of 105 reports), and general appearance (93% of 96

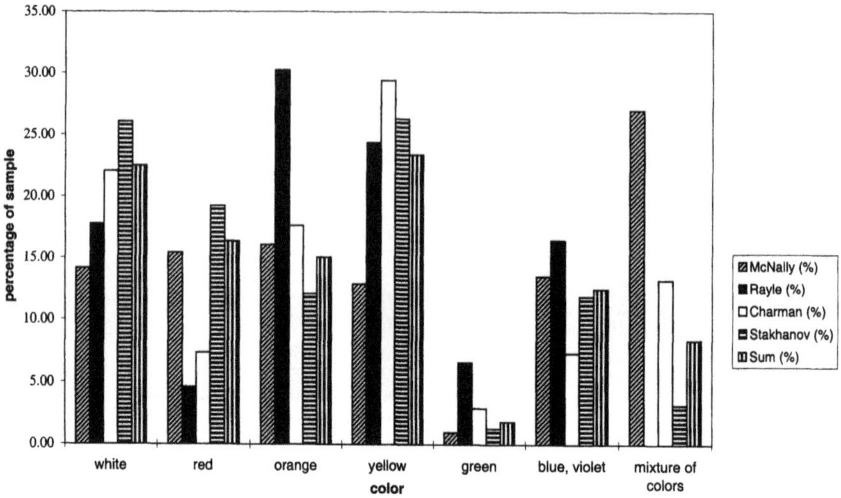

Figure 1.6. Frequency distribution of reported color (several surveys) (Smirnov 1987c).

reports). Of 93 reports, only 12% described a change in behavior or appearance immediately before the ball ended.

1.4.8 Motion

Various kinds of motion of ball lightning have been described, including (1) from clouds to near the earth, or from near the earth to clouds; (2) a spiral, zigzag, or random path above the earth; (3) motionless above the earth; and (4) movement between clouds (Barry 1966). Ball lightning only very rarely seems to convect; that is, it does not usually rise as would heated air in the absence of other forces.

Table 1.2. Spearman Rank Correlation Coefficients for Color Descriptions in Different Surveys[a]

	McNally	Rayle	Charman	Stakhanov
McNally		−0.21	0.19	−0.04
Rayle	−0.21		0.62	0.54
Charman	0.19	0.62		0.79
Stakhanov	−0.04	0.54	0.79	

[a]The only surveys that show signification positive correlation ($\alpha = 0.05 \Rightarrow r_s \geq 0.714$) are those by Charman (1979) and Stakhanov (1991).

Of 108 reports, 54% described predominantly horizontal motion, 19% vertical motion, 19% mixed motion, and 9% no motion (Rayle 1966). This would imply absence of convective behavior in 63% of the reports.

In Rayle's (1966) survey, of 103 reports, 16% said that ball lightning was formed in contact with a metal surface, 14% in contact with the ground, and 12% in contact with a nonmetallic surface. Of 88 reports, Rayle found that 18% said the ball was guided by the ground surface, 14% by power or telephone wires, and 7% by other metal structures (total 39%). A further 3% stated that cloud layers guided the ball, and 19% said it followed another guide. In 39% of reports, it was stated that there was no guide. Of 67 reports in the same survey where ball lightning made contact with any solid object, 37% described surface contact with a metal object. Of 98 reports, Rayle found that 15% stated that the ball lightning was in contact with metal when it disappeared, and 27% that it was contacting nonmetal. Many of these cases might be explained as St. Elmo's fire, and this is discussed further in Chapter 3.

Occasionally ball lightning is said to rotate, roll, or bounce, and it is said, very rarely, to move against the wind. Spinning or rotation was described in 36% of 96 reports (Rayle 1966). Witnesses have described ball lightning passing through small holes, screens (such as insect screens), or solid objects without any apparent effect on the ball. Such behavior was described in 24% of 101 reports (Rayle 1966).

1.4.9 Behavior inside Aircraft Fuselages

Ball lightning seen within aircraft usually traveled inside the fuselage, along the center of the aircraft from front to rear at a moderate speed relative to the aircraft. These reports show remarkable consistency (see Chapter 7).

1.4.10 Odor

Of 98 reports, 23% described odors (Rayle 1966). These odors were often described as sharp or acrid, and observers who are scientists have described the smells as reminiscent of ozone, burning sulfur, or nitric oxide.

1.4.11 Sound

Only a few observers report sounds associated with ball lightning other than at its formation or decay; these witnesses describe hissing, buzzing, or fluttering sounds (Brand 1923). [Of 108 reports, Rayle (1966) found that 23% reported sound, but it was not clear whether this should exclude sound on decay.] Most reports of ball lightning state that the decay was silent, but in some reports the decay is described as explosive. Of 78 reports, 69% described silent decay and 31% explosive decay (Rayle 1966).

1.4.12 Traces and Damage

Many of those who report ball lightning said there were traces or damage remaining after its disappearance. Rayle (1966) found that this was so in 42% of 95 reports. Based on such damage, some scientists have estimated that the energy of some ball lightning is as high as several megajoules (Goodlet 1937). From studies of case histories involving damage, experiments to replicate ball lightning and models to describe it, Barry (1980) claimed a log-normal distribution for the energy density of ball lightning. Further discussion of ball lightning energy and energy density is deferred until Chapter 11, after a review of the physical evidence from which these quantities are derived.

1.4.13 Types of Ball Lightning

Brand (1923) distinguished between two types of ball lightning—*aufsitzende* (attached) and *freischwebende* (floating). The distinction between the former and St. Elmo's fire is unclear, although in many cases Brand found a significant heating effect, which would not be caused by St. Elmo's fire (see Chapter 3).

Rayle's (1966) statistical survey led him to the tentative conclusion that there might be at least two distinct types of ball lightning. Group A consisted of events that followed a lightning flash to ground, where the ball approached within 1 ft (30 cm) of a solid (presumably near the Earth's surface) and where it was seen to end quietly on or near the ground. Group B contained events where the ball was first reported in midair, where it never approached the ground, and where it was not connected with a lightning flash to the ground. In both these cases, the ball lightning was reported at a great distance; its diameter was atypically large and its luminosity atypically low.

A number of authors (e.g., Hubert 1996) have continued to speculate that there may be more than one kind of mechanism responsible for phenomena described as ball lightning.

1.4.14 Log-Normal Distributions for Ball Lightning Parameters

Log-normal distributions are described by straight-line graphs using a cumulative probability scale on the vertical axis and a logarithmic scale on the horizontal axis. Log-normal distributions have been described for diameter, lifetime, energy density, and luminosity of ball lightning, and have also been reported for its velocity and its closest distance from the observer (Fig. 1.7). Dijkhuis (1992a,b) used these log-normal distributions to argue in favor of fractal structures for ball lightning (see Chapter 12).

A number of parameters associated with conventional lightning also follow log-normal distributions. Uman (1987, Appendix B) writes that these include the peak current and the time interval between strokes. Uman indicates that in other

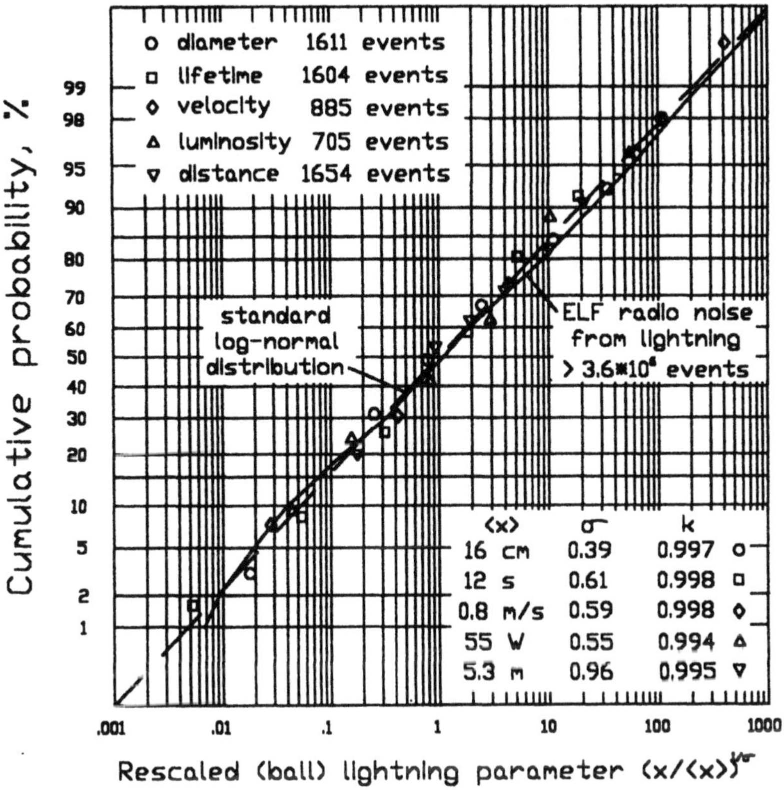

Figure 1.7. Cumulative probability of rescaled ball lightning parameters from Amirov's and Bychkov's global survey, relative to the standard log-normal distribution and extremely low frequency (ELF) radio noise from lightning at 500 Hz according to Fraser-Smith et al. (Dijkhuis 1992a,b). Reproduced by kind permission of Dr. G. C. Dijkhuis.

situations in nature where the log-normal distribution is found, the magnitude of the log-normally distributed variable is physically connected to growth via the "law of proportional effects." He gives the example of cloud growth, where larger elements of cloud, as they convect, are more likely to encounter other elements and hence should grow more rapidly than smaller ones. He gives other examples of attempts to correlate lightning parameters to the log-normal distribution. Uman states that "no physical argument has yet been advanced to account for the fact that many lightning parameters are log-normally distributed."

Amirov and Bychkov (1995c) carried out a statistical analysis which, they claimed, showed that ball lightning diameter, lifetime, radiation power, and the closest distance between the ball lightning and the observer could not be described by a log-normal distribution.

1.4.15 Summary: Typical Reported Characteristics

These are the independent characteristics given in 75% or more of reports and on which a number of surveys agree. Ball lightning reports are associated with thunderstorms of at least average violence. There is a lightning flash just before the observation, which is most likely a flash to ground. Ball lightning is spheroidal, with a diameter between 20 and 50 cm. It is observed for between 2 and 5 s. It is of uniform brightness across its surface. Its size, luminosity, and general appearance are reported to remain fairly constant throughout its lifetime, including its decay. When ball lightning is reported inside an aircraft, it is said to travel along the center of the aircraft from front to rear at a moderate speed relative to the aircraft.

In more than 50% of reports, the following additional independent characteristics are described. There was medium to heavy rainfall just before the observation. Ball lightning is bright enough to be clearly visible in daylight. The motion of ball lightning is predominantly horizontal.

1.5 Developing Models for Ball Lightning

If the signal-to-noise ratio is zero [as Humphreys (1936), Berger (1973), Schönland (1950), Campbell (1992) and others have suggested], the problem of ball lightning would be solved. If signal-to-noise ratio is not zero, then the quest for a valid ball lightning theory is a legitimate goal. This controversy is discussed in Chapter 10.

The suggestion that there is more than one type of ball lightning has not been fully incorporated into the discussion of the merits and demerits of different ball lightning models. Some authors have been very critical of the models proposed by others because they do not explain all the reported characteristics of ball lightning.

Most ball lightning models set out to explain some of the following reported properties:

1. Its apparent association with thunderstorms.
2. Its apparent association with conventional linear lightning, especially flashes to the ground.
3. Its reported shape, diameter, and duration, and the fact that size, luminosity, and appearance are reported to remain fairly constant throughout its life.
4. Ball lightning has been seen in the open air and in enclosed spaces such as buildings or aircraft.
5. When ball lightning is seen inside an aircraft, it is reported to travel along the center of the aircraft from front to rear at a moderate speed relative to the aircraft.

Table 1.3. Classification of Ball Lightning Models[a]

Internal energy source	External energy source
1. Ball lightning is gas or air behaving in an "unusual" way.	1. High-frequency (hundreds of megahertz) electromagnetic field.
2. Ball lightning is a heated sphere of air at atmospheric pressure.	2. Steady current flow from cloud to ground.
3. Ball lightning is a very high-density plasma that exhibits quantum-mechanical properties characteristic of the solid state.	3. Focused cosmic ray particles.
4. Ball lightning is due to a closed-loop current flow contained by its own magnetic field.	
5. Ball lightning is an air vortex that provides containment for luminous gases.	
6. Ball lightning is a microwave radiation field contained within a thin spherical shell of plasma.	

[a]After Uman (1969).

6. A significant number of ball lightning reports describe motion that is inconsistent with simple, convective behavior of a hot gas.
7. Ball lightning has been reported to have caused damage.

Some models also try to explain the following reported characteristics of ball lightning:

1. It may decay silently or explosively.
2. It has been reported to pass unimpeded through small holes, screens, or solid objects.
3. It may sometimes be associated with acrid odors.
4. It may occasionally be associated with hissing, buzzing, or fluttering sounds.
5. It is sometimes said to rotate, roll, or bounce.
6. It is, rarely, said to move against the wind.

Uman (1969) has classified ball lightning models (see Table 1.3), which are reviewed in Chapters 11–13.

Chapter 2

Thunderstorms and Lightning

The purpose of this chapter is to offer a brief review of the physics of thunderstorms and lightning. This is a developing and large area of study in which many questions remain unanswered and there are many differences of opinion. For the sake of brevity, some of these differences have been smoothed over in this chapter, and the explanation here is necessarily simplified. It is intended to be sufficient to enable an understanding of the probable relationship between ball lightning and conventional linear lightning.

For a more thorough understanding, which is essential for all those who study ball lightning, the reader is referred to the several excellent reviews of this topic. For a broad, nontechnical overview, Uman's book, *All About Lightning* (Uman 1971, 1986) is very useful. His excellent academic review, *The Lightning Discharge* (Uman 1987), offers comprehensive coverage up to the date of publication and largely supersedes his earlier book, *Lightning* (Uman 1969, 1984). The two-volume set edited by Golde (1977) is of considerable value even two decades after its publication, but at present it is out of print. The first volume was concerned with the physics of lightning and the second with lightning protection. The latter subject is also covered by Golde's earlier book (1973). Lightning risk assessment and the design of protective systems is based on a clear understanding of the damage ordinary lightning can cause, a topic on which some authors of ball lightning papers have evidently been ignorant.

Other useful reviews discuss electrical discharges; these include Nasser's book, *Fundamentals of Gaseous Ionization and Plasma Electronics* (Nasser 1971), and the collection of papers in *Electrical Breakdown in Gases*, edited by J. A. Rees (1973).

The reader is also referred to the glossary in Appendix B in this volume. Terms included in the glossary are marked with an asterisk in this chapter.

2.1 Introduction

Throughout history, lightning has been one of the greatest mysteries of meteorology and physics, and even now it is not fully understood. Beliefs concerning thunder and lightning abounded in mythology and many religions (Uman 1987). To this day, many superstitions concerning lightning prevail even in modern society.

Most people view thunderstorms and lightning with awe. Certainly, a respect for the dangers of lightning and an awareness of its hazards are prudent. Uman (1986, 1987) points out that every year in the United States alone, lightning causes the deaths of a hundred or so people; injuries to several hundred more; deaths and injuries to livestock and other animals; thousands of forest and brush fires; as well as millions of dollars in damage to buildings, communications systems, power lines, and electrical systems.

The scientific study of lightning began about 250 years ago. In 1752, Benjamin Franklin demonstrated its electric nature. In one famous but exceptionally hazardous experiment, he tied a metal key to the end of a kite string and flew the kite in a thunderstorm. The electric charge in the cloud raised the voltage of the kite string. This high voltage caused a spark to jump from the key to grounded objects, showing that the cloud was electrified (Dibner 1977). Fatalities resulted from attempts to reproduce a number of Franklin's experiments, and he was indeed fortunate that he was not among them.

At any one time, more than 2000 thunderstorms are in progress on the Earth's surface, mostly where vertical convective activity is high, and with most thunderstorms developing around midday in spring and summer (Moore and Vonnegut 1977). Recent whole-earth satellite observations have enabled accurate studies of the geographical and temporal frequency of thunderstorms (Turman and Edgar 1982).

During fair weather, there is a potential difference of about 300 kV between the Earth's surface and the electrosphere, a region that extends from about 50 km above the ground upward (the atmospheric electric field*). In this region, the atmosphere is a good conductor of slowly varying electric signals (Uman 1987). It is widely believed that this potential difference is due to the world-wide distribution of thunderstorms (supply current*, point discharge current*). This potential difference produces a total air–earth conduction current* (or fair-weather discharging current) of about 1 to 2 kA, which, since the surface area of the Earth is about 5×10^{14} m^2, corresponds to a current density of about 2 to 3 pA/m^2 (air–earth current*). Measurements show that an average of nearly 1 A of current flows into the stratosphere during the active phase of a typical thunderstorm. This confirms that one to two thousand thunderstorms must be active at any given time to maintain the fair-weather global electric current flowing to the surface. This whole system is often called the *global electric circuit* (global circuit*).

The Earth and its atmosphere have together been likened to a "leaky, spherical capacitor" (Uman 1974). Thunderstorms are responsible for giving the Earth its net negative charge of about 1 MC, with an equal positive charge being distributed throughout the atmosphere. These charges in turn produce a downward-directed atmospheric electric field, the fair-weather field, with an intensity of about 0.13 kV m^{-1} at the surface (fair-weather electricity*). Markson (1978) has suggested a correlation between variable solar activity and atmospheric electrical mechanisms.

2.2 Thunderstorm Electricity

When moist, warm air is heated, its density decreases and it begins to rise by convection. Both the surrounding air pressure and temperature decrease with increasing altitude. The rising pockets of air expand, which causes the air to cool so that the moisture eventually condenses to form clouds. As the cloud continues to cool, more moisture condenses and the water droplets of which the cloud is composed grow and coalesce. Eventually, some may become too large and heavy for the air currents within the cloud to continue to support them, so they begin to fall as rain.

Thunderclouds, or cumulonimbus (Cb) clouds usually result from atmospheric instability and develop as warm, moist air near the ground convects upward and replaces denser air above it, but the convective activity is much more pronounced than in other cloud forms. They may be regarded as "large atmospheric heat engines with water as the primary heat-transfer agent" (Moore and Vonnegut 1977).

Scientists are uncertain exactly how cumulonimbus clouds become electrically charged (thunderstorm charge*). Air currents in cumulonimbus clouds can be exceptionally powerful. As the particles within a cumulonimbus cloud grow and interact, some become charged, probably through collisions. As a result of these collisions, the heavier particles become negatively charged and the lighter particles acquire a positive charge. The negatively charged particles fall to the bottom of the cloud, and most of the positively charged particles rise to the top. The small, positively charged particles shoot up to the edge of the stratosphere, which acts as a ceiling. They spray out sideways to form the characteristic anvil head of a mature thundercloud. The upper portion of the cloud thus normally acquires a net positive charge and the lower portion of the cloud usually becomes negatively charged. (The electrical distribution of charge within a thundercloud has been inferred from electric field measurements outside the cloud and from soundings.) This separation of charge produces enormous electrical potential differences both within the cloud and between the cloud and ground. These can rise to millions of volts, and eventually the air breaks down and a flash begins. Lightning is thus an electrical discharge between positive and negative regions.

Even when lightning is not produced, pellets of ice may grow through the accumulation of liquid droplets. When the updrafts are very intense, they can suspend the growing ice pellets for long periods, allowing them to become larger. Eventually some may become too large to be supported by the updraft and begin to fall as hail. Typical diameters of hailstones are 5 to 10 mm, although exceptionally a 140-mm hailstone has been recorded.

Usually, therefore, although not always, a thundercloud consists of two major regions of separated charge—the positive or P-region at the top and the negative or N-region near the base of the cloud. The altitudes of the different regions of the cloud and the magnitude of the charges vary in measurements from different parts of the world. The magnitudes of the charge on each of these regions, measured from outside the cloud, are typically about 40 C, but can exceed 100 C (Uman 1987). A simple model of a thundercloud is therefore based on a charge dipole, with the negative end closest to the earth (thunderstorm dipole*). This induces positive charges in the soil; the effective center of this region of induced charge is as far below the soil as the original charge is above it. (The soil surface acts like a mirror and the induced positive charge is like a virtual image.)

Thus this dipole sets up an electric field beneath the cloud which has a direction opposite to that of the fair-weather field. Theory provides a means of estimating the magnitude of this field from the geometry and the charges in the cloud (Uman 1987, Section A.1.2; Brook and Ogawa 1977). Ball lightning models have been proposed in which thunderstorm electric fields focus heavier reactive particles, such as cosmic rays, so that nuclear reactions produce ball lightning (Chapter 12). These models would also suggest a correlation between solar activity and incidence of ball lightning.

When the surface field strength exceeds 1.5 to 2 kV m^{-1}, grounded objects with small radii of curvature will begin to emit ions. This process is called *point discharge**. These positive ions are attracted toward and stream upward to the N-region. Perhaps because of the accumulation of these ions, a small positively charged region is then found in some thunderclouds beneath the N-region; this is called the p-region. Its charge is about 10 C to 24 C (thunderstorm tripole*).

Grounded objects with small radii of curvature may also produce a visible corona discharge* in the form of St. Elmo's fire. St. Elmo's fire may resemble ball lightning. It sometimes appears around airplanes, towers, masts of sailing ships, and treetops. Further discussion of St. Elmo's fire is found in Chapter 3. The intensity of a radial electric field at a charged, spherical surface is inversely proportional to the square of the radius of curvature, which confirms why a corona is observed near sharply pointed objects. Such objects will enhance the local field due to the induced charge.

When the surface field strength reaches about 3 kV m^{-1}, dielectric breakdown and intracloud (IC) lightning flashes (intracloud flash*) occur within the cloud, and these discharges coincide with a large and abrupt change of the surface field back to fine-weather polarity. There is then a gradual return to foul-weather conditions. When these conditions are restored, some discharges within the cloud may initiate cloud-to-ground flashes,* as indicated below.

Much of the precipitation that falls from thunderclouds does so in characteristic, intense bursts of rain or hail and carries with it an electric charge. These bursts are related to changes in the surface electric field, although the exact cause-and-effect relation is not clear (FEAWP*). The first lightning bolt often precedes a sudden downpour of rain by a few minutes.

Through the active phase of the thunderstorm, lightning is followed by a field excursion to fair-weather field conditions; then there is a gradual return to foul-weather field conditions, followed by further lightning, and so the cycle continues. At the end of the storm, the final lightning is often anomalous, lowering a substantial amount of positive charge to Earth (see later discussion) and having associated with it an "end of storm oscillation" (EOSO*) (Moore and Vonnegut 1977).

Some models of ball lightning suggest that, like St. Elmo's fire, it is a form of dc discharge, in which case its behavior and appearance would be strongly influenced by changes in the atmospheric electric field. Others suggest that it is after a particularly intense flash between cloud and ground that a ball is formed. It is therefore essential to try to obtain detailed information concerning other events in the storm before and after a ball lightning event.

2.3 Lightning

Uman (1987) defines lightning* as "a transient, high-current electric discharge whose path length is measured in kilometers." Lightning is a giant electrical spark in the sky formed between charge centers of different polarity. Figure 2.1 contains a classification scheme for forms of lightning. Uman (1987) discusses most of these different forms of lightning and the processes associated with them in detail.

Flashes of lightning vary in length. A flash between a cloud and the ground may be up to 14 km long, while a flash that travels between adjacent clouds may be more than 140 km long. Lightning can occur within a cloud, between a cloud and the air, and between two clouds. These are called *cloud discharges*. Their electrical energy dissipates in the air, and may damage airplanes traveling through it, but it does no harm on the ground. Possible damage to aircraft caused by ball lightning is discussed in Chapter 7. The most common type of lightning, called *intracloud lightning* (I C), is created when charges within a cloud form an electrical spark. This accounts for well over half of all lightning discharges. Electric currents between a cloud and the air cause *cloud-to-air lightning* (CA), and an electrical current between two clouds produces *cloud-to-cloud* or *intercloud lightning* (CC). These are less common than IC lightning or cloud-to-ground lightning. Occasion-

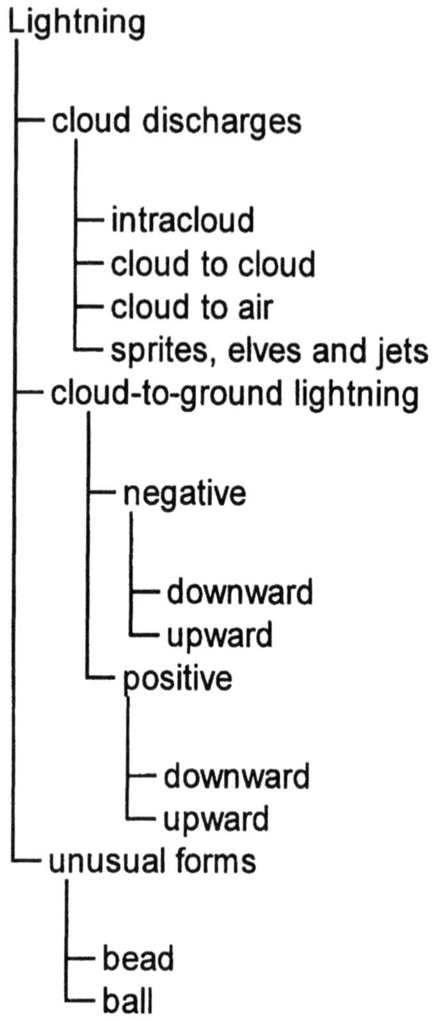

Figure 2.1. Lightning classification.

ally, ball lightning is reported to emanate from a cloud or to form near a cloud following cloud discharges (cloud flash,* cloud-to-cloud flash,* intracloud flash).

Most lightning visible from the ground is between a cloud and the ground, and this is thus known as *cloud-to-ground* lightning (cloud-to-ground flash*). Cloud-to-ground (CG) lightning is responsible for nearly all deaths, injuries, or damage due to lightning. A flash of CG lightning consists of one or more electrical discharges called *strokes*. CG lightning is often called *forked lightning** or *linear lightning*.

Berger (1977a, 1978) has classified CG lightning into four kinds, according to the polarity and direction of the leader that initiates the discharge. Hence CG lightning can be upward or downward, positive (+CG) or negative (–CG). The development and behavior of lightning depends on its type (ground-to-cloud flash,* negative cloud-to-ground lightning,* positive cloud-to-ground lightning*).

The most common form of CG flash, accounting for more than 90% of CG flashes world-wide, is the negative, downward flash, which is from the N-region of the cloud to the ground (Uman 1987). The following scenario describes current theory about such a downward –CG flash. Intracloud discharges between the N- and P-regions of the cloud may initiate CG lightning by creating free electrons and ions. Preliminary breakdown* (electrical breakdown*) within a cloud initiates a stepped leader* (leader*), which in turn initiates the first stroke of a flash. Electrons, being much less massive, are considerably more mobile than ions, so they are accelerated toward the ground by the N-region of the thundercloud. The stepped leader travels downward, directed by the electric field beneath the cloud, but it does so in a series of steps, each about 50 m in length, and between these steps it pauses for about 50 μs and changes its direction (pilot streamer*).

As it descends, the local electric field between its tip and the ground increases. This is because the stepped leader is at approximately the same potential as the base of the cloud since it is so effective a conductor. Theory enables estimates of the relationship between height of the leader and the field (Uman 1987, Section A.1.3). When it is within 10 to 100 m from the soil surface, grounded objects begin to emit filaments of electrically conducting plasma called *streamers** (ground streamer*). If one of these makes contact with the stepped leader, attachment occurs and electrical contact is made between the cloud and ground, so that a return stroke* then propagates along the ionized path in the direction opposite to that of the stepped leader. The return stroke is the bright light that is seen in a flash of lightning. The lightning return stroke is a current-carrying plasma column (lightning channel,* lightning discharge,* spark discharge*). Return strokes travel at about the speed of light. They are thought to have a diameter of several centimeters (Uman 1969, 1987). They discharge very high currents, the highest within the flash, and so heat the air in their paths to temperatures above 30,000 K, more than 5 times the surface temperature of the sun. Ball lightning is frequently reported immediately after a CG return stroke; or occasionally, it is said to precede such a stroke.

Air heated by return strokes expands rapidly (the pressure in the lightning column reaches about 8 atm), producing a shock wave that propagates away from the channel as a sound wave that is heard as thunder. Because sound travels in air at about 330 m s^{-1}, while light travels at about 3×10^8 m s^{-1}, there is a perceptible interval between lightning and thunder unless the flash strikes the ground very close to the observer. Within a few hundred meters from the strike, the observer will hear a single loud bang caused by a return stroke, sometimes preceded by hissing or clicking caused by an upward streamer from the ground to a stepped leader. Over greater distances, the preliminary sounds may be like the tearing of cloth (Uman 1971, 1986; Hill 1977).

The duration of the interval between lightning and thunder enables an estimate of the distance to the return stroke. The approximate relationship is $s = 330t$, where t is the time in seconds and s the distance in meters. Because the return stroke is formed between the ground and a cloud, the distance s from an observer to the return stroke channel varies from x_1, the horizontal distance between the observer and the impact of the stroke on the ground, to approximately $(x_2^2 + y^2)^{1/2}$, where y is the vertical height of the cloud base and x_2 its horizontal distance from the observer (Uman 1971, 1986). This variation may result in the familiar "rumble" of thunder.

In ball lightning reports, descriptions of the time interval between lightning and thunder and a description of the thunder may be useful in establishing the circumstances that prevailed just before the ball appeared and may help to resolve whether conventional linear lightning had struck the ground in the immediate vicinity. A lightning channel perpendicular to the observer's line of sight produces a high-intensity clap of thunder because $x_2 = 0$, while if the channel is along the line of sight, it will be perceived as a rumble. The duration of a rumble of thunder may give some indication of the geometrical arrangement of the observer, the cloud base, and the approximate strike point of the lightning on the ground. Thunder is normally audible only over a distance of up to 24 km. Intracloud and intercloud lightning also produce thunder.

The flash may then end, but if further discharges within the thundercloud make charge available, a dart leader* may then propagate along the residual, ionized first-stroke channel in the same direction as the original stepped leader and initiate a subsequent return stroke. Some models of ball lightning assume that the ball may be formed in this residual stroke. This process may be repeated many times; typically there are about four return strokes, called *subsequent strokes*, in each flash (lightning flash,* lightning stroke*). The length and duration of each lightning stroke vary, but typically, a flash lasts about 0.5 s, and each stroke lasts about 1 ms, with the interval between strokes about 40 to 80 ms. The eye cannot usually resolve the individual strokes, but rather the lightning may seem to flicker. The average peak power per stroke is about 1 kW. Peak currents are about 10 to 20 kA, but occasionally they can range up to hundreds of kiloamperes. Usually, first strokes are associated with more current than subsequent strokes.

If a constant current I is maintained for t seconds in an ohmic conductor of resistance R, the energy transferred through joule heating is given by I^2Rt. If the current i varies, but is unidirectional, and provided the conductor does not melt, the energy dissipated is

$$\int i^2 R \times \mathrm{d}t.$$

The action integral is a measure of the heating effect of lightning. It is defined by

$$\int i^2 \times dt,$$

the time integral of the square of the instantaneous current. In the case of metallic conductors, the temperature rise is quite modest because the value of R is low. Objects that are struck and are poor conductors are most likely to undergo substantial temperature rises. The average value of the absolute energy of a lightning flash was estimated as 200 MJ. Flashes with continuing currents might be expected to give larger values. The action integral of a negative first stroke and flash has a median value of 5.5×10^4 A^2s, while the action integral of the most energetic 5% of positive flashes is 270 times greater (Golde 1977). Thus the energy of the most energetic positive flashes could be more than 55 GJ.

The time between successive strokes may be extended to tenths of a second if continuing current* flows in the channel after the stroke. Continuing currents are on the order of 100 A and last for about 0.1 s, thus lowering about 10 C of charge from a cloud (Uman 1987). They are caused by direct transfer of a charge from cloud to ground. Between 25% and 50% of CG flashes have a continuing current component. Continuing current features in some ball lightning models.

Lightning associated with continuing currents is often called *hot lightning*. Although the peak temperature of all lightning is very high and the currents are very large, these very high temperatures and currents are only maintained for a very short time unless there are continuing currents, and so the heating effect of lightning without such currents is surprisingly rather small. Where there are continuing currents, however, these provide sufficient heating effect for a sufficiently long time to start combustion.

One lightning flash of exceptionally long duration lasted 2 s with 54 component strokes, 26 of which showed leader and return strokes (Workman, Brook, and Kitagawa 1960, cited by Berger 1977a). This must surely set an upper limit of <1 s on the duration of continuing currents for the great majority of lightning discharges. Ball lightning models that require continuing currents to persist for several seconds thus do not receive strong support from experimental studies of linear lightning unless they are based on the ignition of combustible materials.

The rate of change of current di/dt for –CG flashes, which determines the magnitude of their electromagnetic inductive effect, is far greater than for +CG flashes, especially in the case of subsequent strokes. Ball lightning models that are based on electromagnetic induction may be more readily related to –CG than to +CG flashes.

Positive, downward CG lightning may be of particular relevance to ball lightning models (positive discharge*). It is probably initiated from the P-region of a thundercloud, either when the P-region is horizontally separated from the N-region, such as when strong shear winds displace the N-region laterally and the P-region is exposed to the ground (Uman 1971, 1986, 1987), or toward the end of a storm when the charge in the P-region may be discharged to Earth in a single stroke, often of exceptional severity (Berger and Vogelsanger 1965, Golde 1973, p. 12).

One major difference between positive and negative lightning is that +CG lightning flashes more often consist of a single stroke that entirely discharges the P-region of the thundercloud. The median duration of a positive flash is about 7 times greater than for negative single-stroke flashes. Another difference is that +CG flashes are initiated by leaders that are not stepped (Uman 1987). Positive lightning is rather rare in most thunderstorms. Positive flashes are more common in winter thunderstorms, which produce few flashes overall; they are relatively uncommon in summer thunderstorms. This is because in winter the freezing level is at a lower altitude. The proportion of lightning that is +CG increases with increasing latitude and with increasing altitude above sea level, and it is greater in severe storms (Uman 1987). [It is interesting in this context to note that Brand (1923) found that ball lightning was reported to occur most frequently toward the end of a storm and is more often reported in winter than in summer. Rayle's survey (1966) disagreed with both these conclusions. It is important in this context that the data acquired include the time of year of a reported event, time of incidence of ball lightning within a thunderstorm, and geographic location.]

Positive discharges are of particular interest because their maximum current and total charge transfer are usually much greater than that of more common –CG lightning. The largest recorded peak currents for CG lightning (200–300 kA) are due to return strokes of +CG lightning (Uman 1987). Positive flashes consist of a single stroke followed by a period of continuous current that is of a duration comparable to continuing currents in those –CG lightning flashes that have them (Uman 1987). Furthermore, +CG flashes have very broad frequency distributions, so that the highest measured values for these parameters are considerably higher than for –CG strokes or flashes. Berger (1977a) called them "giant flashes" because of their exceptionally high $\int i^2 dt$ values of up to $1.5 \times 10^7\ A^2$ s for the highest 5% of strokes, compared with $5.5 \times 10^5\ A^2$ s for the top 5% of first negative strokes. They are thus responsible for much more severe thermal and mechanical effects than –CG strokes. Their energy may be 10 times higher.

Thus +CG lightning with high values of action integral is associated with a large heating and mechanical stress effect. Ball lightning models requiring large currents or heating effects may describe phenomena that would be expected to be associated with positive lightning. These models include those describing the ablation or puncturing of a solid surface.

Turman (1977) reported satellite detection of a rare and outstandingly energetic form of lightning flash, which he called a lightning superbolt*. These may represent the most energetic "tail" in the frequency distribution of energies of "positive giants." The discovery of superbolts may be relevant to ball lightning models that require greater than typical energy of lightning. If +CG or superbolt flashes generate ball lightning, then this might enable predictions to be made about the relative frequency of different kinds of CG flash and the incidence of ball lightning. This might also explain why not all CG flashes generate ball lightning. If lightning superbolts are extreme forms of +CG discharges, then perhaps they have longer duration, although their optical power vs. time graphs suggest durations only about twice that of ordinary lightning. At the time of writing, National Aeronautics and Space Administration (NASA) measurements with the Optical Transient Detector and the Lightning Imaging Sensor have yet to find optical events as bright as superbolts (E. Williams, personal communication, 1998).

Negative CG upward lightning results from upward leaders that are propagated from tall natural geographical features and manmade structures such as towers, or by artificially triggering lightning by firing rockets attached to grounded wires toward thunderclouds (rocket-triggered lightning*) (Uman 1987). Positive, upward lightning is the rarest kind of CG lightning (Uman 1987). Upward leaders from a ground structure or a topographical feature such as a mountain often enter a thundercloud and result in a more or less continuous current flow of 100–1000 A. This may then be followed by dart-leader and return-stroke sequences of the kind that follow downward –CG lightning first strokes (Uman 1987). Sometimes long, upward-moving negative leaders make contact with positive downward-moving leaders, and upward and downward return strokes then propagate from a junction point. Lightning channels that branch upward, rather than in the usual downward direction, characterize upward lightning of either polarity.

If the current from a lightning stroke passes through a conductor with finite resistance per unit length, a potential gradient will be set up along the conductor. If a second, shorter conductor at or near ground potential is placed alongside the first conductor, the potential difference between the two conductors may exceed the critical breakdown potential of the air (or another insulator) so that an arc or *side flash* occurs, and a large proportion of the current may then be diverted through the second conductor. The first conductor might be a tree whose resistance per unit length is several kΩ m^{-1}. The second conductor might be a human being with a resistance of about 1 to 2 kΩ, a building, an underground water pipe, or a gas or electrical mains supply to a building. The side flash has been recognized in recent years as an important cause of lightning damage. It also offers a mechanism by which energy from a lightning discharge can enter a building without the building suffering a direct hit. The side-flash is the mechanism that makes it especially dangerous to shelter under trees during thunderstorms.

2.4 Unusual Forms of Lightning

We defined ball lightning in Section 1.1. It is immediately evident that this definition shows that ball lightning is not a subset of "true" lightning as defined here. Thus the expression "ball lightning" is a misnomer resulting from the association between ball lightning and conventional linear lightning.

Some descriptive names given to lightning do not describe distinct forms—for example, heat,* sheet,* rocket,* ribbon,* spider,* and streak* lightning are all considered to be various manifestations of intracloud, cloud-to-air, and cloud-to-ground discharges.

Beaded* lightning is the name given to a lightning channel that seems to fragment into a series of luminous regions, perhaps tens of meters in length. This luminosity seems to be more persistent than that of the conventional CG lightning channel, lasting for perhaps 1 to 2 s. Bead lightning occurs in lightning triggered by rockets with especially long continuing currents. It is also observed as a form of intercloud lightning (Uman 1969; Barry 1980). The mechanism for bead lightning is uncertain; proposals have included the magnetic pinch effect (Uman 1962, 1969; Uman and Helstrom 1966), although this has since been called into question (M. A. Uman, personal communication, 1983). The persistence of luminosity has suggested to some authors that there is a relationship between ball and bead lightning (e.g., Uman 1968b, Hubert 1975).

Tornadoes are associated with the most intense manifestations of atmospheric electricity, and continuous lightning, point discharges, and ball lightning have been reported. Charging is probably by triboelectrification.* About 20 lightning discharges occur per second within a tornado, and it has been estimated that an electrical power of at least 20 GW can be associated with a single tornado event. Other less violent vortex motions in the atmosphere, such as dust devils, generate electric fields, but the intensity is insufficient to produce luminous effects. Ball lightning may occur in the spout rather than at the lower tip (dust-devil effect*) (Flora 1954, Tepper 1958, Vonnegut 1960, Fulks 1962, Brook 1967, Colgate 1967).

Lightninglike discharges have been observed above active volcanoes, in the vicinity of nuclear explosions, and in New Mexico gypsum sandstorms (at White Sands National Monument). Transient luminous phenomena that may be electrical in nature have been reported during earthquakes (Uman 1987).

2.5 Atmospherics and Radio Noise

Electrical discharges in thunderstorms behave in essentially the same way as the earliest and most primitive radio transmitter—the spark transmitter designed by H. R. Hertz. A lightning discharge between two charge centers of different polarity short circuits the electric field and causes it to collapse. The field may recover once the discharge is over, unless the flash has discharged the charge centers. These transients generate an electromagnetic wave that radiates from the discharge and propagates through the atmosphere. Such discharges take place, as we have seen, within clouds, between clouds, and between clouds and the ground.

The signals generated may be considered to be produced by many dipoles of different dimensions. Thus the radiofrequency spectrum of lightning is quite broad. In Chapter 13 we discuss Kapitza's model for ball lightning formed by standing waves of electromagnetic radiation in the radiofrequency part of the spectrum.

2.6 Recent Developments

This chapter opened by indicating that the physics of lightning is a developing area of study. A number of recent discoveries in this field have generated much excitement and discussion in the atmospheric physics community.

"Red sprites",* "blue jets",* and "elves" are optical phenomena in the upper atmosphere that have recently been detected using low light-level television technology and that are associated with thunderstorms. However, anecdotal reports of optical emissions above thunderstorms go back more than 100 years (Lyons 1994) and such luminosities have sometimes been reported by pilots (Vaughan and Vonnegut 1989).

Sprites are extensive but faint luminous flashes, predominantly red in color, that appear directly above an active thunderstorm at the same time as cloud-to-ground or intracloud lightning strokes. Sprites usually occur in clusters. There is strong evidence that sprites usually occur in the final stages of thunderstorms and coincide with large +CG lightning flashes. Most observed sprites are located well behind the associated storm fronts, in regions dominated by positive cloud-to-ground lightning. Sprites extend upward from the cloud tops to altitudes of about 95 km; beneath the red region, blue, threadlike filaments often extend downward to about 40 km (Sentman et al. 1995, Sentman and Wescott 1996).

Sprites are infrequently observed with the naked eye because they only occur above active thunderstorm systems, and since they are of low intensity, they can only been seen with the dark-adapted eye. Furthermore, high-speed photometer readings show that they have a duration of only about 3–10 ms and occur randomly with only about 1% of lightning strokes. Their brightness is comparable to that of the aurora borealis, and so they are often obscured by the greater illumination from the lightning with which they are associated. The optical energy is roughly 10–50 kJ per event, with a corresponding optical power of 5–25 MW. The optical energy is assumed to be 0.1% of the total energy of the event.

The first images of a sprite were obtained by chance in 1989. The first color pictures of sprites were obtained in 1994 by a team from the University of Alaska at Fairbanks. Since their first discovery, video sequences and images of well over a thousand events have been obtained from the ground, from aircraft, and from the Space Shuttle (Hampton et al. 1996, Sentman et al. 1995, http://elf.gi.alaska.edu).

Various theories have been proposed to account for red sprites. These include (1) fluorescence or luminescence of middle or upper atmospheric gases, (2) quasi-static dielectric breakdown between electrified clouds and the high atmosphere, (3) quasi-static heating and impact ionization of the middle atmosphere, (4) runaway breakdown triggered by cosmic rays, and (5) radiofrequency breakdown caused by the electromagnetic pulse from lightning (Sentman and Wescott 1996).

Many images have also been obtained from aircraft of blue jets, another formerly unrecorded form of high-altitude optical activity above thunderstorms (Wescott et al. 1995). Blue jets, which are a phenomenon distinct from sprites, are weakly luminous discharges, deep blue in color, that seem to emanate directly from the tops of thunderclouds and shoot upward in narrow cones, about 15° wide, traveling through the stratosphere at a speed of about 100 km s^{-1}. When they reach heights of 40 to 50 km, they fan out and disappear. They have an estimated optical energy of about 4 kJ and a total energy of about 30 MJ. They are not aligned with the local magnetic field. As yet, they remain unexplained, although, like sprites, they are associated with positive lightning.

A group at Stanford University detected a third form of stratospheric lightning in December 1994. These were named "elves" in the same vein as the name "sprite." These are halos of red light that are formed above a CG flash at altitudes of about 90 km and spread into rings up to 300 km across in less than 1 ms. Taranenko and Roussel-Dupré (1995) of Los Alamos National Laboratory, Los Alamos, New Mexico, have developed a mathematical model to explain this. In this model, the electromagnetic pulse produced by lightning expands upward and away from the CG flash. A sufficiently intense pulse will provide energy to the ions and free electrons at the stratospheric–ionospheric border sufficiently to make nitrogen molecules emit red light. Although the electromagnetic pulse expands at the speed of light, the ring of shining particles formed at the lower edge of the ionosphere grows faster. The positive flashes with the largest peak currents (measured by the National Lightning Detection Network), hence expected to be optically the brightest, are associated with elves rather than with sprites (E. Williams, personal communication, 1998).

Two other types of unexpected emissions that appear to originate in thunderstorms and which may be related to sprites, jets, and elves have recently been observed from space. First, the Compton Gamma Ray Observatory has detected millisecond gamma ray bursts of terrestrial origin, with energies in excess of 1 MeV, over thunderstorm regions. They are believed to come from sources at altitudes greater than 30 km. Second, the ALEXIS satellite has observed exceptionally intense pairs of very high frequency (VHF) pulses, called *transionospheric pulse pairs* (TIPPS), which originate from thunderstorm regions, but which are some 10,000 times more intense than the atmospherics produced by normal lightning activity (Massey and Holden 1995, Jacobsen et al. 1998). They appear to be related to very intense VHF pulses detected at ground level (Willett, Bailey, and Krider 1989, Jacobsen et al. 1998). These phenomena are discussed again in Chapter 12.

All these phenomena together suggest that thunderstorms exert a much greater influence on the middle and upper atmospheres than was previously anticipated. Current knowledge suggests that sprites, jets, and elves are a normal feature of every thunderstorm system of average size or greater, and may be essential components of the Earth's global electrical circuit.

Chapter 3

Phenomena that May Be Mistaken for Ball Lightning

3.1 Physical Phenomena

Figure 3.1 is a flow chart showing the process of classifying ball lightning reports. Table 3.1 lists some of the physical phenomena that have been or may be incorrectly described as ball lightning and indicates the diversity of phenomena responsible for reports. Scientists studying reports of unidentified flying objects (UFOs) encountered a similar situation, and much can be learned from their investigations (Condon 1969, Sagan and Page 1972). Readers will note that misidentification is far less likely in the case of ball lightning that is reported to have been seen in an enclosed space such as a room or an aircraft.

3.1.1 Effects of Conventional Linear Lightning

Many of the more obvious effects of lightning were mentioned in Chapter 2. Some more specific effects that may have relevance to ball lightning reports are discussed here.

Humphreys (1936) makes reference to ball lightning reports being explained by "flash at discharge point." He writes, "Many accounts of ball lightning describe a brilliant ball seen on the spot struck at the instant of discharge. This seeming greater flare presumably is real and caused partly by the supply of material by the solid struck and, in many cases, partly by the rich visible spectrum of this material in comparison with that of the air."

The heating effect when lightning strikes open ground is discussed in a quantitative way in Chapter 4. This localized thermal effect may also cause heating or combustion of solid materials on the ground or gases nearby, or could initiate

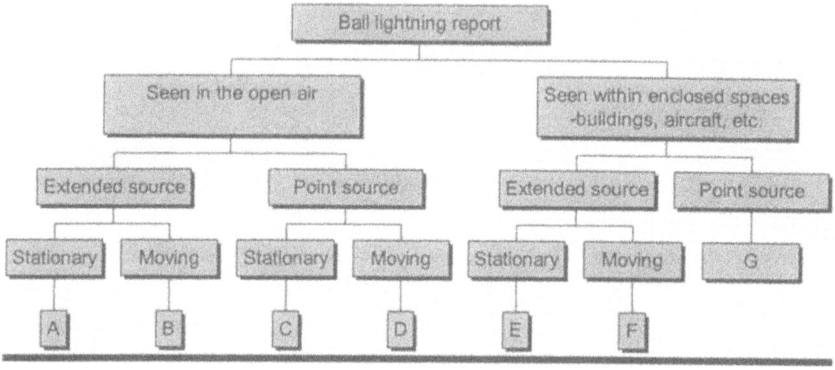

Figure 3.1. Flow chart showing the process of classifying ball lightning reports.

chemical reactions or combustion of any methane present in the atmosphere near, for example, a swamp (see Section 3.1.3). Tentative experimental support for this may be found in the triggered lightning experiments reported by Fieux, Gary, and Hubert (1975), who observed that stationary, roughly spherical luminous regions with a diameter of about 25 cm formed near the ground. They suggested that these might be caused by an outgassing phenomenon involving hot gases, a localized electrical discharge mechanism, or combustion of hydrogen or methane.

The duration of the luminosity of ordinary lightning is significantly less than that reported for ball lightning, so that the further hypothesis of a positive afterimage must be invoked to explain the apparently extended lifetime in the scenarios discussed here.

The step potential is discussed in Chapters 2 and 4. If the electric field due to the step potential exceeds about 100 to 200 kV m^{-1}, the soil will break down. Electric arcs may then be observed on the soil. A related model by Lowke (1996) is discussed in Chapter 13.

It may be that the observer happens to be in the direct line of the lightning channel, in which case it may be perceived as a circular or spherical region, and an afterimage may result (see Section 3.3.5). A localized electric arc may be created by a side-flash, for example, between a branch of a tree and another object, or between two separated conductors within a building.

A side flash may allow current to penetrate buildings via conductors such as the mains cable or water supply pipes. This may cause an arc discharge indoors. An arc may similarly be formed outdoors on the gap in a broken conductor. Such an arc might then generate a positive afterimage (Humphreys 1936).

Table 3.1. Ball Lightning Evaluation Chart

List A: Open air/extended source/stationary	List B: Open air/extended source/moving
St. Elmo's fire	Items in list A + autokinetic effects
End-on lightning/positive afterimage	St. Elmo's fire—must remain in contact with a conductor
Enhanced lightning channel/positive afterimage	End-on lightning/positive afterimage
Side flash/positive afterimage	Enhanced lightning channel/positive afterimage
Side flash/combustion	Side flash/positive afterimage
Sun/moon	Fireball/bolide
Ignis fatuus	Pyrotechnics
Streetlights	Ignis fatuus—small amount of movement only
Insect swarms	Streetlights—apparent motion on photographic film if camera moves
Unusual meteorological phenomena (parhelia, etc.)	Insect swarms
Ball lightning?	Ball lightning?
List C: Open air/point source/stationary	**List D: Open air/point source/moving**
Planets and stars	Items in list C + autokinetic effects
Comets	Meteors
Balloons	Orbiting artificial satellites
Ball lightning?—usually reported as an extended source. According to descriptions, probably too low in intensity to be observed as a point source at great distance.	Satellite reentries
	Balloons
	Ball lightning?—usually reported as an extended source. According to descriptions, probably too low in intensity to be observed as a point source at great distance.
List E: Enclosed space/extended source/stationary	**List F: Enclosed space/extended source/moving**
St. Elmo's fire	Items in list E + autokinetic effects
Side flash/positive afterimage	St. Elmo's fire—must remain in contact with a conductor
Side flash/combustion	Side flash/positive afterimage
Ball lightning?	Ball lightning?
List G: Enclosed space/point source	
Nil	

3.1.2 Corona Discharge (St. Elmo's Fire)

St. Elmo's fire is the visible manifestation of corona discharge or corona point discharge, and has also been called *brush discharge* or *glow discharge*. It is often blue or bluish white in color, roughly spherical, with a typical diameter of 10 cm, although significantly larger diameters have been reported. Sometimes hissing or buzzing sounds have been described. Its lifetime may be several minutes because it derives its energy from the atmospheric electric field. It occurs when the atmospheric electric field is intensified in a thunderstorm. It is formed by corona discharge from a grounded object with a small radius of curvature. Examples include the masts of ships, points near electric pylons or metallic cables, or, if wet, near wooden posts or lengths of cord or rope. It has been seen around the heads and hands of human beings. It may also be observed at the extremities of aircraft in flight since as the intensity of corona discharge increases with wind speed. A similar, luminous phenomenon may sometimes be seen along the length of high-tension power lines.

Unlike ball lightning, St. Elmo's fire cannot move independently through the air. If it moves, it remains in contact with a conductor throughout its lifetime. Currents of air or variations in electric field pattern may move it. Hubert (1996) equates St. Elmo's fire with the steady glow region seen in electrical discharge experiments. Corona discharge often precedes a lightning strike to the region in which it is seen. St. Elmo's fire is one of the most common causes of ball lightning reports, and consequently it is important that eyewitness descriptions be studied carefully with this in mind.

Hubert (1996) has indicated that coronal discharge may be sensitive to gas composition and to traces of impurities, and consequently a wide range of colors and other departures from classical behavior, such as formation at lower field intensities, may be reported. He suggests that the appearance of a persistent, luminous ball of any color in contact with a conducting object, even if not sharply pointed, may be a form of coronal effect in the presence of impurities, probably of organic origin.

In Rayle's (1966) survey, of 103 reports, 16% said that ball lightning was formed in contact with a metal surface, 14% in contact with the ground, and 12% in contact with a nonmetallic surface. Of 88 reports, Rayle found that 18% said the ground surface, 14% power or telephone wires, and 7% other metal structures (total 39%) guided the ball. Hubert (1996) found that 39 out of 253 reports bore a significant similarity to St. Elmo's fire. In another survey, it was found that up to 25% of 111 ball lightning reports could thus be explained (Stenhoff 1988b). The presence of an overvoltage on telephone and power lines or electrostatic induction may allow a corona to be observed inside buildings (Hubert 1996), even on conductors that are electrically insulated from their surroundings. Hubert has also suggested that continuing currents in the ground or in components of a building may produce secondary discharges, but presumably the duration of such continuing currents would necessarily be much greater than those measured in experimental studies.

A report that may be an example of St. Elmo's fire seen indoors was received from a research scientist. During a summer thunderstorm while there was a light rainfall, he saw a

> bright luminous [opaque] ball, perhaps about 15 cm in diameter about 2.5 m away and close to or directly beneath a steel beam that is one of the structural members supporting the overlying floors. The circumference of the ball was irregular and seemingly made up of individual sparks which came and went relatively rapidly compared to the duration of the ball itself. This duration was about 2–3 seconds, during which time the ball was either stationary or moved only slightly. At the end of that time the ball disappeared completely and instantaneously. . . . [My wife, who was in another room] did not see the ball but had been alarmed by a loud noise (she thinks perhaps a thunderclap) and then a "sizzling" or "crackling" noise like a wood fire burning or bacon frying [which] moved downward along the outside wall. She believes as well that she saw a flash outside the house as the sizzling/crackling sound was going on. . . . We searched the house both inside and outside for any signs of damage after the incident but found none. (Lennox, D. H., personal communication, 1983).

3.1.3 Ignis fatuus (will-o'-the-wisp)

In marshes and swamps, methane (CH_4) and phosphine (PH_3) are produced by anaerobic decay of organic matter. In such an environment, luminous globes, often ellipsoidal in form, have occasionally been reported to float above the swamp (Minnaert 1954, 1993). They are occasionally seen over freshly dug soil. These are called *ignis fatuus, jack-o'-lantern, will-o'-the-wisp, specter light,* or *swamp or marsh gas*. They are sometimes likened to flames about 1 to 10 cm high and up to 5 cm broad. The boundary is ill defined. The color is usually blue, but occasionally it is said to be more yellow and luminous near the center. This phenomenon is more often reported in warm weather. Sometimes they are seen on the ground, and sometimes they float about 10 cm above it. They may be blown along by the wind before being extinguished. They have occasionally been observed to burn for several hours. There may be a sound like a "pop" on formation, and they may be associated with faint odors.

Ignis fatuus is not generally associated with any perceptible warmth. There are many suggested explanations of ignis fatuus. That favored in the review by Mills (1980) is the chemiluminescent oxidation of some volatile, unstable, organic compound formed by anaerobic fermentation and entrained in the methane. Barry's (1967, 1968) experiments on "cool-flame" combustion of hydrocarbon–air mixtures at concentrations below the combustion limit, which were intended to provide an experimental analog to ball lightning, duplicated some properties of ignis fatuus (see Chapter 14).

Ignis fatuus is itself not a fully accepted phenomenon. Humphreys (1936) gives another interpretation: He suggests that hunting owls covered with luminous decaying matter from their nests are seen as will-o'-the-wisps.

3.1.4 Other Meteorological Phenomena

It appears unlikely that ball lightning reports will result from misidentification of many other meteorological phenomena since most other luminous meteorological sources are quite subtle and difficult to observe.

Noctilucent clouds are unusual forms of cloud, probably caused by ice crystals and meteoric dust forming at about 82 km above the Earth's surface, and therefore higher than any other clouds. They are occasionally visible during midsummer nights at high altitudes, long after sunset, when they are illuminated by light from the sun, which is well below the observer's horizon (Minnaert 1993).

A large number of other, rare meteorological effects are described in detail by Minnaert (1993). These are beyond the scope of this book, and readers are encouraged to consult Minnaert's fascinating and comprehensive account.

3.1.5 Astronomical Phenomena

There is clearly no link between astronomical objects and thunderstorm conditions. However, adverse atmospheric conditions during thunderstorms may affect (or prevent) normal observations of astronomical bodies. Other atmospheric effects, such as mirages caused by temperature inversions, may distort optical images. These are discussed in the following paragraphs. A very useful reference on astronomical phenomena and their visual characteristics is Norton (1978).

Meteors, fireballs, and bolides: *Meteoroids* are small bodies that are mostly dust particles or debris from comets. These are sometimes captured by the Earth's gravity. All objects but the smallest dust particles entering the atmosphere at high speed will experience heating by friction. This heating produces light. *Meteors* (or "shooting stars") may range in appearance from faint objects observable only with telescopes and visible for less than 1 s to bright *fireballs* that last several seconds. If the object is large enough, it will produce an exceptionally bright fireball or *bolide* with an apparent diameter comparable to that of the sun or moon. The object may then not be completely vaporized before striking the ground. Any surviving rocks are called *meteorites.* Meteorites may produce craters. Some bolides explode with a loud detonation. Singer (1971) points out that the trajectory of meteors almost invariably appears to be straight, while ball lightning is often reported to move in a curved path.

The popular vocabulary associated with ordinary and ball lightning encourages confusion between ball lightning, fireballs, and meteorites. A flash of lightning, and sometimes ball lightning, is often called a *thunderbolt* (see glossary), which implies a solid falling object. Ball lightning is sometimes called a *fireball*. Photographs of some fireballs show similarity to many ball lightning photographs. Photographic evidence of ball lightning should be checked with this in mind.

The sun and moon: The sun and the moon each subtend approximately 9 mrad (0.5°) when observed from the Earth. If an observer incorrectly identifies either as a nearby object at distance r in meters, its assumed diameter in centimeters would be approximately given by $0.9r$. Reports of essentially stationary objects that subtend approximately the above angle should be carefully checked to establish whether they might be caused by misidentifications of sun or moon, perhaps seen through broken cloud.

Planets and stars: Only five planets—Mercury, Venus, Mars, Jupiter, and Saturn—are visible with the naked eye. These planets remain close to the ecliptic, and their positions in the sky can be checked using an astronomical ephemeris. Venus is the brightest planet at maximum, with a maximum magnitude of –4.4 and is often visible in daylight. After dark, it can cast shadows when at its brightest. The disc of Venus subtends a maximum angle of over 1′ of arc, but may appear larger through atmospheric distortion. When at its brightest, it is best seen by Southern Hemisphere observers in the morning. Venus has often been reported as an unidentified flying object (Condon 1969). Again, reports of essentially stationary objects should be checked to determine whether planets might be responsible. Eyewitness descriptions of altitude are notoriously unreliable, so the first check should be whether at the time of the reported event there was a bright planet visible.

Most *comets* are not visible with the naked eye. Very bright comets have been uncommon in the twentieth century. Comets usually appear as small, cloudy objects. They move very slowly against the background of stars. Although the classic description of ball lightning is superficially similar to that of a comet, the low brightness and very slow motion of the latter makes it an unlikely candidate for generating a ball lightning report.

Aurorae are luminous displays visible at night at high latitudes, occurring most frequently near the earth's geomagnetic poles. Auroral displays are most commonly visible with the naked eye in either hemisphere at latitudes between 60° and 70° magnetic. They are caused by the interaction of charged particles streaming from the sun, attracted to the auroral regions by the Earth's magnetic field, with the atoms (mainly atomic oxygen) and molecules in the upper atmosphere (above about 100 km). They have colors characteristic of atomic and molecular nitrogen and oxygen. They appear as colored arcs, rays, bands, streamers, and curtains, usually green or red. However, the glows produced by aurorae are not localized and therefore are unlikely ever to be misidentified as ball lightning.

3.1.6 Artificial Satellites and Satellite Reentries

Artificial satellites appear as starlike point sources of light moving slowly against the star background. Their brightness varies considerably, but is probably too low to be misidentified as ball lightning. However, when satellites reenter the earth's atmosphere, they produce very bright displays similar to fireballs (see earlier discussion). Atmospheric friction causes the satellite to become incandescent and it may break up into many fragments, each moving at a different speed because of variable drag. Satellite reentries may be confirmed with astronomical observatories.

3.1.7 Streetlights

Streetlights have been the cause of images photographed and recorded on videotape and misidentified as ball lightning (see Chapter 9). Fortunately, streetlights are supplied with mains electricity, and "still" images that result from a combination of streetlights and camera motion are likely to show fairly regular modulations in exposure along the length of the luminous trace. Ball lightning photographs should be treated with caution when the photographer claims to have seen nothing unusual at the time.

3.1.8 Insect Swarms

Callahan and Mankin (1978) have suggested that corona discharge may be observed from nocturnal swarms of insects moving through intense electric fields such as those in thunderstorms. This was confirmed by experimental studies. The corona discharge occurs as a result of their body structure. It appears that the insect body shell forms a dielectric conductor, with the electric field intensity enhanced at sharp points, while body fluids form an electrolyte. A corona is observed near the sharp points when the insect is placed in an electric field.

3.1.9 Pyrotechnics (Fireworks and Flares)

Reports that describe pyrotechnic appearance or behavior should be carefully studied with fireworks and military flares in mind. Chapter 1 indicated that the photographs by Jensen (1933) were thought by some to have been a result of a prank using fireworks.

3.1.10 Blimps and Weather Balloons

Slowly moving spherical sources of luminosity may be weather balloons or other dirigibles. The sun below the observer's horizon may illuminate those at high altitude. The article by V. E. Lally (1969) is a useful source for detailed further reference.

3.1.11 Miscellaneous Phenomena

Airplane landing lights: In addition to the familiar landing lights used by commercial aircraft, unusual navigation lights may be found on military aircraft. Some aircraft, such as helicopters and Harrier jump jets, may hover for prolonged periods of time. Helicopters may carry searchlights.

Other possible sources of misidentification include vehicle headlights, falling molten metal (perhaps from a metal object that has been struck by lightning, or from an electric arc in power cables that has been initiated by a lightning strike), soap bubbles (which may reflect light), and airborne debris (which some have suggested may produce corona discharge in thunderstorm conditions).

3.2 Physical Effects that Cause Distortion

3.2.1 Atmospheric Refraction

All distant, luminous sources may undergo significant optical distortion or displacement owing to the variable refractive index of the atmosphere. There is an excellent review of this subject by Viezee (1969). There are also several references to optical mirages in Minnaert (1993).

3.2.2 Miscellaneous Other Optical Effects

Humphreys (1936) suggests that some people have been startled by the reflection of a brilliant streak of lightning off a polished convex metal surface, e.g., that of a kitchen pan or doorknob.

Many ball lightning reports describe observations made or photographs or videotapes taken from indoors through window glass. It is likely during a storm that the glass will be wet. Water droplets are roughly hemispherical and act as a planoconvex lens. They are capable of producing unusual, circular images. These may be responsible for unusual images found on photographic film or videotape.

3.3 Psychological Aspects of Reports

3.3.1 Perception and the Limitations of Eyewitnesses

Since there are so few photographs alleged to be of ball lightning, and since ball lightning is rarely reported to have been detected by instruments, much of the information we have about the phenomenon results directly from the descriptions of eyewitnesses, which we call ball lightning reports. In this section, we consider only ball lightning reports that result from *distal physical events*, so we are not concerned here with hoaxes or hallucinations. The distal physical event concerned may be a ball lightning event (if these exist), or may be another physical phenomenon, as discussed earlier. Linking the event to the report involves a series of physiological and psychological processes, all of which are subject to distortions. The final report may thus bear little resemblance to the distal physical event.

Ninety-four percent of ball lightning reports describe thunderstorm conditions, and 64% describe the formation of ball lightning immediately after a cloud-to-ground lightning flash (Rayle 1966). These are very special and unusual conditions when one considers the role of perception in determining the accuracy of the reports. Many people are extremely anxious during thunderstorms, and the powerful visual and auditory stimuli during a storm may also have a significant effect on the accuracy of perception.

3.3.2 Perception as the Link between Distal Event and Report

Wertheimer (1969) and Hartmann (1969b) describe the sequence of events linking distal events to reports. These may be summarized thus:

1. The sequence begins with a *distal physical event*—an energy change or source some distance from the observer.
2. Energy is transmitted to the sense organs of the observer; the energy received is called the *proximal stimulus*. The energy leaving the source is subject to attenuation and distortion during transmission. The degree of this is determined by the medium through which transmission occurs. Thus the proximal stimulus is neither a faithful copy of the distal physical event nor may it be complete. The proximal stimulus may be insufficiently informative about important properties such as size, distance, or speed.
3. The proximal stimulus is encoded in *neural events*, which produce *sensations*. The conversion of *proximal stimulus* to *sensations* is also subject to distortions. These occur in the cerebral cortex and are determined by fatigue or the degree of alertness of the observer. Sensation can be modified by anomalies such as colorblindness. There are various kinds of *afterimages* (see later discussion) that persist after cessation of the original stimulus.
4. These sensations are combined into *percepts* and finally into *cognition*. *Perception* is the process by which the original stimulus is recognized. When the perceptual framework is impoverished, the interpretation necessary to reach a perception is much more difficult. However, perception can still operate in conditions of incomplete information. In the case of unfamiliar events, however, perception tends to "fill in" the missing information, perhaps incorrectly. *Cognition*—the judgment or conviction about the actual identity of the stimulus—depends on previous experience,

opinions, and beliefs, and this influence is particularly strong when perception is impoverished.

5. The production of the *report* itself is a source of further distortion. Responses may depend heavily on the way questions are asked and on the level of fluency or literacy of the observer. Like cognition, *memory* of the percept is subject to the distorting effect of mental set, expectation, and suggestion. The question of whether an event will lead to the production of a report depends on the motivation and attitude of the witness.[1]

3.3.3 Events Observed during Thunderstorms

The physiological consequences of a nearby lightning strike include a number that may have a direct effect on the human perceptual system. These include loss of consciousness, loss of memory, afterimages, and temporary or permanent blindness or deafness.

Rayle (1966) collected some interesting statistics about thunderstorms. The target for his survey consisted of 1764 NASA employees in the United States, and thus was an unusually well qualified, and, one would assume, naturally curious sample. Forty-two percent estimated that they had only experienced zero or one thunderstorm per year while outdoors. Twenty-three percent had seen the point of impact of ordinary lightning, 13% on more than one occasion. Ten percent of the sample reported that they had seen ball lightning, 4% on more than one occasion. Six percent said that they preferred not to watch lightning displays. We therefore conclude that most people experience storms infrequently, that close observations of lightning flashes are rare, and that many people are anxious during thunderstorms. These factors are all relevant to the accuracy of reports of unusual events during thunderstorms.

3.3.4 Perception and Memory of Unusual Events

There are two contrary lines of argument here. One view suggests that the unusual nature of the experience of a ball lightning observer impresses itself clearly on his or her memory. Certainly, the detailed and fairly consistent descriptions received of events lasting only a few seconds tend to support this. The other suggests that the lack of any conceptual framework by which to assess this unusual experience introduces a greater degree of distortion than would be expected in a description of a more commonplace but unexpected event, such as a road accident or a robbery. Many who report ball lightning say they had no prior knowledge of the phenomenon. It is well known that even descriptions of familiar events can be very

[1]Hartmann (1969b) estimated that just over 1 man/year was required to generate one report of an event lasting 2 min, and visible over a region inhabited by about 23 million people. This was the spectacular reentry of the Zond IV satellite.

unreliable. Hubert (1996) has suggested that the unusual nature of a ball lightning event could produce a psychological "time dilation" effect that leads to exaggerated estimates of duration.

Two anecdotes are useful here. The first concerns the experience of astronomers from the National Radio Astronomy Observatory at Socorro, New Mexico, who accumulated eyewitness descriptions of two very bright meteors with the aim of collecting fragments. The astronomers had the advantage of prior knowledge of the observational properties of the phenomenon being reported. They discovered that memory faded very quickly after such an unusual event. One day after the event, about half the reports already contained substantial errors, and this fraction increased to about 75% after 3 days and 90% after 4. After 5 days, people reported more imagination than truth, and were evidently reconstructing in their imagination an event based on a vague recollection of what had happened.

Although estimates of the duration of the event were mostly very accurate, perception of the color of the meteors was very unreliable, with almost all colors being reported. Speeds were generally grossly overestimated, and altazimuth positions were very poorly judged. As often happens in reports of meteors, many witnesses claimed to have heard a loud sizzling noise as the meteors passed over, but there is no known physical reason why this would be possible (Drake 1972).

The second concerned the reentry of the Zond-IV satellite in 1968. Many witnesses to this event interpreted and reported their observations in terms of UFOs, and there were descriptions of cigar, rocket, or saucer shapes "flying in formation," with windows, exhausts, noise, vertical descent, estimated altitude or distance below 20 miles, and reactions of fear exhibited by animals (Hartmann 1969b).

In support of reports of ball lightning, however, there appears to be a significant contrast between ball lightning and UFOs in the degree of consistency of the reports. Whereas UFO reports appear to describe heterogeneous properties (Condon 1969), there is a much greater level of consistency among reports of ball lightning (Rayle 1966), at least for the most basic characteristics such as duration, motion, and size.

3.3.5 Afterimages and Persistence of Vision

An observer who is directly in the path of a lightning flash may see the flash end-on. Alternatively, the observer may see the point of contact of a cloud-to-ground lightning flash, or an arc resulting from a side flash. In each case, an afterimage, or, more correctly, an aftersensation may result. *Positive* afterimages are those where the sensations are the same as those in the inducing stimulus, whereas they are reversed in the more commonplace *negative* afterimage. A positive afterimage may be invoked to explain the perception of a region of an apparent luminosity similar to ball lightning. These result from exposure of the retina to a bright source of illumination that is intense relative to the background. This kind of afterimage is formed on the cone portion of the eye that is close to the center of the retina. The image subtends the same size at the retina as the original stimulus. In attempting to center the image, the observer will seem to see it drift across his visual field, an effect called *autokinesis*. This movement may be smooth or erratic, depending on eye movement. These images persist for 2 to 60 s and can be caused to disappear by sudden eye movement. The image fades with time.

The apparent size of the afterimage will increase in direct proportion to the distance of the surface onto which it is projected. (This is known as Emmert's law.) The color may change as the image fades, and can range through the entire spectrum, typically alternating between the color of the original stimulus and its complement (Barry 1980, Wertheimer 1969). In 60 of 106 reports of ball lightning, eyewitnesses reported that they did *not* see the ball originate (Rayle 1966). In these circumstances, the positive afterimage is not feasible. Furthermore, the appearance of ball lightning, including its luminosity, is generally reported to remain constant. Humphreys (1936) suggests that imagination may be coupled with the afterimage to produce other, false sensations such as heat or odor.

Charman (1979) makes the interesting observation that persistence of vision prevents pulsations on a time scale below about 0.04 s to go unnoticed. Together with the phi-phenomenon, which is the impression of movement given by presenting two objects in quick succession in two adjacent but different positions, the sequential occurrence of a series of spatially separated, short-lived spheres could generate the impression of the continuous motion of a single, long-lived ball. This possibility had previously been tentatively implied by Humphreys (1936).

The following may be an example of a positive afterimage, perhaps with combustion of substances on the ground, following an enhanced lightning channel. A physicist was attending a conference at Boulder, Colorado, in June 1969. He was at his hotel at about noon. Just before the event there was rainfall of moderate intensity.

> The Rocky Mountains rise abruptly from the plane to the west boundary of Boulder. A sudden storm came over from the mountains and I happened to be watching a boulder-strewn grassy area leading to the foothills when a blue-white flash of lightning came down to hit this area. It left behind a ball of the same color, stationary and apparently just above the grass. It was very intense, rather like a huge (very high-pressure) mercury vapor lamp. It extinguished almost as abruptly as such a lamp being switched off, leaving only a small whisp of smoke from the ground. (Froome, K. D., personal communication, 1982)

The estimated distance from the observer to the ball was about 1/2 mi (1 km), and the ball was in sight for 5–10 s.

Some scientists have interpreted their own ball lightning experiences as the result of optical illusion. Dr. R. D. Smith, who observed possible ball lightning at about 3 p.m. one afternoon in 1956 or 1957, adopted this hypothesis, but later rejected it. He wrote

> I was standing at the window of my office in UKAEA [United Kingdom Atomic Energy Agency] Harwell looking south during a summer thunderstorm. There was a nearby lightning strike, probably on the BEPO reactor chimney. A [yellow] ball then traveled from east to west at about the level of my first floor window between me and the adjacent buildings and disappeared from my view. I estimate the speed at about 60 mph [30 m s^{-1}], but since it was not possible to gauge exact distances and since the phenomenon was so unexpected, it can only be regarded as a very approximate estimate. At the time, I didn't know about ball lightning. I decided it was probably an optical effect in the eye rather than real. However, I now believe it was ball lightning. (R. D. Smith, personal communication, 1983)

The ball was visible for 2–3 s.

3.3.6 Magnetic Phosphenes

The main biological effect associated with alternating and interrupted magnetic fields is that of *magnetic phosphenes*. A phosphene is defined as a sensation of light produced by physical stimuli other than visible light. Magnetic phosphenes are reported to be produced by the making and breaking of the contacts of a large, solenoid-type coil, or by the application of alternating fields of frequency 10 to 100 Hz, to the temporal areas of the human head. The intensity is greatest (for any given field strength) between 20 and 30 Hz. Magnetic phosphenes appear as colorless or occasionally light-blue-tinted shimmering luminosities in the peripheries of the visual field (Becker 1963). Although a lightning discharge produces a magnetic field and there is a high value of di/dt (i.e., the current rapidly changes), magnetic phosphenes invariably appear near the borders of the visual field and it seems unlikely that they would be mistaken for ball lightning. The effect on the retina of electric currents may also produce phosphenes (Charman 1979).

However, one observer of ball lightning, himself a university professor in a science department and a member of the Brain Research Association (Great Britain), used his own observation to support a similar interpretation:

> The event occurred when a house party was in progress and about forty people were present in the house. Those of us in the kitchen saw the ball in the middle of a group of several people, roughly in the direction we were looking when the lightning struck. The people in the lounge similarly saw the ball near the fireplace. The girl in the bath saw it towards the

window. I feel sure it could not wander all round the house, and no one has ever seen two at once to my knowledge.... As a physicist, I interpreted the phenomenon of ball lightning as the effect of a *very* strong electromagnetic pulse on the brain. (Swithenbank, J., personal communication, 1977)

3.3.7 Miscellaneous Perceptual Phenomena

Estimates of time: Although it was noted earlier that eyewitness estimates of the time taken for bright meteors to traverse the sky were very accurate, the perception of time may be strongly influenced by fear or fright. Witnesses of road accidents may experience the sensation that time has slowed down just before impact. Since many people are anxious during thunderstorms, it may be that estimates of the duration of ball lightning are exaggerated.

Size: The apparent size of an unfamiliar object is estimated by the angle it subtends at the retina, which is, of course, a function of distance as well as its actual physical dimensions. Estimates are thus far more accurate when lower and upper limits can be placed on distance; for example, if the stimulus is seen to pass in front of and behind objects whose distance is known (Rock 1975, Vernon 1971). Clearly, estimates of size and other quantities derived from it, such as distance and speed, will be far more reliable in a well-lit environment with foreground and background. This is one reason why reports of ball lightning seen within enclosures such as buildings and aircraft are so much more reliable.

Distance: There are various perceptual clues that are used to estimate distance. These include binocular vision (the disparity between the images formed by the two eyes), the degree of convergence of the eyes in viewing the stimulus, and the degree of accommodation of the lens of the eye. Where there is motion of the stimulus relative to the observers, parallax also offers a clue to distance (Rock 1975, Vernon 1971).

Motion: Objective perception of the movement of a stimulus is known as *real movement perception*. There are conditions where illusions of movement may be perceived, for example, the phenomenon of autokinesis. A single, stationary point of light seen otherwise in darkness may be seen to move, and this is thought to occur because of the absence of a point of reference against which to judge the position of the light. Naked-eye observers often see autokinetic movement of astronomical objects. *Autostasis* is the term that has been coined to describe the related illusion in which a moving object may be perceived as stationary. Errors in estimation of size, distance, and time will cause uncertainty in estimated speed (Rock 1975, Vernon 1971).

The perception and memory of colors also has inherent uncertainties. Memory of color is often more unreliable than recollection of shape. Defects of color vision are commonplace—about 10% of people, about 20 times as many men as women, are colorblind. We noted in Chapter 1 the poor correlation among various surveys concerning the reported color of ball lightning.

3.3.8 Hoaxes and Hallucinations

Most of those who report ball lightning show no indication of intending to gain publicity from it. There is the risk of a reporter of ball lightning occasionally being the victim of a hoax. Occasionally, one hears from those who report ball lightning alongside a range of so-called "paranormal" experiences, and in these circumstances perhaps suspicion should be aroused. Walker (1968) provides useful advice for these circumstances.

3.4 Limitations of Reports based on Visual Observation Alone

The foregoing discussion shows that there are many phenomena that are likely to be mistaken for ball lightning, such as the effects of conventional linear lightning and St. Elmo's fire, and many phenomena that will be an occasional source of confusion. The distorting effects of perception must also be taken into account. A particular problem is the degradation of memory with time. To improve matters, it is essential that ball lightning reports be obtained and followed up as quickly as possible after an alleged event.

There are, however, significant levels of consistency in many reports of ball lightning and this homogeneity, when compared with the descriptions of many other unusual phenomena, may indicate that there is a distinctive and interesting phenomenon or phenomena at the root of some of these reports.

3.5 The Importance of Physical Evidence

Even the most detailed and best-authenticated report of a visual observation of ball lightning can yield only limited quantitative data. Physics can only progress so far with limited qualitative information. Many mundane explanations for ball lightning reports could be immediately eliminated by the presence of convincing physical evidence. It is therefore essential that high priority be given to the investigation of ball lightning reports that include traces or damage. This is the subject of Chapters 5 to 9.

Chapter 4

Assessment of Electrical, Thermal, and Mechanical Risks

The dramatic nature of many ball lightning reports suggests at face value that like conventional linear lightning, it is potentially a very serious hazard that can cause injury, death, or damage on a scale comparable with that of conventional linear lightning.

The chapters that follow assess the possible risks associated with ball lightning. The approach in each is to present representative case history materials without much comment, and then to evaluate cases individually and collectively with the intention of making generalizations applicable to many future reports. In accordance with the principle of Ockham's razor, wherever possible traces and damage are explained in terms of well-understood phenomena, particularly by the effects of conventional linear lightning. In many cases, eyewitness reports are quoted verbatim to attempt to give the reader some idea of the raw material with which an investigator of ball lightning works, and to avoid the "sanitizing" effect of rewriting accounts for a scientific readership. Reports cannot always be conveniently categorized, so some are relevant to more than one chapter. It is therefore essential that individual chapters not be read in isolation.

It is useful to consider some general effects of lightning, also attributed to ball lightning, which will have relevance in many situations. Ordinary lightning can cause electrical, thermal, and mechanical damage; similar damage has also been attributed to ball lightning. Chapter 5 reviews some of the effects of a direct or indirect lightning strike to the human body or to animals. Among the injuries that have been attributed to ball lightning are electric shocks and burns, which have ranged from superficial to fatal. The chapters that follow also consider evidence that ball lightning can also damage buildings, aircraft, and trees. Some damage to aircraft is evidently thermal in origin, and this is discussed in Chapter 7. Some of the case histories that follow refer to thermal, electrical, or mechanical damage to buildings or trees; this is discussed in more detail in Chapters 6 and 8.

4.1 Electrical Effects

4.1.1 Electrical Effects Attributed to Ball Lightning

With the increased use of electrical and electronic devices through the twentieth century, electrical damage by ordinary lightning has become an increasing problem. Electrical damage has been recorded in ball lightning reports throughout the period in which electrical power has been in use.

Mrs. G. Pumfrey (personal communication, 1984) recalled an experience some 70 years previously. As a child of six, she was playing in the nursery of her parents' home at Gainsborough, Lincolnshire, England. Ball lightning appeared to come through an open window and traveled past her. The ball was translucent, about the size of a soccer ball, and bright enough to be clearly visible in daylight. It seemed to be spinning or rotating. She was blinded by its intensity, which increased as she watched it, so she ran through an open door into the hall, where she found her mother collapsed on a sofa with shock. A domestic who was cleaning the front entrance steps said she had seen the ball pass over her hair; she was also momentarily dazzled, but otherwise no one was affected. However, the fuse for the electric lighting circuit in the house was melted. Mrs. Pumfrey was not certain that this event was related to a thunderstorm or lightning.

Sometimes ball lightning is blamed for quite extensive damage. This is a report of an event at Yarralumla, Australia (date unknown):

> During a severe thunderstorm several years ago, ball lightning hit my backyard and subsequently my neighbor's. While the only evidence in my yard was the death of a 100 ft. [30 m] tall gum tree [which died within 48 hours], my neighbor's yard and house were substantially damaged. The ball lightning bounced over my boundary fence and landed on a steel reinforced concrete post, destroying the post and buckling the steel reinforcement. It then jumped onto my neighbor's verandah, where it fused the wrought iron tables and chairs into a molten mass, and entered the house through closed glass doors. The glass did not break, but the wooden door runners split open. The ball then hit the colored television, rendering the picture half [monochrome] and half color . . . The ball left through the wall of the house, leaving a one-foot diameter hole framed with shredded wallpaper. It then bounced across the road into an adjacent property, where

> I assume it "died" as no further damage was reported. (A. McEwin, personal communication, 1983)

There were overhead electricity wires near the tree.

At about 5 a.m. one morning in 1972, during a thunderstorm with heavy rain at Ivybridge in Devon, England, a woman saw a ball the size of a soccer ball, float past her window. There was then a tremendous explosion and "feeling of pressure" which apparently distressed a dog. From the aftermath of the explosion, the witness deduced that

> the ball [had] touched the ground by a wire and concrete post fence. Pieces of concrete post went through roofs the other side of our house, some 60 yards [55 m] away. The wire disappeared, then the discharged electricity, after traveling down the wire, leapt to our metal framed garage door, from there through a car (full of petrol) from by the petrol cap to the near-side bumper; it then leapt to the deep freeze (a write-off) and from there into the house electricity supply. It completely destroyed the fuse boxes and blew off the meter cupboard door which was solid sapele [a species of mahogany]. All the underground telephone cables to about 15 houses were welded into a solid mass outside our house and our telephone box in the house was on fire. The six houses in the cul-de-sac lost many panes of glass and the force of the explosion moved our 14 ft. [4.3 m] bay window in the sitting room about 2 in. [1 cm]. . . .
>
> Someone living at Catisfield, the other side of the valley, said, "It's a pity you couldn't see it—your house had an orange corona around it for several seconds after the bang." . . . The extraordinary thing *was* that the two children aged 5 and 3 slept through the whole [event] and only woke when carted off by a helpful neighbor. However, every drawing they did after that for about six months had a large black or purple mass in one corner, which was called "it." (A. Estyn-Jones, personal communication, 1976)

In summer 1975, one afternoon at about 5 p.m., a yellow "fireball" about the size of a football was reported to have exploded at Iffley Lock on the Thames near Oxford (G. W. Scott Blair, personal communication, 1976). The ball appeared while a thunderstorm was in progress some miles north. There was no thunder or lightning in Iffley, but the clouds were very black. The ball was round, with no protrusions. It was in sight for several seconds as it fell vertically toward the lock, where it hit a tall post carrying telephone wires and exploded loudly "like a bomb." There was a smell, probably of burned insulation. The underground and overhead electrical cables to the lock were put out of action. Telephone fuses in several neighboring houses were melted.

Anderson and Freier (1972) reported an incident dated September 1, 1971, in Minneapolis, Minnesota. A ball was not observed directly, but several neighbors described a red-orange glow seen near an oak tree, lasting for 2 or 3 s, following thunder that almost coincided with a lightning flash. The tree was the tallest object in the area. There was some indication that leaves were singed. There was a path of burned grass across the lawn. The soil had been moist, and there was rainfall at the time of the event. The trail of burned grass meandered some 0.46 m from the tree and split into two. The main branch of the trail, which was crooked and passed through flowerbeds, ended at the house, terminating at a metal downspout from the eaves. A ¼-in (64 mm) hole was burned in the base of a light bulb in the yard. Power in the house failed as the red glow ceased. The other, subsidiary branch ended at a picket fence. The interpretation of the authors was that the ball had fallen through the tree and burned a path in the grass, splitting in two partway along the trail.

On July 3, 1982, just before 5 p.m., two oak trees some 50 yd (46 m) from a house at Woodlands St. Mary, Berkshire, England were struck and severely damaged by ordinary lightning. Fragments of wood and bark, some as large as 4 lb (1.8 kg) were scattered over a wide area (see Chapter 8). The lightning current dug a trench 2 ft (60 cm) wide, 16 in. (41 cm) deep, and 11 ft (3.4 m) long, and apparently grounded by a steel crash barrier in the road. There was damage to a number of television sets, telephones, and other circuits.

Mr. Frank Bell was watching television in a room with closed windows when, just before the lightning flash, he saw a ball about 2 m away above the television (F. Bell, personal communication, 1983). He described it as "a glowing silvery-gold color, about 5 to 6 in. [13 to 15 cm] across, and rotating slowly." He said that "It flew around the room, and out through the open doors, right through the house into the bathroom, where there was the biggest bang of all. At the same time as this bang, the TV set exploded as I was watching it" (Pike 1982). He said the ball appeared to be rotating "like a ball of wool or string," was "as bright as an ordinary lightning stroke" and was uniformly bright all over. Its appearance did not change throughout the observation, which lasted about 1 to 2 s. No heat or odor were noted, but the ball made a "swishing sound." Mrs. Bell was in the kitchen when she heard the swishing sound, although she did not actually see the ball as it passed within about 6 ft (1.8 m) of her. "She came running into the sitting room in a state of shock" (F. Bell, personal communication, 1983). Pike's (1982) account is slightly at variance with the later communication from Bell since he quotes Mrs. Bell as saying she saw "this glowing thing go through."

The windowpane in the bathroom had a hole in its bottom, right-hand corner, and most of the fragments of glass were on the outside. The hole was 12.5 in. (32 cm) × 8.5 in. (22 cm). There was no evidence of burning, nor any sign of any tree fragments that might have broken the window. There was also a ¼-in. (0.6 cm)-wide crack along the full length of the double glazing in the sitting room, alongside the telephone cable, although the telephone still functioned normally. Other damage included an immersion heater, which was beyond repair, and a refrigerator whose fuse melted.

On July 12, 1868 at Guildford, England, a 3-in. (8 cm) ball lightning was considered to be the cause of the excavation of a trench 10 ft long (Corliss 1982, p. 60).

Two reports of ball lightning are related to the same thunderstorm in Staffordshire, England on March 21, 1983. One of these is discussed later. In the other, more than twelve houses in Knutton, Newcastle-under-Lyme, suffered power cuts following a sighting of an orange ball above the rooftops; this was said to have made a 2-ft (60 cm) hole in one roof and then bounced before disappearing explosively. Television sets and junction boxes in some houses were damaged (R. W. Maddison, personal communication, 1983).

4.1.2 Electrical Effects Due to Conventional Linear Lightning

Median lightning currents (crests, exceeding 2 kA) are 30 kA for negative first strokes and flashes, 12 kA for negative subsequent strokes, and 35 kA for positive flashes. The top 5% of lightning currents are 80 kA for negative first strokes and flashes, 30 kA for negative subsequent strokes, and 250 kA for positive flashes (Berger, Anderson, and Kröninger 1975, Golde 1977b). Some lightning damage is caused by direct strikes in which the entire lightning current is discharged through an object or structure. Sometimes a branch of the main lightning channel will strike an object or structure, with most of the current being discharged elsewhere. Sometimes the rapidly changing magnetic field due to a nearby CG flash or side flash (see following discussion) induces substantial currents in nearby conductors. Secondary effects may be thermal or mechanical, and these are discussed later.

An important mechanism by which lightning current can be transferred from one conductor to another apparently insulated from it, or by which lightning can pass through a building (see Chapter 6), is the side flash. One scenario in which this can apply is if a conductor in which the lightning current is traveling is inefficiently grounded. A path through the air or another insulator, if it breaks down, may offer less resistance than the path through the conductor, and it can traverse a gap. Thus current passes through the air or through another imperfect insulator such as soil (Uman 1986).

Such a path through soil can produce a trench in the ground, presumably by rapid vaporization and expansion of water in the soil. An event of this sort was described by Müller-Hillebrand (1957). A pine tree 70 m from a house was struck and suffered only superficial damage, but a side flash produced a trench 50 m long, leading to a metallic fence through which the current discharged to the electrical system of the house. Golde (1973) shows that this will occur when the product of the lightning current and the effective resistance of the root system of the tree exceeds the breakdown potential of the soil.

Golde (1973) also gives another scenario in which a side flash may be important. Chapter 5 gives several examples of the deaths of farm animals in sheds. If a shed has a corrugated iron roof and there is a return stroke nearby, the return stroke, roof, and ground act as a pair of series capacitors. If the return stroke is at a potential u_1, the capacitance between the return stroke and the roof is C_1, and the capacitance between the roof and ground is C_2, then the roof will be raised to a potential u_2, which is given by

$$u_2 = u_1 \frac{C_1}{C_1 + C_2}. \tag{4.1}$$

A person or animal inside the shed may be considered to be at ground potential. If the potential u_2 exceeds the breakdown potential of the air gap, a side flash may cause electrocution without the roof even being struck.

In Chapter 6, it is shown that even for buildings protected by lightning conductors, there is a danger of side flash if there is insufficient clearance between the lightning conductor and a grounded object in the vicinity.

The side flash has also been responsible for the deaths of people sheltering beneath trees. The resistance of a tree between head height and ground is a few thousand ohms. A person standing nearby is effectively at ground potential, so if the potential difference between the two is sufficient to cause breakdown, a side flash will arc to the head of the victim.

Uman (1986, p. 46) mentions that a lightning current can initiate an arc from the electrical mains, which provides an artificial continuing current whose duration far exceeds that of a lightning flash. This would happen if the mains cables are raised by a direct strike or a side flash to a sufficient potential to cause breakdown across an air gap. The arc would constitute a low-resistance plasma path that may allow fairly large currents of long duration to continue to flow between electrical wires or between a cable and a grounded object. The effective action integral $\int i^2 \mathrm{d}t$ of such a discharge may be sufficient to cause much thermal damage. Such an arc may cause progressive breakdown of insulators (dielectrics) so that a short circuit may propagate fairly slowly along a pair of conductors. A similar breakdown mechanism can cause a discharge between overhead power lines. Hermant and Hubert (1983) describe such an event, which is supported with a photograph (Fig. 4.1).

When lightning strikes open ground, the current is dissipated into the earth. If the ground is isotropic, the current will distribute evenly in all directions. The spreading resistance at the point of contact can be shown by integration to be given by $(2\pi\sigma r)^{-1}$, where σ is the conductivity of the soil and r is the radial distance from

Figure 4.1. Short-exposure photograph taken during the extinction phase of a lightning flash to a powerline tower at a distance of 180 m. The luminosity observed was interpreted as the consequence of an arc resulting from the overvoltage produced by the lightning stroke, short-circuiting the chain of insulators. The luminosity persisted for about 2 s (Hermant and Hubert 1988). [Photograph: A. Hermant. Reproduced by kind permission of the authors.]

the strike point. In practice, the situation is likely to be much more complex, as discussed by Lowke (1996) (see Chapter 12).

The potential difference between two points at distances d and $d + s$ from the strike point when the instantaneous current is i is given by

$$V = \frac{i}{2\pi\sigma} \times \frac{s}{d(d+s)}. \tag{4.2}$$

This potential difference V, known as the step voltage, may be responsible for driving injurious or fatal currents through the body (see Chapter 5), for heating the ground in the vicinity of a strike (see Chapter 3 and later discussion), or, if it exceeds the breakdown potential of the soil or of the air above the soil, for a side flash or flashover. When subjected to an impulsive voltage, the breakdown potential of soil is about 200 to 1000 kV m^{-1} (Saraoja 1977), while the electrical breakdown strength of an air gap several meters wide is comparable at about 500 kV m^{-1} (Golde 1973, Berger 1977). Thus it must generally be retrospectively a matter of conjecture whether flashover is more likely through the air, through the soil, or through both because much will depend on local conditions.

In the immediate vicinity of a strike, flashover through the soil or through the air above the soil would be expected once the current increases to a sufficient value. Given the preceding values, this would begin to occur when the current reaches about 50 A, and flashover would be expected to take place across a region approximately 1.4 m in radius once the current reaches i_{max}. The resulting arc within the air would be intensely luminous. Lane (1945) published a photograph that shows Lichtenberg-like traces (treelike images) on grass, emanating from a golf course flagstick that was struck by lightning presumably generated by such a mechanism.

Saraoja (1977) also provides resistivities of different kinds of soil. For example, sandy clay has a resistivity between 40 and 300 Ωm, which gives conductivity between 3.3 and 25 mS m^{-1}. If the maximum lightning current i_{max} is 100 kA and a person stands on sandy clay 10 m from the point of impact of lightning with his feet separated by a radial distance of 0.5 m, Eq. 4.2 shows that the maximum potential difference between the feet is approximately 5 kV. If the resistance between the feet is 1 kΩ, then the maximum current through the legs will be 5 A, which is potentially fatal.

4.1.3 Discussion of Case Histories

In many of these reports, there is only a vague description of a ball, or even a rumor that ball lightning had been seen, although all the damage is attributed to the ball rather than to ordinary lightning, notwithstanding the fact that a thunderstorm was in progress at the time. It has proved difficult to obtain clear, firsthand descriptions of a ball. Examples of this given earlier include the reports from Yarralumla, Ivybridge, Iffley, Minneapolis, and Knutton.

In many cases, such as the reports from Yarralumla and Ivybridge, extensive damage attributed to ball lightning is more readily explained as the path of an ordinary lightning current, perhaps including a side flash. In the Yarralumla case, it is possible that lightning caused an arc from the electrical power lines near the tree, continuing current then being supplied by the electrical mains. Uman (personal communication, 1997) has suggested that this may have been the mechanism responsible for the vaporization of a large volume of water in the Dorstone, Hereford case (Morris 1936, Goodlet 1937; see Chapters 1 and 11).

The trench leading away from the oak trees that were struck in the Woodlands St. Mary case can be explained by a side flash in which the return stroke current passed from the trees through the soil to the grounded steel crash barrier. This would occur if the product of the return stroke current and the effective resistance of the root system of the tree exceeded the electrical breakdown strength of the soil between the tree and the steel barrier. The side flash presumably also contacted underground electric and telephone cables and was responsible for all the electrical damage described. The trace described in the Minneapolis case (Anderson and

Freier 1972) can similarly be explained as a side flash from the tree to the electrical mains in the house. Flashover may have occurred above the surface of the grass or within the soil, or current may have passed through the soil, causing a local heating effect.

4.2 Thermal Effects

4.2.1 Thermal Effects Attributed to Ball Lightning

It is often impossible to separate thermal from electrical effects. Thermal effects on the ground have ranged from slight burns on grass to craters in asphalt.

On Monday, March 28, 1977 at 9 p.m., a woman at Newgate Street Village, Hertford, England, saw a white light motionless just above a flower bed in the garden of her house, no more than 20 ft (6 m) away from her as she stood in her kitchen. The light was between the size of a bicycle lamp and a car headlamp, perhaps about 4 in. (10 cm) in diameter. There was a wire perimeter fence just beyond the flower bed, and this passed close to a substation transformer adjacent to the property. The light disappeared after 3 to 5 s. There was nothing in the kitchen that could have caused a reflection. The following morning there was a heavy frost. She discovered a round patch of burnt, blackened grass, partly reduced to ash, exactly where the light had appeared. It was about the same size as the light, and the grass in front of it was somewhat yellow-brown. The black ash smelled of fresh burning (K. Lloyd, personal communication, 1977). A weather report confirmed the presence of cumulonimbus cloud and elevated atmospheric pressure, although there were no reports of lightning or thunder (K. Grayling, personal communication, 1977).

Similar traces on grass were found in connection with a ball lightning report from a scientist who worked at an atmospheric electricity research institute. This occurred in 1951 in Germany during a storm of average intensity. Following a number of flashes of conventional linear lightning, one of which was apparently to a mast carrying a 5-kV power line, a ball reminiscent of a car headlight appeared, moving obliquely and rapidly downward at about 50 to 100 m s^{-1}. The ball disappeared behind some trees and there was an immediate, loud bang. There was a cloud of blue smoke. Another observer located in a different house saw bright rays emitted from the ball as it exploded prior to the production of smoke. About 5 m from the mast, a burn spot 50 cm in diameter was found in the meadow grass. Some insulators on the mast were damaged (Dolezalek 1951).

Dr. A. Wittmann (1971) of the University of Sternwarte, Gottingen, Germany, reported a personal observation of ball lightning during a thunderstorm with heavy rain at Neustadt near Coburg, Germany. He saw a bright, yellowish-white "plasma ball" approximately 24 m away at a height of about 16 m. The diameter was estimated as 0.5 to 1.0 m. It fell with a speed of about 4 m s^{-1} into the top branches of a tree at a height of about 9 m. It then immediately disintegrated on contact with the branches, fragmenting into eight to twelve smaller spheres of the same color and a diameter of about 12 to 15 cm. These smaller spheres fell to the ground, where they disappeared silently. Three to five minutes later the same phenomenon recurred. In the place where the smaller spheres had contacted the road surface, Wittmann found circular patches of melted asphalt of a diameter of 12 to 15 cm (Barry 1980).

Another event involving damage to a road surface took place in the early summer of 1931. A very severe thunderstorm began at about 8 p.m. at Portrewydd, Gwent, Wales. At midnight or later, two boys sat in a window watching the storm. About 300 yards (270 m) from their house was a paddock surrounded by a brick wall. Suddenly they saw a ball about the size of a football strike the wall and rebound onto the road. There was a shower of sparks and it disappeared. The ball had been in sight for more than 5 s. The next morning they found a hole in the wall and a small crater in the road (R. L. Hooper, personal communication, 1977).

One afternoon at about 3 p.m. in July 1936 or 1937, a man was in his workshop, a wooden building, at Stavely, near Kendal, Cumbria, England. He heard a fearful rending and tearing sound and ducked instinctively. He thought for a moment that the roof was being torn off. The sound was loud enough to be heard ¼ mile (0.4 km) away. Looking out of the window, he saw a luminous, football-sized ball drifting lazily away from him. Twenty yards (18 m) further on it passed within a few yards of the kitchen window of the adjacent house in which two women were watching the storm. Its velocity was so low that they had plenty of time to observe it. The ball carried on for a further 30 yards (27 m) or so, crossing a main road and grounding itself at a speed limit post set some 3 or 4 ft (0.9 to 1.2 m) from the edge of the road in a sloping grass bank. This was at the northwest end of the village. The grounding must have been imperfect, since the current tore a zigzag path in the turf between the metal post and the edge of the road. The current passed under the road, which was about 20 to 24 feet (6 to 7 m) wide. It caused an explosion in the center, probably due to expansion of steam, creating a conical hole about 2 ft (60 cm) in diameter and scattering tarmac for about 150 yards (140 m) around (J. C. Braithwaite, personal communication, 1983).

Three people were in a room in a house in Moss, overlooking the Oslo Fjord in Norway late one evening. The date was August 1956, and there was no storm at the time and no rainfall, although the sky was overcast and distant lightning was visible on the horizon across the fjord. A white-blue ball, slightly larger than a tennis ball, entered through the window about 20 ft (6 m) away, following a rapid, predominantly horizontal trajectory. The ball was surrounded by an orange-blue corona, which increased it to the size of a football. It was bright enough to be clearly visible in daylight and was limb-darkened. At its closest, it was about 7 ft (2.1 m) away. It bounced several times off the walls and also off a picture. Its size and brightness were constant, but it seemed to flicker as it struck the walls. Having been in the room for about 15 s, it departed via the same window, breaking the pane with a crash as it left. There was a smell of ozone. The ball left a scorch on a wall and a scar on the picture. The marks on the picture were described as "crinkly" (T. H. Phillips, personal communication, 1982). A very similar earlier event was reported in the United States in which ball lightning entered a room through a window, moved around the room, scorching objects as it did so, and departed through the same window (Corliss 1982, p. 63).

In extreme cases, ball lightning is reported to have caused combustion. From Flammarion (1905, p. 63) we have the following report:

> On July 12, 1872, a . . . fireball made its appearance in the Commune of Hécourt, Oise. It was of the size of an egg, and it was seen burning upon a bed. Efforts were made in vain to extinguish it, and presently the entire house, together with the neighboring dwellings and barns, became a prey to the flames.

Cade and Davis (1969) interpret the following report, also from Flammarion (1905, p. 230), as ball lightning.

> The ship *Dayfield* from Liverpool was struck by lightning November 25, 1845. Instantly the deck was seen covered with globes of fire and large sparks that set fire to the vessel. As it threatened the powder magazine, the captain decided to abandon the ship. A rush was made for the boats, but as only thirty pounds of bread could be saved, many perished of hunger and thirst.

At about 1 p.m. on a dull and overcast day in February 1981 a woman was cooking lunch on an electric stove in the kitchen of her ground-floor flat near Warminster, Wiltshire, England. Suddenly, out of the corner of her eye, she saw in an adjacent room a small but brilliant blue ball of light about the size of a large marble. Its color was similar to that of a match that had just been struck, but it was much brighter. The windows of the room were probably open but covered with net curtains. The ball traveled swiftly at about 1 m s^{-1} toward her, about 1.2 m above the ground, passing in front of her face, then it veered off toward one of the electric rings at the rear of the stove as if attracted to it. She quickly removed a saucepan that was on the ring. Both this ring and the ring in front of it were glowing red hot. The ball attached itself to the ring, crackling loudly and ejecting sparks. She was very alarmed and turned off the main switch to the stove, whereupon the ball burned itself out. The entire event lasted only a few seconds, and the ball spent most of its lifetime in contact with the ring. At the same time as its appearance, she noticed loud crackling on the radio in the adjacent room, although the radio had an intermittent fault, which produced a similar noise. The electric ring no longer worked, but other parts of the stove circuit were undamaged. The ring consisted of an element spiral surrounded by insulation powder, encased in a heat-resistant metal sheath. There was a furrow melted in the sheath of the electric ring 0.14 cm in width near the surface and 0.045 cm in width at its deepest point, surrounded by a yellow deposit. Some of the melted metal formed two roughly hemispherical globules, solidified at one end of the furrow (Fig. 4.2). The sheath was constructed of a nickel–chromium–iron alloy with a melting point of 1628 to 1658 K. About 1.46×10^{-7} m^3 of metal was melted, requiring energy somewhat in excess of 600 J (Stenhoff 1981).

Bychkov et al. (1996) have published details of an event in Russia in 1994. Following a conventional lightning flash to a neighboring cottage sharing the same electrical circuit, holes and sooty deposits were found on a metal pot on an electric stove in a kitchen. There was also damage to a metal washing unit. There was no one in the kitchen at the time of the event, and windows and doors were closed. The authors rejected the suggestion that a pulse from the electric connections to the stove had caused the damage because the electric circuit of the stove was not damaged and because there were no traces on the base of the pot. They therefore suggested that the damage was caused by ball lightning.

4.2.2 Thermal Effects Due to Conventional Linear Lightning

Although the peak temperature of lightning is about 30,000 K, this is maintained for only a few microseconds, so a relatively small amount of heat is

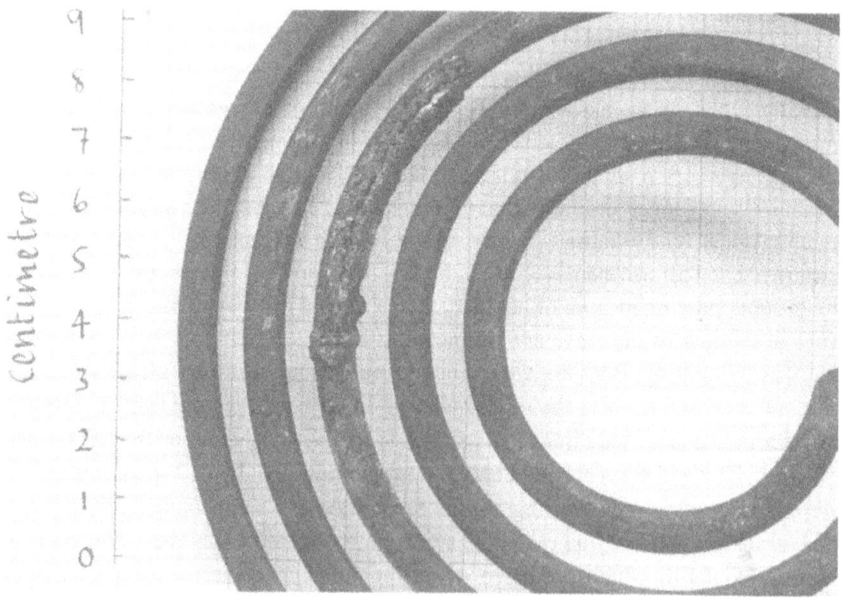

Figure 4.2. Electric stove ring allegedly damaged by ball lightning in Warminster.

transferred during this time. Some idea of the heating effect of a lightning flash is given by its action integral (see Chapter 2) and by the resistivity of the material struck. Thus the heating effect depends on the magnitude of the lightning current, the duration of that current, and the resistivity of the material struck. Hot lightning has a high action integral, largely due to the long duration of continuing current between flashes. As indicated in Chapter 2, positive flashes usually consist of a single stroke associated with large currents and long current tails. Thermal damage is most likely to be caused by positive giants and by lightning with continuing currents. Materials undergoing combustion initiated by lightning may cause further damage on contact with the soil or other surfaces.

The temperature increase of a lightning conductor when it is struck by lightning is relatively modest because it is made of metal and has low resistance (Golde 1973). Materials with higher resistivity such as wood, on the other hand, may be raised to very high temperatures and, if they are combustible, they may catch on fire. Thus buildings constructed of timber or with thatched roofs are particularly vulnerable to lightning, which may set them on fire. A poor electrical joint between two conductors may also be raised to a substantial temperature by lightning current and by the sparking that will take place across the junction.

Hot lightning and positive flashes may produce ablation pits in metal sheets or even cause penetration. This is of particular significance in the protection of aircraft. If a current i dissipates through a metal surface, the thermal energy at the point of contact is given by

$$\text{thermal energy} = u\int i\,dt = uq, \tag{4.3}$$

where q is the charge delivered by the lightning and u is the anode voltage drop. Hence the mass of metal melted should be proportional to q. Experimental data have been collected relating the area of the hole to the charge and thickness of the sheet (Hagenguth 1949, Golde 1973).

Under appropriate circumstances, an ablation plasma might be produced where the lightning channel makes contact with the earth. The spreading resistance $(2\pi\sigma r)^{-1}$ at the point of contact may be calculated by assuming that the lightning channel has a radius r on the order of 1 cm making contact with the earth. This will have an electrical conductivity σ of between 10^{-5} S m^{-1} (dry magnetite) and 10^{-1} S m^{-1} (wet quartz). Resistances of between 10^2 and 10^6 Ω indicate that a current of 1 kA would dissipate between 10^5 and 10^9 J in 1 ms. Even the lower energy is sufficient to dissociate fully 1 g of calcium carbonate, which would have a noticeable effect over at least 10 cm of the lightning channel. Chemical reactions resulting from this event might continue for some time. It could be that the resulting region of enhanced luminosity is itself interpreted as ball lightning, in which case it will appear stationary on the ground, or it could generate a positive afterimage that may appear mobile (see Section 3.3.5) (Stenhoff and Wooding 1977). Alternatively, flashover might occur. Lowke (1996) proposed a discharge theory of ball lightning relating to corona discharge sustained by the electric fields associated with charges from a lightning strike following lower-resistance paths in the earth. This model is discussed further in Chapter 13.

4.2.3 Discussion of Case Histories

In reports such as those from Hécourt or the ship *Bayfield*, there is no clear indication that ordinary lightning, perhaps hot lightning or a positive flash, could not be the direct or indirect cause of the damage. However, present knowledge about the thermal effects of ordinary lightning is perhaps not entirely successful in explaining all the damage described in Section 4.2.1.

It is possible that an ordinary lightning strike to ground, especially if it is hot lightning, will burn the grass or even create a crater in an asphalt road surface. However, the Newgate Street Village report might perhaps be more readily explained as ignis fatuus since lightning was not seen just before the event.

The report from Germany in 1951 seems to describe a lightning-initiated arc from a 5-kV power line, with the grass burned at the point at which the arc struck the ground.

It is possible that asphalt could be melted by hot lightning or by a positive flash, as in the Neustadt report. Perhaps it is more probable that burning material, ignited by lightning, fell to the ground and melted the asphalt. The craters in the Portrewydd and Staveley reports may have been caused by vaporization and expansion of water beneath the road surface following a flash to the ground. In the latter report it appears that ball lightning began and ended with a CG flash. The zigzag path in the turf between the metal post and the edge of the road seems to be a clear indication of a side flash.

It is suggested that the damage to the metal pot described in the Russian incident of 1994 consisted of ablation damage resulting from an electric arc originating elsewhere in the room. If the electric arc reached earth via the external metal body of the stove, the electric circuit of the stove would not necessarily have been damaged.

The manufacturer of the electric ring damaged in the Warminster (1981) case commented that testing the electric rings to destruction produced a brilliant light similar to an arc welder and a crackling sound. The damage to the electric ring could be attributed to normal failure. Failure of the electric ring might have been initiated by an overvoltage from a lightning flash, but the witness reported no thunderstorm. A possible hypothesis would be that normal electrical failure occurred first and that the brilliant light produced a positive afterimage, but this is inconsistent with the witness' description of the sequence of events.

I find it difficult to explain the damage described in the Moss report (or the similar report from the United States). Damage was evidently thermal, but even if the heat was from the ball, it is difficult to see how it transferred this energy sufficiently quickly to the surfaces in the room to produce the effects described. Rapid heating of picture glass might be expected to cause differential expansion and cracking.

4.3 Mechanical Effects

4.3.1 Mechanical Effects Attributed to Ball Lightning

Just as the thermal effects of ordinary lightning are really a consequence of electrical joule heating, so effects that appear to be mechanical are in some cases thermal or electrical in origin. Given these limitations in the classification scheme used here, we now consider damage by ball lightning that appears to be mechanical. This includes pressure-wave damage related to reports of explosive decay of ball lightning, puncturing or fracturing of efficient but brittle dielectrics such as glass, and fractures and fissures in poor dielectrics such as concrete, brick, or mortar. These effects are also discussed in Chapter 6, which considers structural damage to buildings.

A number of reports describe fracturing of poor dielectrics. In the Yarralumla report (see Section 4.1.1), there is a description of ball lightning landing on a steel-reinforced concrete post, destroying the post and buckling the steel reinforcement. In the Ivybridge report (see Section 4.1.1), ball lightning was said to have touched the ground by a wire and concrete post fence. The concrete post shattered and fragments went through roofs some 55 m away. In the Portrewydd case (see Section 4.2.1), ball lightning was reported to have caused a hole in a brick wall.

Other reports describe what may be pressure-wave damage. In the Yarralumla report (see Section 4.1.1), a ball was said to have entered the house through closed glass doors. The glass did not break, but the wooden door runners split open. In the Ivybridge report (see Section 4.1.1), nearby houses in the cul-de-sac lost many panes of glass and the force of the explosion moved a 4.3-m bay window about 1 cm. In the Woodlands St. Mary report (see Section 4.1.1), a window pane had a 32 × 22 cm hole in its bottom, right-hand corner, and most of the fragments of glass were on the outside. In another room there was also a 0.6-cm wide crack along the full length of the double glazing in the sitting room.

Tomlinson (1895) and Cade and Davis (1969) describe an incident that occurred on board the ship *Chichester* on March 7, 1840 near Galway, Ireland during a thunderstorm. Ball lightning descended from the masthead and broke through the deck. As it did so, it knocked over several crew members. It passed through the captain's cabin while he and his daughters were eating dinner, and passed above the table, shattering glasses and dishes. There were no injuries. The center of the deck of the ship was raised, the patent lights were extinguished, and the skylights were thrown up. Magnetic compasses no longer worked and the watches of those on board stopped. There have been other reports of breaking of drinking glasses and plates, for example, at Selsdon in 1980 (Chapter 8).

A farmer was driving his van along the A40 road near the Haverfordwest Golf Club, Pembrokeshire, Wales on April 4 or 5, 1985 at about 12:20 p.m. There was apparently no storm present and no thunder and lightning. He said, "It was raining heavily when in front of me I saw a ball of fire the size of the front wheel of a tractor. The van [was] lit up, then there was a big bang and the windscreen shattered." He first saw the ball in midair some way ahead above the road. The ball was spherical and red, and bright enough to be clearly visible in daylight. It was uniformly illuminated across its surface, and its appearance did not change much throughout the event. The van windshield was completely shattered by the explosion (see Fig. 4.3) (Hughes, D., personal communication, 1985). The event was brought to my attention by Mr. Ian Jones, a meteorologist operating a station about 14 miles (23 km) away. Mr. Jones reports that there were no days with thundery conditions around April 4 or 5. He kindly provided a barogram showing that the pressure on April 4 fell from 1008 mb at midnight to 994 mb at 3 p.m., remaining fairly steady throughout April 5.

Figure 4.3. Car windshield allegedly damaged by ball lightning. [Photograph: Chris Warren.]

Still other reports describe objects apparently displaced by ball lightning. In Chapter 5 we give a report dated February 1767 from Presbourg, France. A blue, conical flame escaped suddenly with a detonating noise from a brazier. It carried up the chimney with it some hams that had been hung in the chimney piece (Flammarion 1905).

When a child was killed in 1865 in Germany after he allegedly touched ball lightning with his foot, his toys were tossed around by the explosion (von Haidinger 1868, Singer 1971) (see Chapter 5).

4.3.2 Mechanical Effects Due to Conventional Linear Lightning

The rapid expansion of a lightning channel as it reaches its peak temperature of about 30,000 K produces a pressure wave that is initially a supersonic shock wave. The pressure wave is heard as thunder, and may also be responsible for widespread disturbance of tiles on a building that has been subjected to a direct lightning strike (Golde 1973, p. 55), or for breaking brittle materials such as glass. The spectrum of thunder contains a wide range of frequencies (Hill 1977) and some of these may occasionally cause damaging resonances.

Electrical conductors that carry parallel lightning currents will experience an attractive force due to the magnetic fields surrounding them. This may cause displacement or distortion of the conductors, the collapse of hollow conductors, or the compression of stranded electrical cables (Golde 1973, p. 56). The magnitude of the force F is given by

$$F = \left(\frac{\mu_0}{2\pi}\right) \times \frac{I_1 I_2}{a} \times l, \tag{4.4}$$

where $\mu_0 = 4\pi \times 10^{-7}$ H m^{-1}, I_1 and I_2 are the currents in the conductors, a is their separation, and l is their length. Two conductors carrying a current of 100 kA and separated by 1 m will experience an attractive force of 2 kN per meter length. Field visits following lightning damage must ascertain whether two parallel conductors may have behaved in this way.

A much weaker magnetic force is responsible for straightening out bent conductors carrying a current, for example, a 90-degree bend in the gutter on a building (Golde 1973). This force, which, for a current of 200 kA is only a few tens of thousands of newtons, may distort a conductor or cause it to be detached from its fixings.

4.3.3 Discussion of Case Histories

Several examples have been given of trenches produced in soil. These were explained in Section 4.1.2 as the effect of vaporization of water in soil by a side flash from conventional linear lightning. In some cases (e.g., Linguy, 1897; Chapter

6) where objects have been moved from one place to another, there is no clear indication that ball lightning was seen. Some of these displacements may have been caused by tornadoes, which are often said to be associated with luminous regions reminiscent of ball lightning. An example may be a description dated 1731 or 1732, interpreted by Mendenhall (1890) and Corliss (1982, p. 56) as ball lightning, in which a phenomenon violently threw up a piece of wood and a large stake. Others may be caused by pressure waves or by magnetic forces of the kind described in Section 4.3.2.

Chapter 5

Assessment of Risk of Death or Injury by Ball Lightning

5.1 Deaths of Humans Attributed to Ball Lightning

There are many early accounts of fatalities attributed to ball lightning. I am especially indebted to Adrian James, director of the Ball Lightning Division of the Tornado and Storm Research Organisation, who compiled much of the information in this chapter (James 1992).

Sir Thomas Atkyns' *Ancient and present state of Gloucestershire* gives the following account:

> In the year 1556, in less than two months dyed Maurice Walsh, together with seven of his children, occasioned by a fiery sulphureous globe rolling in at the parlour door at Dinner time, which struck one dead at table, and occasioned the death of the rest. It made its passage through a window on the other side of the room. (Atkyns 1712)

James (1992) points out that a later account makes no reference to the fiery globe:

> A dreadful thunderstorm fell on this seat [Little Sodbury] in the year 1556, when Maurice Walsh, Esq., was lord of the manor. The lightning entered at the parlour door, and forced its way out at the window at the opposite side of the room, when Mr. Walsh and his family were at dinner. One of the children was killed on the spot, and six others, with the unfortunate father, were so much hurt, that they all died of shock in less than two months afterwards. (Rudder 1779)

Congregations in churches have several times fallen victim to incidents apparently involving ball lightning (and, indeed, ordinary lightning).

For example, James (1992) discusses such an event, which occurred at Withycombe, Devon, England on October 21, 1638. A tornado struck the church there and a ball lightning description seems to have been associated with it. To quote from a contemporary account, *A true relation of those sad and lamentable accidents which happened in and about the Parish of Withycombe in the Dartmoors, on Sunday 21st October last, 1638* (quoted by James 1992):

> the extraordinarie lightning came into the church so flaming, that the Church was presently filled with fire and smoke, the smell whereof was very loathsome, much like unto the scent of brimstone; some said they saw at first a great ball of fire come in at the window and pass through the Church, which so much affrighted the congregation that the most part of them fell upon their seats, and some upon their knees, and some upon one another, with a great cry of burning and scalding, they all giving themselves up for dead. [The lightning] seized upon [the Minister's wife], fired her ruffe and linen next to her body, and her cloaths, to the burning of many parts of her body in a very sad and pitifull manner. [One woman] had her flesh so torne and her body so grievously burnt, that she died the same night. . . . Another woman adventuring to runn out of the Church, had her cloaths set on fire, and was not only strangely burnt and scorched, but had her flesh torne about her back almost to the very bones.

Here is a similar report from the same period:

> The last Lord's Day (July 2 1665), as Mr. Hobbs was preaching in his Parish Church of Erpingham [Norfolk, England], in the afternoon, there did arise a great storm, and there descended the appearance of a great grey ball, which as was supposed did beat down the southwest corner of the steeple, towards the top of it; for at that instant, it fell and carried along with it the south porch; And as soon as the said ball was come into the Church, it turned upon the south side, on which were Men's seats; Mr. Hobbs being in the pulpit, saw the men fall some one way, and some another, in such manner that he thought they had all been struck dead. It passed towards the chancel and brake. Upon which the Church was as if it had been all of a fire. It left a great smoke and stink behind it, and upon the breaking there was a great and hideous outcry in the Church, and in the confusion there was one man found stark dead and many others lamed, who yet continue so. One woman who sat in the porch is so weak as 'tis thought she will not live. It did raze and tear the Church wall in two places on the inside. One Mr. How who sat above the chancel is lamed and about the top of his thigh in the groin, is [a] round red place and down from that about the breadth of a finger, a red streak to his foot which is very painful and his stocking on the inside is seared, but not without. From *Newes*, July 13, 1665, quoted by Schove (1951).

Camille Flammarion's book *Thunder and Lightning* (1905) abounds with examples of dramatic ball lightning reports involving death or injury. He quotes a report from Muschenbroek of an incident at a church in Duytsbourg, Solingen, Germany, in 1711. During the sermon and in the middle of a storm, a fireball fell into the church through the clock tower and exploded. The sanctuary was set on fire and became thick with smoke. Three people were killed and more than a hundred injured.

Flammarion (1905) also tells us that on July 11, 1809 at about 11 a.m., a fireball entered the church of Chateauneuf-les-Moustiers, Basses-Alpes, France, just as the bell was ringing and a large congregation had taken their seats. Nine persons were immediately killed and eighty-two others were wounded. A woman in a hut on a neighboring hill saw three fireballs descend that day and was convinced they would reduce the village to ashes.

One of the most famous reports of a ball lightning-related fatality concerns the death in St. Petersburg of the Swedish scientist, Professor Georg Wilhelm Richmann (b. 1711). The exact date is uncertain—Dibner gives both August 6, 1752 and July 26, 1753. In 1751, Benjamin Franklin had made the suggestion that lightning was an electrical phenomenon. The famous experiment in which Franklin flew a kite into a thundercloud probably took place in June 1752. Richmann had had some success in repeating Franklin's observations with a lightning rod, so, accompanied by Sokolov, an engraver from the Imperial Academy, he hurried home on seeing an approaching thunderstorm. Once inside the house, Richmann approached the lightning rod. Sokolov said that when Richmann was

> a foot away from the iron rod, [he] looked at the electrical indicator again; just then a palish blue ball of fire, as big as a fist, came out of the rod without any contact whatsoever. It went right to the forehead of the professor, who in that instant fell back without uttering a sound .... (Dibner 1977)

Two engravings have been published showing Richmann's death. In a contemporary report on the circumstances of death (Anon. 1755 quoted by Lee 1977) we read:

> There appeared a red spot from the forehead from which spirted some drops of blood through the pores, without wounding the surrounding skin. The shoe belonging to the left foot was burst open. Uncovering the foot at that place they found a blue mark, by which it is concluded that the electrical force of the thunder, having forced into the head, made its way out again at the foot.

In Sussex in 1780, several balls of fire fell from a large black cloud into the sea. The witness was then struck by conventional lightning while others in the house were killed (Singer 1971, Brereton 1781). On July 27, 1789, at about 3 p.m. at Feltri, Marche Trevisane, France, ball lightning about the size of a cannonball fell in a great hall in which six hundred people were seated, wounding seventy and killing ten. All the lights went out (Flammarion 1905).

Ball lightning is sometimes reported near ships and boats and here, too, fatalities have been reported. James (1992) gives the following report from Borlase (1753):

On Tuesday, August 2, 1757, between one and two of the clock, as James Tillie, Esq. . . . with his neighbors and servants, were lying aground on a boat in a sand-bank in the River Tamar [Cornwall, England] . . . waiting for the tide to throw a net for salmon, a sudden clap of Thunder broke over their heads. In a field adjoining the grass seemed on fire and the whole field in a flame, and a fire-ball was observed just to pass over a hedge at the top of a very steeped wood which hangs over the Tamar. The fire-ball fell on the boat; and passing from the South-West in a direct line from the larboard bow to the stern, James Widear . . . in the bow, had a violent blow on his right shoulder and head; Mr. Samble, who sat next to him, was struck deaf for some time; Mr. Tillie sat next in the middle of the boat, and plainly perceived the fire-ball, about five inches in diameter, somewhat sharp, and pointed in the fore-part, to pass by him at about three feet distance: he was violently struck on the back part of his head by the current of air attending the ball; his eyes were shut, and he leaped from his seat about two or three feet high: and on return of his senses, opening his eyes, was surprised to find himself standing, for before the shock he was sitting on the oar. The right side of his face continued very warm for two hours, and the corner of his hat was carried away, as if half of a small bullet had been shot through it. Robert Atkins . . . was near the stern, with his face to the South-West, but not in the direct line of the fire-ball: he was struck speechless, thrown on his back upon the fishing net, remained insensible for two or three hours afterwards, his face was black, as if the priming of a gun had been blown by accident over it; his left eye weak, and contracted a fortnight after; with a great numbness in all his limbs, until a brief circulation of the blood ensued, and then he had violent pains, which wore off slowly, leaving him weak and low, with complaints of frequent head-ache, but no visible mark of hurt. Mr. Pethen . . . was standing three feet from Atkins, upon the seat of the stern, with his face to the South-West, and had scarce done speaking when the fire-ball fell on his left temple, and struck him dead into the river. He was immediately taken up by some of Mr. Tillie's servants upon the shore. His peruke on fire, and smoaking in the stern of the boat, Mr. Tillie took up and extinguished: it had a hole burnt in it as large as a crown piece, and smelt strongly of sulphur. His hat was blown into the water, rent three or four inches long, the lining only ript. His cloaths were but little rent: on his hip there was a black scar about three or four inches long, and discolored as if filled with gunpowder: his neck and left breast were soon after changed to a claret colour; but where the fire-ball fell, it left no wound more visible than the puncture of a pin, neither did it discolour his temple; but the third day the lower part of his face altered a little, and the other parts of his body discoloured more and more, till his burial. All his cloaths smelt like gunpowder newly discharged. There were three persons standing on the adjoining sand; one was violently struck on the head, the second had his eye singed, and the fire-ball fell between the legs of the third into the sand, from whence he only perceived a sudden warmth. The day had been showry, neither hot nor cold; and the sun shone, though faintly, about ten minutes before the explosion. No one had his face turned towards the South-West but Mr. Pethen and R. Atkins.

Arago (1854) writes:

In September, 1780, before the thunderbolt which threw him to the ground and killed two of his servants, Mr. James Adair of Eastbourne, Sussex, saw several balls of fire fall from a large black cloud into the sea.

Arago (1854) also includes the following report:

> . . . on the 13th of July, 1798, the ship, the Good Hope, belonging to the East India Company, located at 35° 40′ S. latitude and 44° 20′ E. longitude from Greenwich, was struck by lightning of globular form, which produced the most violent detonation, killed a sailor instantaneously, and seriously injured another.

From *The Gentleman's Magazine* dated February 14, 1809, we have the following report of several maritime fatalities related to ball lightning (Arago 1854, Anon. 1909e, Corliss 1982, James 1992):

> A remarkable occurrence took place on board the Warren Hastings, moored at the Motherbank (Portsmouth). The morning being fine, it was deemed necessary to get up the top-gallant masts. About 3 in the afternoon, the atmosphere to the westward indicating a violent storm, several sailors were sent aloft to strike the top-gallant masts; but, when lowering them, the wind blew tremendously, and the rain fell in torrents, accompanied by heavy claps of thunder. Three distinct balls of fire were emitted from the heavens; one fell into the main top-mast cross-trees, killed a man on the spot, and set the main-mast on fire, which continued to blaze for 5 minutes. A few hands ran up the shrouds to bring down their dead companion, when the second ball struck one of them, and he fell down upon the guard-iron in the top, from which he bounced off into the cross-jack braces. His arm was much shattered and burnt, and it was expected he must undergo amputation. The third ball came in contact with a Chinese, killed him, and wounded the main-mast in several places; the force of the air, from the velocity of the ball, knocked down the chief mate, who fell below but was not much hurt. For some time after there continued a sulphureous smell.

Chapter 4 referred to a report in which a child who apparently touched ball lightning with his foot was killed in 1865 in Germany, and the explosion tossed around his toys (Von Haidinger 1868c, Singer 1971). In June 1866, a large fireball, described as "a mass of fire rolling along the hill towards the building in which the party had taken shelter," was said to have killed two young women in the Malvern Hills, Worcestershire, England (Tomlinson 1895, Cade and Davis 1969). The death of a woman in Palestine, Texas in 1876 was attributed to ball lightning or perhaps a meteorite (Anon. 1876, Corliss 1982).

At Baguières-de-Bigorre, France, sometime around the 1870s, a plum tree was shattered "to atoms" by a 6-ft (1.8 m) diameter lightning ball. The surrounding garden was also left in much disarray, with the ground covered with several inches of large hailstones. Nearby, nine shepherds were sheltering in a cabin. Four of them were killed instantly and three died soon afterward (Anon. 1887u).

The two reports that follow, both from Flammarion, are rather similar:

> In 1890, a young farmer was working on a plot of ground, two or three miles from Montfort-l'Amaury. A storm breaking out, he stood up against his horses to take refuge from the rain; moving away a few yards in order to get his whip, there was seen, when he returned, a ball of fire almost touching the ear of one of his horses. A moment later it exploded with a deafening noise. The two horses fell—one of them unable to get up again. The farmer himself was dashed to pieces. . . .

> In the middle of a violent thunderstorm, a Dr. Gardons saw several fireballs moving close to the ground in different directions, making a "crackling" sort of noise. One of them was seen to have come out of an excavation full of stagnant water. The fireballs killed one man and several animals, and did much damage to trees and houses nearby. (Flammarion 1905)

Flammarion (1905) also quotes a report by Mlle. N. de Soubbotine (1901/1902?) from the *Bulletin of the Société Astronomique de France*:

> A terrible storm broke out at Ouralsk [now Oral, Western Kazakstan] on May 22, 1901. It was a fête day and the streets were thronged with people. Towards five in the afternoon some young men and girls, twenty-one in all, had taken refuge in the vestibule of a house, and a girl of seventeen, Mlle. K., had sat down on the threshold, her back turned towards the street. Suddenly there was a violent clap of thunder, and in front of the door there appeared a dazzlingly brilliant ball of fire, gradually descending towards where they were all grouped. After touching Mlle. K.'s head, who bowed down at once, the fireball fell on the ground in the middle of the party, made a circuit of it, then forcing its way into the room of the master of the house, whose boots it touched and singed, it wreaked havoc with the apartment, broke through the wall into a stove in the adjoining room, smashed the stove-pipe, and carried it off with such violence that it was dashed against the opposite wall, and went out through the broken window.
>
> After the first feeling of fright, this is what transpired. The door near which Mlle. K. was seated had been thrown back into the court, and in the ceiling there were two holes of about 18 cm each.
>
> The young girl, still seated with her head bowed down, looked as though she was asleep. Some of the people were walking in the courtyard, having seen and heard nothing, and the others were all lying in the vestibule in a dead faint. Mlle. K. was dead. The fireball had struck her on the nape of her neck and had proceeded down her back and left hip, leaving a black mark all along. There was a sore on one hand, with some blood on it, and one of her shoes was torn completely off, and there was a small hole in one of the stockings.
>
> All the victims became deaf.

Ball lightning has sometimes been used as an explanation for reported cases of the highly controversial phenomenon called "spontaneous human combustion." For example, Cade and Davis (1969) implied that the death in the following incident might have been caused by ball lightning. A man was awoken at about 5 a.m. by the screams of his wife, who was in their living room. He ran to her and found her lying on the floor, burning fiercely, while a blazing ball hovered above her. He was badly burned while he attempted to put out the fire. He called for help, and those who came to assist him poured buckets of water over the woman, but sadly this was to no avail because she died soon afterward in hospital.

A report of a similar fatality, but with a clearer link to a ball lightning description, dates from 1886:

> At Crawforth, Indiana, on August 9th, during the fall of a slight shower of rain, but in the absence of any indication of a thunderstorm, a ball of fire was seen to enter the window of a house occupied by one of the most prominent citizens of the town. Shortly afterwards, Mr. Riley was observed lying on the floor, his body, according to the American account from

> which we quote, burnt almost to a cinder and unrecognizable. A black streak was traced upon the carpet from the window to the fireplace, in which line the body was found. The family, who were sitting outside the house, witnessed the ball of fire enter the window and apparently disappear up the chimney. . . . It is difficult to understand how [the ball lightning] could continue its course having discharged sufficient energy to carbonize a human body. (Anon. 1886b, James 1992)

Flammarion (1905) describes yet another fatality attributed to ball lightning. During a violent thunderstorm, several glowing spheres were seen moving close to the ground in apparently random directions, making a sound described as "crackling." One of them emerged from an excavation full of stagnant water. (This may suggest the presence of methane, which is flammable. Refer to comments on ignis fatuus in Chapter 3.) The deaths of one person and several animals were attributed to the fireballs, and many trees and houses were damaged.

There are very few reports of fatalities attributed to ball lightning during the modern era. In August 1978, an amorphous yellow "blob" made its way into a tent occupied by climbers. This occurred at an altitude of some 3900 m in the Caucasus Mountains. The "blob," floating 1 m above the floor, suddenly dived into a sleeping bag, whose occupant screamed out in pain. The ball jumped out and went on to circle over the other bags, entering first one and then another, accompanied by anguished howls from the victims. When the climbers were flown to hospital, their injuries were described as "worse than burns," with pieces of muscle ripped from the bones. Indeed, one climber had been killed instantaneously; it was speculated that this was because his sleeping bag was on a rubber mattress that had insulated it from the ground (Anon. 1984b, James 1992).

Lloyd's List for July 1, 1995 includes the following account:

> Kiev, June 29—Lightning has killed 10 people in storms throughout Ukraine in the past week, officials said today. . . . Officials had earlier reported three deaths near Cherkassy, east of Kiev. They included a 68-year-old woman, who was bathing her granddaughter when she and the girl's brother were struck dead by what meteorologists call "ball lightning." (Anon. 1995)

5.2 Deaths of Animals Attributed to Ball Lightning

An earlier case indicated that animals as well as humans may apparently be victims. Corliss (1982) describes an undated report from Iowa in which a yellowish-white ball "the size of a washtub" bounced along a road, making a rushing noise (Uman 1971). It was said to have demolished a shed and killed a horse. Uman also mentions an incident in Hendon, England, in September 1880 where ball lightning fell into a pond, killing over a hundred fish (Anon. 1880b).

Flammarion (1905) writes

> On September 10, 1845, at about two in the afternoon, in the course of a violent storm, a fireball came down the chimney into a room in a house in the village of Salagnac (Creuse). A child and three women who were in the room suffered no harm from it. Then it rolled into the middle of the kitchen, and passed near the feet of a young peasant who was standing in it. After which it went into an adjoining room, and disappeared without leaving any trace. The women tried to persuade the man to go in and see whether he could not stamp it out, but he had once allowed himself to be electrified in Paris, and thought it prudent to refrain. In a little stable hard by, it was found afterwards that the fireball had killed a pig. It had gone through the straw without setting fire to it.

He also gives a more dramatic account that illustrates the apparently capricious reported behavior of some ball lightning:

> On February 16, 1866, a thunderstorm descended upon a farm in the Commune of Chapelle-Largeau (Deux-Sèvres), and the circumstances attending its explosion are too remarkable to be overlooked. After a tremendous thunderclap, a young man who was standing near the farm saw an immense fireball touch the ground at his feet, but it did him no damage, but passed, still harmlessly, through a room in the farmhouse in which there were nine persons. The only effect it produced was the flaring up of some matches upon the chimney-piece.
>
> It proceeded towards the stables, which were divided into two compartments. In one there were two cows and two oxen: the first cow, to the right of the entrance, was killed, the second was uninjured; the first ox was killed, the second was uninjured.
>
> The same effect was found to have been produced in the other compartment, in which there were four cows; the first and third were killed, the second and fourth were spared: the odd numbers taken and the even numbers left. (Flammarion 1905)

The Gentleman's Magazine for July 27, 1809 wrote:

> At Boston, this night, was experienced a most alarming tempest; it began about half past eight, and was not over till nearly eleven. During a great part of that interval, rain descended in torrents, and the thunder and lightning were of the most awful kind. About ten, four houses were struck by fire-balls; a window in one of the houses was driven in, the bed-clothes were torn off a bed, and a cat was killed. (James 1992)

In another case, Flammarion (1905) wrote:

> On August 24, 1895, at about ten in the morning, in the midst of a storm of wind and rain, several persons saw descending to the ground a whitish-colored globe of about an inch and a half in diameter, which, on touching the ground, split into two smaller globes. These rose at once to the height of the chimneys on the houses close by and disappeared. One went down a chimney, crossed a room in which were a man and a child, without harming them, and went through the floor, perforating a brick with a clean round hole of about the size of a franc. Under this room there was a sheep-fold. The shepherd's son, seated at the doorway, suddenly saw a bright light shining over the flock of sheep, while the lambs were jumping about in alarm. When he went up to them, he was startled to discover that five sheep had been killed. They bore no trace of burning, or of wound of any kind, but about their lips was a sort of foam, slightly pink in colour.

In the adjoining house, the second fireball had also gone down a chimney, and had exploded in the kitchen, causing great damage.

5.3 Injuries Attributed to Ball Lightning

Some of those who have alleged close encounters with ball lightning have sustained injuries but survived. Arago (1854) describes an incident that occurred on June 20, 1772:

> while a thunderstorm rumbled over the parish of Steeple Ashton, in Wiltshire, a globe of fire was seen to hang in the air for a considerable time, and afterwards to fall perpendicularly onto houses where it caused much damage. . . .
>
> the Reverends Messrs. Wainhouse and Pitcairn, who were in a room in the presbytery, suddenly saw appear, at the height of their faces and about a foot [30 cm] away, a globe of fire of the size of a fist. This globe was surrounded by a black smoke. In exploding it made a noise comparable to that of the simultaneous discharge of several pieces of ordnance. Immediately afterwards a strongly sulfurous vapor spread throughout the house. Mr. Pitcairn was dangerously injured; his body, his clothes, his shoes and his watch presented all the same appearances as those resulting from a stroke of lightning of the more usual kind. Different colored lights filled the apartment, and underwent vigorous oscillation.

On October 9, 1883, a fireball that exploded into flying sparks was said to have injured three seamen (Anon. 1883, Corliss 1982).

At about 4 p.m. on August 24, 1895, a boy in his garden at Culdaff, Donegal, Ireland was allegedly injured by ball lightning. It was a stormy day with heavy showers but no thunder. The large ball, of brilliant intensity, approached him and burst into several fragments with a report like that of a gun, although it continued in its path along the ground somewhat diminished in size. He said it was the size of the table in his bedroom. He flung his hands over his face to protect himself and his hand was badly mangled. The thumb and first two fingers of his left hand were shattered and required amputation. His right hand and left cheek were scratched, cut, and blackened (Minchin 1895b, Cade and Davis 1969, Corliss 1982).

Injuries attributed to ball lightning include burns. On March 3, 1557, Diane of France, illegitimate daughter of Henri II (who was then the Dauphin), was married to François de Montmorency. On their wedding night, an "oscillating flame" came in through the open bedroom window, glided round the room from corner to corner, and finally moved toward the bed. It singed Diane's hair and burned her nightdress (Bellay 1886, Flammarion 1905).

We also have this report from Flammarion (1905):

> A party of five women took refuge during a storm in the entrance to a house in order to escape from the rain and the lightning.

> They had scarcely gained the doorway when there was a tremendous thunderclap which sent them flying backwards—and the two girls who had joined them—knocked senseless by lightning in the form of a fireball. One of the girls remained unconscious for a long time; all the others were more or less seriously injured, but all recovered. The strangest circumstance in connection with this affair, however, still remains to be told.
>
> On the same side of the street as the passage, in a neighboring house, nine or ten yards away, in a ground-floor room of which the door was shut, a young woman was working at a sewing-machine. At the moment of the thunderclap, she experienced a violent shock throughout her whole body, and a fierce burning sensation in the hollow of her neck. It was found afterwards that between the shoulder-blades and also on her leg, she had been badly scorched, but the wounds quickly healed. Now, in the room of this victim, no trace was to be found of the passing of the fireball, neither on the ceiling, nor on the floor, nor on the walls. There was absolutely nothing to show how the electric fluid could have made its way in from the spot in which the fireball had exploded in the neighboring house, separated from it by two thick walls.

James (1992) gives another report of ball lightning entering a church from Lewis Turner's *History of the ancient town and borough of Hertford* (1830):

> 23 July 1763: During the time of Divine Service a fireball fell upon the Church of All Saints in Hertford, which penetrated the same and greatly terrified the congregation. It burst in the Blue Coat Boys' gallery with a terrible explosion, that was heard in every house in the town, but did no damage other than singeing a boy's hair.

Another report from Flammarion (1905) describes burns suffered during an observation apparently of ball lightning:

> a fireball fell upon the door of a house, pushed it violently open, and made its way into the kitchen.
>
> At the sight of this strange visitor, the cook bolted from the room. A [garment worker], who was at work near the window, received a small burn on her forehead, of about the size of half a franc, with a slight weal a couple of inches long—like the tail of a comet.
>
> After bursting, the fireball made its way up the chimney, from which it removed a mass of soot, smelling somewhat of sulphur.

He also describes a very similar event:

> In February, 1767, at Presbourg, a blue, conical flame escaped suddenly with a detonating noise from a brazier, breaking it to pieces, and scattering the glowing cinders all around. It then went twisting about the room, burnt the face and hands of a child, escaped partly through a window, partly through the door, broke into a thousand pieces a second brazier in another room, and disappeared finally up a chimney, carrying up with it and discharging from the chimney-top into the street several hams which had been hung up the chimney-piece. For several days afterwards the atmosphere of the house retained a smell of sulphur. (Flammarion 1905)

A house at New Harmony, Indiana was extensively damaged during a lightning incident on August 19, 1886. Ball lightning rolled down some steps and along a carpet (without scorching it) and grazed the toe of a girl. This caused severe inflammation as if by burning. Others present were blinded and "suffocated" for several minutes (Anon. 1887).

Mr. J. Durward, a former deputy director of the Meteorological Office in Great Britain, was flying to Iraq in the summer of 1938. The captain of the aircraft reported that just after passing the Toulouse gap, while flying through dense nimbostratus cloud at 2500 m, ball lightning came through an open window in the cockpit and burned off his eyebrows and some of his hair. A hole was also burned in his safety belt and a dispatch case. The ball then passed through into the rear cabin, where Mr. Durward saw it disappear explosively (Gold 1952, Cade and Davis 1969, Singer 1971).

The present author reported an event that occurred at Warley, Smethwick, West Midlands, England on August 8, 1975 (Stenhoff 1976, 1988). A severe thunderstorm began about 6 p.m. that evening. Many buildings were damaged. At about 7:45 p.m., a lady was in her ground-floor kitchen. Both a louvred window and a door leading to a porch were open. Figure 5.1 shows the front exterior of the house. Outside the porch was a conifer tree about 10 ft high. The woman had just filled a kettle with water (see further comments) and turned toward the gas stove. She saw a ball of light appear between and above the stove and a refrigerator, both of which were next to the open door. Above the stove was a flue leading to an open chimney, sealed with a metal plate inside the kitchen, and above the refrigerator was an air vent.

The ball was round, about the size of the wooden balls used on bowling greens (diameter about 4 in.), with a bright blue to purple core, surrounded by a flame-colored halo. The woman felt burning heat that made her "glow all over" and heard a sound "something like a rattle."

The ball moved toward her at a height of about 39–40 in. Instinctively, she brushed the ball away from her with her left hand. It may have moved downward about 2 or 3 in. as she struck it. She said, "I can't really say whether it exploded as soon as I touched it, it all happened so fast. I saw it heading for me. As I went to put the kettle on, it seemed as if [the ball] exploded or went off with a bang. I went to knock it away and screamed all at the same time." She estimated that the lifetime of the ball was about 1 s; during this time there was no change in its appearance. The explosion was so loud that her next-door neighbor heard it. She smelled singeing.

After the event, she found a hole melted in her dress and tights. Her legs were not burned, but were numbed and reddened. There was a redness and swelling on her left hand and it seemed as if her gold wedding ring was burning into her finger. The inflammation seemed to appear more quickly than would a scald, and the discomfort from her wedding ring was such that she had to force it off under running cold water. The skin was unbroken. She was only able to wear the ring again after 48 h.

Figure 5.1. The house in which the Smethwick event occurred (Stenhoff 1976).

There was some confusion in her memory of the exact sequence of events. She was not sure whether the ball appeared after or before a clap of thunder, but she thought it more probable that thunder preceded the ball. Her early statement, written 18 days after the event, was as indicated above. However, in correspondence some 6 months later, she suggested that the ball appeared about 5 min *after* she had placed the kettle on the gas ring of the stove. Given the unreliability of memory, it seems more probable that the earlier statement is correct.

Figures. 5.2 and 5.3 show the damage to the dress. There is a central area of primary damage where the fiber has disappeared altogether. The shape and position of this hole are consistent with her statement that the ball was at a height of 39 in. and that she brushed it downward. This hole is about 11 × 4 cm. Around its perimeter, the material has shrivelled, but is not charred. There is a still larger area of secondary damage where the fiber has shrivelled and the pattern of the dress has faded noticeably.

Dr. R. S. Higginbotham, senior finishing adviser of the Shirley Institute, Manchester, England, examined a specimen of the dress fiber. He said it was probably made from polyester fiber (polyethylene terephthalate), with a melting point of ≈ 260°C, but other less likely possibilities were polyamide (m.p. ≈ 210 to 250°C) or cellulose triacetate (m.p. ≈ 290°C) (Higginbotham 1975, Wooding 1976).

Because some models of ball lightning suggest that it is associated with nuclear processes, Dr. T. D. MacMahon at the University of London Reactor Centre examined a sample of the damaged portion of dress material. About 10 weeks after the event, MacMahon examined the gamma spectrum between zero and about 3.6

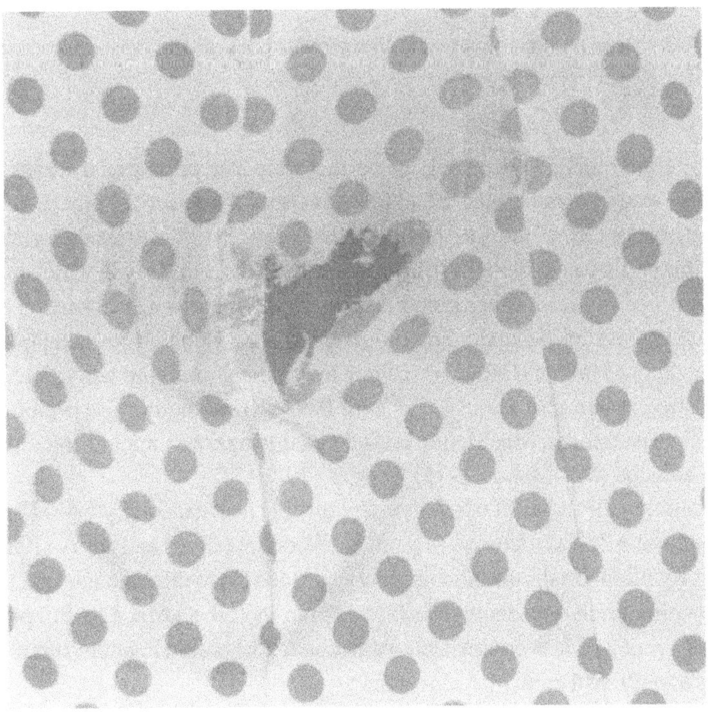

Figure 5.2. Damage to dress fabric in the ball lightning report from Smethwick (Stenhoff 1976). [Photograph: Brian Tate.]

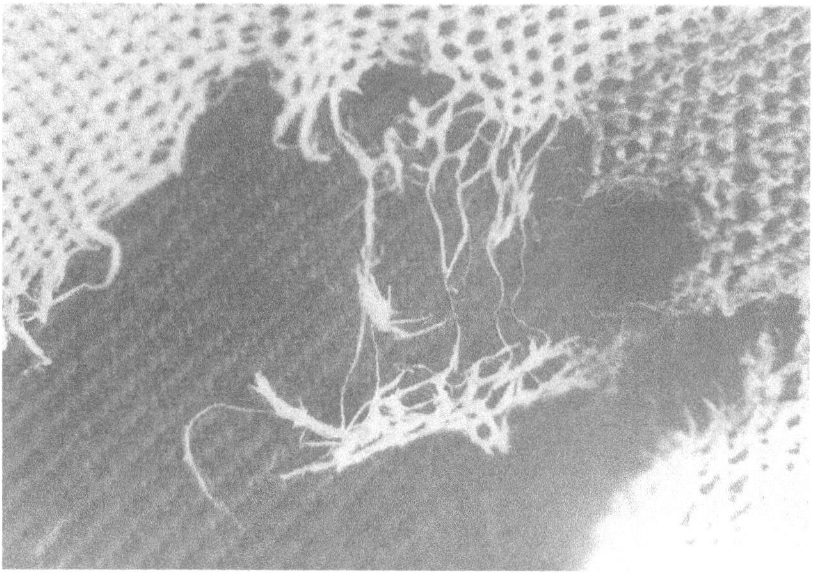

Figure 5.3. Detail of damage to dress fabric in the ball lightning report from Smethwick (Stenhoff 1976). [Photograph: Brian Tate.]

MeV for 1000 min and found only peaks that were also present in the background. However, when he integrated all 400 channels of the spectrum, there was a net increase of about 2.2 counts per minute in the spectrum of the dress material. About half of this could be attributed to fluctuations in the threshold level, which left about 1.1 counts per minute coming from the dress. For comparison, MacMahon counted a 0.39 μCi source of ^{88}Y under approximately the same conditions and obtained an integrated count rate of 2.68×10^5 counts per minute, implying a possible activity for the dress material of about 1.6 pCi (or 0.06 Bq). Without more prolonged and detailed measurement, MacMahon considered this figure an upper limit rather than a definite value (MacMahon 1975, Wooding 1976).

Assuming the material was polyester (specific heat capacity ≈ 1200 J kg^{-1} K^{-1}, melting point ≈ 250°C, density ≈ 1 g cm^{-3}), Wooding (1976) and Barry (1980) used the heat equation to estimate that the energy needed (1) to raise the missing material to its melting point was about 200 J, and (2) to melt the surface to a depth of 0.1 mm was about 240 J. Wooding stated that about 3 kJ would heat the surface of the hand to a depth of 1 mm.

Barry assumed that the ring was heated by thermal contact, and again using the heat equation with values for gold, an estimated ring volume of 1 cm^3 and an estimated temperature rise of 80 °C, he estimated the heat input to the ring as about

200 J. Wooding considered the possibility that the ring was heated by electromagnetic induction. He calculated that a current of 1 kA and an electric component of the electromagnetic field of 10 kV cm^{-1} would be needed to heat the ring to 100°C. If the inductance of the ring were 1 nH and the capacitance 10 pF, its resonant frequency would be about 1 GHz.

Wooding estimated the energy flux of the ball from the witness' statement that she "seemed to glow all over." He compared this sensation to that of a 1-kW radiant heater at a distance of 0.5 m, giving a flux of 60 W m^{-2}, so that a body area of 0.6 m^2 would receive 400 J in 1 s. An equivalent power would be radiated by a spherical blackbody of 0.1 m diameter at 950°C, or at 3500°C for one of a diameter of 1 cm. The thermal energy content of such a sphere would be 2 kJ. Radiation trapping by the opacity of a dense plasma would reduce the radiant flux, so this could be an underestimate of the energy.

Berger (personal communication, 1978) suggested that the observer had seen a short-circuit arc near the stove, caused by a direct flash or a side flash to the house. He suggested that her alarm at seeing the arc was sufficient to cause her to push her hand toward it, and in so doing to tear her dress with her thumbnail. He pointed out that an electric arc would be an unfamiliar sight for most people. However, the damaged fabric shows clear signs of exposure to localized heat rather than of tearing. Fibers were melted rather than sheared.

Stenhoff (1986) discussed whether such an arc might have melted the fabric, but considered that more severe and extensive damage would have resulted. Had a side flash jumped from the stove to the witness' lower body, heating could have occurred at the point of entry of the current, but the witness showed no signs of electric shock. Assuming the stove to have been properly grounded (as safety standards require), it would have offered a lower resistance path to ground than the observer's legs.

Stenhoff (1986) also discussed whether a small volume of methane from the gas stove might have been ignited. A side flash to the gas main could cause heating and expansion of methane, and ignition where it was in contact with the air. The minimum combustion limit for methane is 5.4% by volume. The complete combustion of a methane–air mixture of the dimensions of ball lightning would liberate between 100 J and 1 MJ. The equation for a spherical flame is $\Delta Q = d^3 s\rho\pi\Delta T/6$, where ΔQ is the heat added to a sphere of diameter d by combustion which raises the initial temperature of the gas by ΔT (Singer 1971). The relevant mixture of gases would have a specific heat capacity $s \sim 1.2$ kJ kg^{-1} K^{-1} and density $\rho \sim 0.2$ kg m^{-3}. Stenhoff (1986) argued that the energy content would be directly proportional to the volume and hence the cube of the diameter. He assumed blackbody radiation (an approximation) to deduce from Wien's law that a blue ball would have a temperature of 6200 K, and thus estimated the energy content of a sphere with a 10-cm diameter to be about 700 J. This energy is plausibly close to values obtained by Barry (1980) and the lower estimates of Wooding (1976).

Stenhoff argued, however, that it is difficult to imagine how a volume of escaping gas forming under such circumstances could assume near-perfect symmetry for a period of 1 s. Campbell (1998b) disagreed and referred to experimental work by Barry (1968a, 1980) (see Chapter 12). In these experiments, igniting propane in air by an electric spark produced luminous balls. The concentrations were about 50% of that normally required for normal combustion. The ball thus formed was yellow-green, brightly luminous, several centimeters in diameter, exhibited rapid random motion, and persisted for 1 to 2 s after the ignition source was removed.

Campbell proposed a scenario in which the ball was unrelated to the thunderstorm. According to his interpretation, the witness turned on the gas to the stove, but was slow to ignite it. At that moment or soon afterward, the unburned methane at low concentration was ignited, either by the source of ignition used by the observer or by another, already burning ring. He suggests that the observer did not notice this until she turned back toward the stove, having filled the kettle. However, the luminosities in Barry's experiment were formed inside a Plexiglas tube and one might expect more vigorous convective mixing in burning gases formed in the open air.

A very similar account in which dress fabric was damaged was published at the turn of this century (Elliot 1904). It is interesting that in this case the witness submitted her account in response to a suggestion previously published in *Symons' Meteorological Magazine* that ball lightning is a "subjective impression of some luminous phenomenon." We take this to refer to the idea of positive afterimages (see later discussion).

> As a child of about 12, I was in a house which was struck by lightning and still remember seeing what appeared like a great ball of fire poised above my bird cages for a few seconds.
> . . .
> On that occasion the electricity behaved with its usual eccentricity, deserting the pipe down which it was travelling to make the circuit of a small box-room, where it twisted a wire bonnet into a strange shape, and riddled a cotton dress with small holes such as might be made with the glowing end of a match; it then returned to the pipe and travelled safely to the ground.

Dijkhuis (1993a) reports an event on July 3, 1992, which occurred in Amsterdam. A person was struck on the forehead by a ball. At 11:30 p.m., he was standing in a second-floor room near an open balcony door that was covered with a nonmetallic mosquito screen. The weather was humid, with heavy rainfall, and there was a nearby lightning flash. Immediately thereafter, he saw a vivid green sphere of the size of a soccer ball above a tree adjacent to a nearby office building. The sphere, which was visible for only a fraction of a second, struck him on the forehead, throwing him backward and onto the floor. He was dazed but apparently did not lose consciousness. When his wife, hearing the noise, entered the room, they both detected a smell of ozone. There was no damage to anything in the room, but the victim developed a red spot on his neck. After medical consultation, he was referred for tests. Magnetic resonance image scanning of the brain indicated damage on one side of the frontal lobe.

We have discussed burns evidently sustained through contact with ball lightning. Many witnesses describe symptoms or sensations that might be attributed to electric shock.

On July 28, 1883, a fireball the size of a person's head paralyzed a man in Hartford, Connecticut (Tomlinson 1888, Anon. 1883, Corliss 1982).

Several interesting reports of a lightning incident with evidence of electric shocks were received from Conwy, Gwynedd, North Wales (Stenhoff 1992a). The events occurred on the evening of August 12, 1992. Mrs. Aline Owen was indoors at her home when "a great blue light" entered through the window and struck her on the ankle, throwing her across the room. Mrs. Pam Wignall, who lived in a house opposite, was struck on the chest and was also thrown across her room. Both women had to go to the hospital later for treatment for shock, and Mrs. Wignall suffered burns. Both the electrical supply and the plumbing in Mrs. Wignall's home were damaged, and there was damage to telephones (see Chapter 6).

At about 6:15 p.m., another neighbor, Mrs. Pat Stafford, was looking through the closed window of her bungalow when she saw what she first thought was a "ball of white fire," larger than a football, about 20 to 30 ft (6 to 9 m) from her, traveling horizontally toward her. There was no lightning or thunder, but very heavy rainfall and there may have been some hail. She saw the ball against the background of other houses, and her view of it was uninterrupted. The ball, round, opaque, and white-yellow, was surrounded by a blue, iridescent halo and appeared to be spinning or rotating. It reminded her of a meteor or comet. The light from it was like that of a fluorescent tube, and it was bright enough to be clearly visible in daylight. It was in sight for about 10 to 15 s. It hit an oak tree some 12 to 13 ft (4 m) away, whereupon it became smaller, rolled down the tree, and disappeared with a terrific crack and explosion, dispersing in flashes. There seemed to be "waves of lightning" passing from it into the ground and radial sparks streaming out of it in all directions. It had struck the tree about halfway up. The bark and trunk were split, showering splinters of wood over a distance of about 50 yd (45 m). (see Chapter 8). Although Mrs. Stafford thought the ball disappeared, her husband thought he saw the ball, now smaller in size, cross the lawn.

Adrian James (1992) found the following accounts:

> On July 20, 1783 "at Norwich, in the hamlet of Pockthorpe, a ball of fire fell on a dwelling-house, and passed through it without doing any material injury" (Anon. 1783a), while elsewhere "a ball of fire fell down the chimney of Mr. Hind, glazier; his wife, himself and a woman were sitting in the corner, and it made a great hissing as it came down, giving them some notice of its approach, but not sufficient for them to get away; the man was struck down, and lay as if dead on the floor for at least ten minutes, one of the women had her face

much scorched, and the other (who was not much injured) says she saw the fire run along the house as though a train of gunpowder had been lighted; from thence it turned up the stairs, made a large hole in the plastering, went through a wainscot partition at the top, and turning again passed through a large crack in another, and damaged the chamber ceiling." (Anon., 1783b)

Flammarion (1905) also provides this more recent report:

On March 6, 1894, M. Dandois, professor of surgery at the University of Louvain, went to the neighbouring town of Linden, by railway, to see a patient. On his return, on foot, the sky suddenly so darkened over that he made for the nearest dwelling-place, avoiding, as he did so, the telegraph poles along the road. Suddenly a ball of fire came against him and threw him over a ditch into a field, where he lay unconscious.

A quarter of an hour later, having regained his senses and finding himself undamaged save for a numbness in one arm and one leg, the doctor set out again, congratulating himself on the fact that his umbrella had acted as a sort of portable lightning conductor, for the steels were all twisted, and showed signs of having borne the brunt of the fray. Had the handle been of steel also, the electric current would have run down into his hand, doubtless, and killed him.

Arago (1854) gives the following account, which seems to describe very mild electric shock:

In a letter . . . dated 10th of September, 1713 . . . Maffei reports that . . . having stopped a short time earlier at the Castle Fosdinovo,[1] in the region of Massa-Carrara, during a thunderstorm accompanied by torrential rain, he was received by the lady of the castle in a room on the ground floor; and that there, he and the Marquis of Malaspina saw suddenly appear, at the surface of the flagstones, a bright flame or fire (fuoco), partly white, partly blue, that this fire appeared greatly agitated, but without any progressive movement, and that it vanished as it had appeared, that is, quite suddenly, but after having acquired considerable volume.

At this final moment, Maffei felt a peculiar kind of tickling sensation running down from behind of his shoulder; pieces of plaster detached from the vaulted ceiling of the hall fell on his head; and, lastly, he heard a crack, a sound, different, however, from the usual rumbling of thunder.

Electric shock may be very slight but still have long-term consequences.

A violent storm was raging near Marseilles, when seven persons, seated together in the ground-floor drawing room of a country house, saw a fireball as big as a plate appear in their midst.

It directed its course towards a young girl of eighteen, who, frightened out of her life, had fallen on her knees. Touching her shoes, it rebounded to the ceiling, then came down to her feet again, and so on two or three times, with mysterious regularity, the girl experiencing, it seems, no other sensation than that of a slight cramp in her legs. Eventually, the fireball made its exit from the room through a keyhole!

[1]Incorrectly given by Flammarion (1905) and Cade and Davis (1969) as "Fosdinaro."

> The girl could not get up after it had gone. For a fortnight or so she could not walk without assistance, and it was two years before she got over a liability to sudden weakness in her legs, causing her suddenly to fall. (Flammarion 1905)

5.4 Summary

To summarize, we note the following characteristics of reports of deaths attributed to ball lightning. A large number of these reports are from Flammarion (1905) and so the majority relate to the eighteenth and nineteenth centuries. A large proportion of these events occurred inside churches. Very few reports of similar events have been published in the twentieth century.

In many cases, the cause of death is unclear. In some cases, the cause seems to be electrocution or burning. There are some bizarre descriptions, such as victims being "dashed to pieces." Sometimes there are reports of multiple casualties.

Injuries attributed to ball lightning have included electric shock; paralysis; loss of consciousness; burns; singeing of hair; shattered limbs, fingers, and thumbs (sometimes requiring amputation); loss of hearing and loss of vision.

Corpses and survivors alike have sometimes been left with a localized area of inflammation or other damage (often near the head), sometimes with a red or black line leading to one foot, where further injury may be present; and shoes may be torn or even ripped off.

Animals that have allegedly been killed by ball lightning include farmyard animals such as horses, cows, sheep, and pigs and domestic animals such as cats and fish.

5.5 Discussion

Let us now compare these effects with those that are known to occur when people or animals are struck by ordinary lightning. Golde (1973) and Lee (1977) published excellent reviews of deaths and injuries caused by ordinary lightning.

The relatively minor injuries found among those who survive being struck by lightning include loss of consciousness, loss of memory, burns, temporary paralysis, and Lichtenberg figures on the skin (see following discussion).

Loss of consciousness may last from a few minutes to a few hours, with occasional relapses after apparent recovery. Loss of memory is of the type known as "retrograde amnesia," where the person may not remember having been struck, but only have a recollection of sensations of heat and light. This apart, memory is usually restored within a few hours. (Retrograde amnesia may sometimes be relevant to the accuracy of some reports of ball lightning.)

Paralysis is usually of the limbs, especially the legs. Death is uncommon where electric shock is through the legs and lower trunk. Paralysis usually lasts no more than a few hours to 1 day. Victims may suffer further injury from falls caused by paralysis or loss of consciousness. They may experience numbness or even a total loss of sensation from the waist downward.

Burns occur either at the point of entry of the current, or where the skin was in contact with metal objects because these objects are raised to a high temperature or may even be melted by the lightning current. There have been cases in which a metal object such as a necklace seems to have diverted the lightning current and saved a person from death. Burns are more likely to be caused by positive flashes (see Chapter 2). This is probably because positive flashes have significantly larger values of action integral and therefore cause much more joule heating. Human skin offers a greater resistance to electric current than other parts of the body, so a large voltage drop can occur across the skin, and this may also cause localized burns. (The defining equation for resistance is $R = V/I$, where V is potential difference in volts, I is current in amperes and R is resistance in ohms. Where R is greater for a given I, V will also be greater. This increased V value is known as the "contact voltage" or "touch potential.") Burns may also be found at the exit point of the lightning current.

The electrical resistance of the human body is about 500 to 1000 Ω, so with lightning currents of, say, 1 kA, the voltage between head and feet will be 1 MV, which enables flashover, i.e., much of the current bypasses the body and arcs through the air (see Chapter 4). Consequently many persons have survived even direct lightning strikes.

Lichtenberg figures on the surface of the body are thought to be caused by flashover. They consist of treelike discolorations of the skin where current entered the body. They are caused by electrical sparking on the surface of the skin and disappear after a few hours.

Flashover may cause rapid evaporation of surface moisture on the body, and tight-fitting clothing may be torn or ripped off. This is thought to result from vaporization of surface moisture (Lee 1977). Some victims of lightning strikes have had their shoes ripped or thrown off, and there have even been reports of victims of lightning being found naked. Frank W. Lane (Lane 1968) has published a photograph showing a pair of shoes "blasted off" by lightning. (The wearer apparently suffered only badly cut feet.)

The intense flash of light associated with a nearby lightning flash can cause blindness that may be permanent due to corneal burns, or witnesses may simply be dazzled. Pilots of aircraft whose cockpits are struck by lightning often suffer temporary blindness. Bright light sources produce afterimages. For lower levels of illumination, these are the complementary color of the original source (so a bright source produces an afterimage that appears dark). For higher levels of illumination, people may experience a positive afterimage that appears bright compared with the background.

The shock wave associated with thunder from a nearby lightning stroke can be sufficiently powerful to cause temporary or permanent deafness—in the latter case, the eardrum may be ruptured.

Most deaths by ordinary lightning are caused by direct strikes and side flashes to the head. It appears that Professor Richmann was killed during his lightning experiment by a direct lightning flash to the apparatus (Uman 1986).

Electrocution by lightning may cause ventricular fibrillation and/or respiratory arrest, which may in turn cause death. Ventricular fibrillation or asphyxia may be caused by part of the current being passed through the heart. In cases of respiratory failure, it may be instead that part of the lightning current has passed through the brain and, more especially, through the respiratory center. The latter is located near the base of the skull. Heating effects sometimes cause brain damage. The corpses of those who have been electrocuted often appear cyanosed (blue in color) because blood has ceased to absorb oxygen. Autopsies may reveal damage to heart muscles and lungs, but it is not always possible to ascertain the cause of death. Many people have even survived direct strikes. It is thought that the duration of current flow may sometimes be too short to cause fibrillation.

Lightning that strikes, for example, a nearby tree or the open ground may kill animals, especially quadrupeds. In a flash between cloud and ground, the lightning current passes into the earth. Soil has a finite resistance, so a radial potential gradient is set up around the point of impact. This can create a potential difference between two points on the ground. The further apart these two are, the greater the potential difference. (This is what makes quadrupeds particularly vulnerable.) The voltage may be sufficient to drive a fatal current through the hapless animal. This mechanism is called the *step potential* (see Chapter 4). Lane (1945) published a dramatic photograph of part of a flock of 504 sheep killed by lightning on July 22, 1918, in the Wasatch National Forest, Utah (Lane 1945, plate 28).

It is fortunate that flashover often occurs, preventing current from passing through the body, because humans and other mammals are very sensitive to electric currents. Relatively modest currents of about 40 to 60 mA can cause asphyxia, and currents of several amperes will cause ventricular fibrillation. Lightning currents are typically in the range of 10 to 100 kA, i.e., several powers of ten greater!

5.6 Conclusions

Many of the reports discussed here provide insufficient information concerning the cause of death or injury. Many that attribute death or injury to the effect of ball lightning do not give a clear indication that the ball made contact with the victim. Most reports that provide more than superficial information, however, describe injuries or deaths that could readily be explained by the effects of ordinary lightning. Adrian James endorses this conclusion. In his excellent and critical review of fatalities attributed to ball lightning (James 1992), he points out the difficulty of interpreting historical accounts and highlights the confusion between linear and ball lightning. He also indicates discrepancies of interpretation in comparing different sources, exemplified by the Little Sodbury case in 1556. He indicates that some historical accounts, such as that from Cornwall in 1757, are less ambiguous, but that the cause of death in the latter case was electrocution consistent with ordinary lightning.

Historical accounts of injuries are similarly ambiguous. Most injuries can readily be explained as the effects of ordinary lightning. Only a minority of cases involving superficial burns may be more difficult to explain in this way.

The reports discussed in this chapter thus lead to the conclusion that ball lightning is not in itself a particularly hazardous phenomenon. The same is true of St. Elmo's fire. However, both phenomena are closely associated with ordinary lightning and may precede a nearby lightning flash to the ground. Observers of either phenomenon would thus do well to adopt the same precautions that are generally associated with nearby thunderstorms.

Chapter 6

Assessment of Risk to Buildings

6.1 Case Histories

A number of examples have already been quoted in which ball lightning has been considered the cause of damage to buildings, including churches and houses. This chapter presents further accounts of such damage.

6.1.1 Structural Damage and Fires

Because of their structure and tall spires, churches are vulnerable to lightning strikes and so an affinity of ball lightning for churches might be expected.

> In the parish of Samford-Courtney in Devon on October 7, 1811, a sudden darkness came on, and a fire ball fell in the vicinity of a church. The ringers in the belfry, ringing at the time, declared that they never knew the bells go so heavy, and were obliged to desist ringing. Looking down from the belfry into the church they perceived four fire balls, which suddenly burst, and the church was filled with fire and smoke, some of which ascended to the tower, where a large beam, on which one of the bells was hung, was broken, and the gudgeon breaking, the bell fell to the floor. (Nichols 1928, Blair 1973)

Tomlinson (1859) quotes a very similar event. At about 5 p.m. on August 1, 1846 during an exceptionally violent storm with torrential rain over Leicester that caused much flooding, there were several reports of "fireballs" as well as ordinary lightning. At 8:05 p.m., St. George's Church was struck by lightning and there was a loud explosion. The sexton, tolling the bell, saw the clapper strike the bell twice on its own. Two observers, one the rector of the parish, described a vivid stream of light, followed by a globular mass of fire, red in color, which struck the upper part of the spire. The spire shattered in a tremendous explosion and masonry fell from it and damaged the tower and the church roof (Tomlinson 1859).

On August 7, 1892, after a flash of lightning, a large globe of fire shaped like a teardrop exploded in a church in Germany, causing extensive damage and leaving a sulfurlike odor (Anon. 1892j).

A number of authors have noted the apparent affinity of ball lightning for chimneys. These reports are not always associated with damage. In the summer of 1935 or 1936 during a thunderstorm, a child in a house in Moseley Village, Wolverhampton, West Midlands, England, saw a yellow or orange ball about 12 to 18 in. in diameter come down the chimney, bounce across the room against two of the walls, and disappear silently. It was in sight for some 8 to 10 s. In this case, there were no signs of damage (G. W. Garner, personal communication, 1976).

Arago (1854) describes a more damaging event on June 2, 1843[1] transmitted by M. Babinet to the Académie des Sciences. This report described ball lightning the size of a child's head that entered a room of a house in Paris via the chimney some while after a fairly loud thunderclap. As it did so, it knocked over a frame covered with paper in front of the fireplace. The ball moved slowly just above the tiled floor and approached the feet of the witness, whereupon he jumped backward; it remained there for several seconds, during which time the witness examined it carefully, before the ball moved up to his face, making him fall back into a chair. The ball was bright and shining, but he felt no heat. The ball then elongated somewhat and disappeared via a small, paper-covered hole in the chimney above the mantelpiece, exploding near the top and scattering fragments of brick over the surrounding area. The ball had "unpasted the paper without hurting it."

Flammarion (1905) reports a very similar event:

> In October 1898, [ball lightning] made its appearance in a room and advanced towards a young girl who was seated at a table, her feet hanging down without touching the floor. The luminous globe moved along the floor in the girl's direction, began to rise quite near her and then round and round her, spiral fashion, darted off towards a hole in the chimney—a hole made for the stove-pipe, and closed up with glued paper—made its way up the chimney, and, on emerging into the open air, gave out upon the roof an appalling crash which shook the entire house. It was a case of coming in like a lamb and going out like a lion!

Flammarion (1905) also provides the following dramatic account, which was apparently interpreted by Cade and Davis (1969) as a description of ball lightning:

> In 1897, at Linguy (Eure-et-Loire), a man and his wife were sleeping quietly, when suddenly a terrible crash made them jump out of bed. They thought their last hour had come. The chimney, broken to pieces, had fallen in and its wreckage filled the room, the gable-end was put out and the roof threatened to come down. The effects of the thunderbolt in the room were less alarming than the effects outside, but were very curious. For instance, bricks from one wall had been dashed horizontally against the wall opposite, with such extraordinary force that they were to be seen imbedded in it up above a dresser upon which pots and pans,

[1]Not July 1, 1852 as stated by Flammarion (1905) or Cade and Davis (1969).

> etc., were ranged, and within a few inches of the ceiling, while the windows of the room had been smashed into bits, and a looking-glass, detached from the wall, stood on end whole and entire upon the floor, delicately balanced. A chair near the bed, upon which articles of clothing had been placed, had been spirited away to a spot near the door. A small lamp and a box of matches were to be found undamaged upon the floor. An old gun, suspended from a beam, was violently shaken and had lost its ramrod.
>
> The thunderbolt actually frolicked over the bed, leaving its occupants more dead than alive from terror but quite unhurt. It passed within a few inches of their heads and passed through a fissure in a partition into an adjoining dairy, where it carried a whole row of milk-cans, full of milk, from one side to another, breaking the lids but not upsetting a single can. It broke four plates out of a dozen, leaving the remaining eight intact. It carried away the tap from a small barrel of wine, which emptied itself in consequence.
>
> It ended by passing out through the window without further breakage, leaving the husband and wife unscathed but panic-stricken.

It is interesting to note that Cade and Davis (1969) wrote, ". . . they were terrified to see a dazzling ball of fire which passed over the bed within inches of their heads . . . ," although they quote Flammarion as their source and he does not provide so specific a description.

On July 17, 1964, there were two ball lightning reports from London. The first of these was described as follows:

> About twenty years ago one very sultry mid-morning I opened our upstairs window and to my amazement saw an orange ball pass in front of my eyes. It was a tightly packed ball of fire. I could see the packed flame very clearly for a brief moment. Then there was a loud crack. I ran into the street and saw my neighbor's iron gutter smashed, the neighbor's chimney to their right was smashed and fell to the ground. My own gutter was intact except for two bolts which lay on the ground with scorch marks on them. I have often wondered how the damage was caused because it seemed to me that the fireball was just passing at the moment of the loud crack. (N. Swain, personal communication, 1984)

Cade and Davis (1969) describe what is evidently the same event. Their sources were the London *Evening News* and *Evening Standard* for July 17, 1964. Cade and Davis say there was a violent thunderstorm at the time, but Mrs. Swain states that there was no storm or rain before the appearance of the ball. She saw it at about 11 a.m. and it was in sight for about a second. It passed within only 45 cm of her. It was the size of a medium-sized orange. It was opaque and bright enough to be clearly visible in daylight. It seemed brighter near its outer surface and appeared to spin. Its appearance did not change significantly during the observation. Its movement was mostly horizontal below the iron gutters, and Mrs. Swain wondered whether the gutters guided it. Cade and Davis describe further damage to a neighboring house—the ball apparently fell through the roof and started a serious fire. The garden was strewn with tiles and rubble, and the television aerial was thrown into the street.

The second report was from a school in Shoreditch, where a lightning conductor was damaged, masonry fell into the playground, and windows in the school and a neighboring house were broken. There was no clear description of a ball (Cade and Davis 1969, *Evening News* and *Evening Standard*, June 17, 1964).

Some reports describe damage to brickwork. Some examples were given in earlier chapters. Chapter 5 contains a report of ball lightning seen on August 24, 1895 which descended toward the ground and split into two smaller globes that rose immediately to the height of the chimneys on nearby houses and disappeared. One descended through a chimney, crossed a room, where it was seen by a man and a child, and went through the floor, "perforating a brick with a clean round hole of about the size of a franc" (Flammarion 1905).

During a storm in Karachi, Pakistan, also in 1895, three people in a completely closed room saw a fireball 6 in. (15 cm) in diameter. There was then a thunderous report. One person felt a sharp pain on his face, another had a trembling arm. A rifle holder in an adjacent room was broken, and there were three holes in the wall, one of them where the rifle holder was attached to the wall. There was a powerful odor of sulfur (Ryan 1895).

In Alpena, Michigan, on August 1, 1907, there was a report of ball lightning that fell to the floor, then circled around a room, smashing holes in walls (Anon. 1930a).

On February 12, 1983 at 4:30 p.m. in Maidenhead, Berkshire, England, it started to snow very lightly. There was no thunderstorm. A woman saw a bright blue flash go past her glass front door. A second or so later there was a loud bang that shook the house. It sounded like a loud bomb exploding and was heard for several miles around. There were television and citizens' band antennae on the roof. In her house there was extensive damage—all electrical devices were destroyed. All the light bulbs smashed, plugs were blown from wall sockets and broken, plaster was ejected from the walls near any wiring, and the telephone and answering machine were destroyed (see Fig. 6.1). Some components of a videocassette recorder disappeared altogether. There was a serious fire in the upper part of the house that destroyed most of the contents and the roof. Firemen who attended the scene were adamant that a fireball "which had been seen to pass over Maidenhead" was the cause.

There was a very unusual hole in the brickwork between the back bedrooms. It appeared as if someone had drilled a hole 2 in. (5 cm) in diameter, and the outer bricks seemed to have been blasted away. There were no burn marks, however, but burn marks were seen near the citizens' band antenna, so it was assumed that this was the point of the strike. Several large holes were blown through the house walls. Paving stones in the garden had been lifted up and smashed in half. Many people in the area suffered damage to televisions and telephones. A woman using a

Figure 6.1. House damaged by lightning at Maidenhead on February 12, 1983. [Photograph reproduced courtesy of the *Maidenhead Advertiser*, © 1983.]

telephone nearby suffered minor facial burns (*Maidenhead Advertiser*, February 18, 1983; E. P. Folland, personal communication, 1983).

Some reports of ball lightning make it quite clear that electrical damage is not attributed to the ball. An agricultural research scientist, who requested anonymity, described an event that occurred early in a monsoon storm of average violence on January 1, 1973 at about 9 p.m. at Banting, Selangor, Malaysia. The level of rainfall was typical for a monsoon. The storm was moving in a southerly direction parallel to the long axis of the house, which was of ferroconcrete construction. At the northerly end was a living room (with a bedroom and bathroom above it upstairs), then a passage with a toilet adjacent to it, leading to a study at the far end of the passage.

> My wife and I were in the living room when the storm broke. My wife looks upon lightning as a kind of directed malevolence and went to take refuge in the study, the wooden floor of which [she] considered to be "safer." All the rest of the floors were tiled. Meantime a bolt of lightning struck the ventilator pipe from the bathroom at the north end of the house and blew off some surrounding roof tiles, as we later discovered. Then as my wife was nearing the study doorway she turned to call to me to follow.
>
> At that moment the ball lightning appeared in the passage, so we both saw it from different directions. It was bluish, about one foot (30 cm) in diameter. It moved down the passage into the toilet and thence presumably into the plumbing with a hissing noise. There was no heat or odor.
>
> Shortly afterwards another bolt of lightning struck at or near the south end of the house: the nearness [could] be judged by the viciousness of the zip before the explosion. (Anon., personal communication, 1983)

The ball appeared about 4 s after the first lightning stroke, and was about 30 ft (9 m) from the probable point of contact of the CG stroke. This point was behind the witness as he faced the ball, and out of the line of sight of him or his wife. The ball was formed about 20 ft (6 m) from him. It was translucent and uniformly bright across its surface, but only bright enough to be barely visible in daylight. Its appearance did not change significantly throughout its lifetime of 3 to 4 s. Its speed was about 3 ft/s (0.9 m s^{-1}).

6.1.2 Damage to Glass, Asbestos, etc.

There are many reports of ball lightning passing through windows. Grigor'ev, Grigor'eva, and Shiraeava (1992) have collected many such accounts. In some of these reports, the ball lightning is not alleged to have caused any damage; in others, the ball squeezes through a small crack; while in still others the ball is reported to leave a hole of about its own size or somewhat smaller. In these latter cases, unfortunately, the holes have mostly been found after thunderstorms without ball lightning being seen, and it has been assumed that ball lightning was their cause.

Turner (1997) discussed three cases in impressive detail and used them in support of his ball lightning model (see Chapter 12). He examined damaged window panes with the help of experienced research scientists from the glass industry. He argued that all three windows fractured because of exceptionally even radial stress. He concluded that the most likely cause was ball lightning.

The first report Turner considered was from the Department of Meteorology at the University of Edinburgh, Scotland. McIntosh (1973) attributed the cause of this to ball lightning, while Campbell (1981) subsequently claimed that it was caused by a glass marble that had been projected at the window. The second was found in a window in Stockholm following a thunderstorm in 1944. This was one of two such examples discussed by Müller-Hillebrand (1965), who attributed both to conventional linear lightning. The third, also from Sweden, was discovered in a window in Uppsala after a storm on August 6, 1994.

Turner also mentioned an elliptical hole found in a window in New South Wales, Australia following a reported nearby sighting of ball lightning (Needham 1993).

Kolosovskii (1981) published a report that more satisfactorily correlated ball lightning with damage to a window. A teacher and a class of children at a school in Moskovskaya Oblast, Russia in summer 1977 saw a red, "hairy" sphere about 5 cm in diameter approach the window from the outside. It formed a small round aperture with glowing red edges, which enlarged to 3 to 4 cm. This continued for about 5 s. After that, the lightning flared and disappeared noisily. At that moment, the teacher was holding an appliance connected to the electric main and he experienced an electric shock. The hole was 5 cm in diameter and conical, with the diameter on the outer side about 1 mm greater than inside, and the glass was 2.5 mm thick.

Experiments were carried out with a continuous-wave carbon dioxide laser to try to replicate the effect. The authors concluded that these experiments suggested that the heating of the glass was a rapid surface effect rather than being uniformly distributed throughout its thickness.

Sometimes glass or other brittle materials are reportedly shattered. Chapter 4 referred to a ball lightning report from Knutton, Staffordshire, England, dated March 21, 1983. During the same storm, Mr. Geoffrey Millward, a machine supervisor, was working alone in a mixing shed at Alfa Aggregates, Kingsley, near Cheadle, Staffordshire. This is his account written immediately after the event:

> *Monday 21.3.83.* At about 10:45 a.m., the storm came over—dark, rain, sleet and snow, then this terrific flash of lightning—and this thunderbolt came in under the shed where the two box feeders are, [came] up to the first floor landing [where it] tried to get out through the glass sheets . . . then [turned] down and went crashing through the [corrugated asbestos]

sheets at the end of [the] shed blowing them as far as the china clay. . . .
(R. Maddison, personal communication, 1983)

Dr. Ron Maddison of the Department of Physics at Keele University visited the site and interviewed Mr. Millward (R. Maddison, personal communication, 1983). The ball was believed to have entered the building through a door, made a clean crack in a wire-reinforced glass skylight, and bounced off walls. The hole in the asbestos sheeting was 6 × 10 ft (1.8 to 3.0 m) and sections of the asbestos wall were projected over distances up to 40 ft (12 m) (*Staffordshire Evening Sentinel*, March 23, 1983). The ball had traveled the full length of the building. When first seen, it was about 12 ft (3.7 m) away; it was red, and its diameter was about 12 in. (30 cm). It was bright enough to be clearly visible in daylight and was uniformly bright all over. Its appearance did not change throughout the observation, which lasted about 5 s. It was about 15 ft (4.6 m) away when it disappeared (G. Millward, personal communication, 1986).

6.2 Interpretation

6.2.1 Side Flashes and Buildings

In earlier chapters, we discussed the mechanism known as the side flash and we now consider how this may apply to lightning strikes to buildings. When lightning strikes a conductor grounded at one end, there are two factors that together determine the relationship between the potential u at the top of the conductor and the current i—these are the resistance R and the inductance L of the conductor. The relationship is (Golde 1973)

$$u = iR + L\mathrm{d}i/\mathrm{d}t. \tag{6.1}$$

Median values of maximum di/dt are 12 kA μs^{-1} for negative first strokes, 40 kA μs^{-1} for negative subsequent strokes, and 2.4 kA μs^{-1} for positive flashes. The top 5% values of maximum di/dt in each case are 32 kA μs^{-1} for negative first strokes, 120 kA μs^{-1} for negative subsequent strokes, and 32 kA μs^{-1} for positive flashes (Golde 1977b). The resistive term in Eq. 6.1 is thus dominant for negative first strokes and for positive flashes, while the inductive term is dominant for negative subsequent flashes.

Golde (1973) gives an estimate based on a lightning conductor on the chimney of a house, 10 m above the ground, with a resistance between the chimney and ground of 10 Ω, which is struck by lightning with a current crest value of i = 100 kA and a rate of rise of current di/dt of 50 kA μs^{-1}. He gives the inductance L of a single vertical conductor as approximately 1.6 $\mu H\ m^{-1}$. By way of simplification he ignores any phase relationship between the two terms, and thus uses Eq. 6.1 to show that the top of the lightning conductor is raised to a potential of 1.8 MV. If there is a nearby water tank at a distance D from the lightning conductor, connected by water pipes to the ground, then this will be at ground potential. If the potential difference exceeds the electric breakdown strength of the gap, breakdown will occur between the lightning conductor and the tank, and an arc will be produced. This is another form of side flash. Few houses have lightning conductors, but the argument could apply equally well to a television antenna on the roof.

Side flashes may be responsible for the electrocution of people indoors, especially if they are using the telephone or standing near bathtubs, sinks, or connected electrical appliances (Uman 1971, 1986). The electrical breakdown strength of an air gap of several meters width when subjected to an impulsive voltage is about 0.5 MV m^{-1} (Golde 1973). The foregoing rather simplified calculation implies, therefore, that given the estimated values, a side flash can traverse a distance of up to 3.6 m in this way.

Using the values for the British Code on lightning protection, $i_{max} = 150$ kA and $di_{max}/dt = 40$ kA μs^{-1} (Golde 1973), this gives a distance over which the arc may strike of 4.28 m. The largest distance over which the arc can strike is obtained by substituting values for the top 5% of i and di/dt values for positive flashes; this gives an estimate of about 6 m. It is thus quite possible for a side flash to produce an arc of considerably greater length than the height of a room. This would probably be associated with extensive thermal and mechanical damage, including shock-wave damage (see later discussion) reminiscent of the effect of a lightning return stroke, for in effect the side flash has allowed the return stroke to enter the building. This could explain the breaking of glass in the Woodlands St. Mary, Berkshire case in 1982 (Chapter 4), and the fact that glass fragments were found outside the building. Perhaps an arc was produced inside the bathroom from the immersion heater.

The arc will be exceptionally intense, and although its duration will be comparable to that of the return stroke, it might perhaps sometimes be a source of positive afterimages, particularly when the gap traversed between conductors is a few centimeters. Of course, the arc might also be a mechanism for generation of ball lightning. It is essential to make a field visit to buildings in which incidents are reported in order to establish whether these are possibilities.

We must not underestimate the spectacular nature of the experience of an observer inside a building penetrated by a side flash. The route taken by a side flash may often be interpreted as the path followed by ball lightning, damaging electrical and other appliances as it passes them. In many cases, damage is attributed to the ball retrospectively. As this section was being written, there was a report on the television news describing a serious house fire caused by lightning during a severe thunderstorm in Britain on August 12, 1997. One of the residents of the house said, "I thought we had been hit by a bomb—the noise was absolutely horrendous. We were hit by a thunderbolt which came through the back of the house, out through the front of the house, arced across to this [street] lamp here, and then [went] across to that house over there."

6.2.2 Overvoltages and Flashover

An overvoltage caused by a side flash may cause dielectric breakdown through the insulator surrounding cables carrying mains electricity. The consequent arc may persist after the lightning current has ceased. The arc may burn through further insulators and move along the conductors. This may appear as a moving region of luminosity on a wall if cables are buried within the wall. The speed of propagation of this region may depend on the rate at which the insulator breaks down or melts.

6.2.3 Glass

Golde (1973) writes that conventional linear lightning has punched holes of 1 or 2 cm diameter in glass. He indicates that when lightning strikes an insulator or poor conductor, the point of contact undergoes significant heating. He refers to work by Uman (1964) which deduces the diameter of lightning from evidence such as this.

Slow heating of glass might be expected to produce melting, while rapid heating could cause differential expansion and fracture owing to the relatively low thermal conductivity of window glass. Whether the holes discussed by Turner and other authors could be explained in this way is a matter for debate. Certainly, reports in which ball lightning was not seen to contact a window offer much weaker evidence than those (e.g., Kolosovskii 1981) in which contact was observed. However, glass heated to red heat by ordinary lightning might appear as a glowing, red sphere, and the glow would persist as the glass gradually cooled. Glass and other brittle substances such as asbestos may also simply be smashed by the pressure wave resulting from the rapid expansion of the lightning channel, or perhaps due to resonance with certain frequencies in the spectrum of thunder.

Charman (1971) has pointed out that reports of ball lightning seeming to pass through glass windows and leaving them intact might sometimes be explained by perceptual error resulting from an actual increase in the size of the ball outside the window.

6.2.4 Concrete, Brick, and Mortar

The dielectric strength of porous insulators such as concrete, brick, and mortar, like the dielectric strength of soil, depends heavily on their water content. When breakdown occurs, rapid vaporization of any water held in the insulator will often cause explosive fracture (Golde 1973).

Sometimes lightning punches holes in brickwork. On occasions, the path of lightning through a building may partly be marked by removal of plaster around mortar between bricks, probably because of the greater moisture content of the mortar.

6.3 Postscript: A Field Study

A field investigation by the present author (Stenhoff 1988a) shows the extent and variety of damage that may be caused by an ordinary ground flash (see Figs. 6.2 to 6.4). A 20-m cedar tree whose trunk was only about 2.5 m from a house, and whose branches overhung the roof of the house, was struck during a violent thunderstorm. There was evidently a side flash to the chimney of the house, which traveled to the ground through the domestic electricity system. Bark from the tree was stripped over a 1-m length at the approximate height of the chimney stack. The chimney stack was split and hundreds of tiles were removed, probably by the pressure wave from the return stroke.

In a second-floor attic beneath the chimney, the current followed the seam between bricks, removing a furrow in the plaster that led to an electric clock (Figs. 6.2 and 6.3). The clock was thrown off the wall with such force that as it disintegrated on contact with the wall opposite, its internal components made clear imprints in the plaster. They had fallen through a height of less than 20 cm in traveling a horizontal distance of 3 m. Hence the horizontal velocity of the components was about 15 m s^{-1}. An electric fire unit beneath the clock was projected from the wall. Windows were broken in the attic. The current path was from the clock to the fire and down to a fuse box on the first floor. As the fuses melted, the glass door of the box was shattered.

A second path of the lightning current was outside the house, through a cast-iron downpipe that was fractured at the bottom, where arcing occurred to an outside electric cable. Large pieces of stucco were ripped from the front of the house above the downpipe (Fig. 6.4). On the ground floor, the motor of the central heating boiler was damaged.

The house was unoccupied at the time. Let us for a moment imagine the kind of description that might have appeared in a local newspaper had there been people in the house to witness the event.

> A house in Egham was struck by a fireball during a violent thunderstorm. The thunderbolt fell onto a 20-m tree, ripping bark from it, then bounced onto the chimney, which it demolished. The ball passed through the roof, wrenching off tiles as it went, into an attic room where a woman saw it move horizontally along a wall, removing a furrow of plaster as it did so. The ball was a brilliant, luminous, sparking mass. There was a strong smell of ozone and burning. It made its way to an electric clock, which immediately disintegrated, hurling its parts dangerously across the room. It

Figure 6.2. Damage in the attic of a house in Egham struck by ordinary lightning showing a furrow in the plaster and the electric fire that was torn from the wall. The electric clock had been located between the two. [Photograph: Dr. John Gordon.]

> moved down the wall to an electric fire, which it ripped from the wall with a massive explosion. It sent such a large charge into the electric mains that the fuses blew and the central heating boiler was damaged beyond repair.
>
> Although no-one saw it, a second ball fell onto a cast-iron downpipe, fracturing it as it exploded. The explosion also broke upstairs windows and tore large pieces of stucco from the front of the house.

The evidence that ball lightning presents a risk to buildings is thus very weak. Nonetheless, there are some very interesting and compelling visual reports of ball

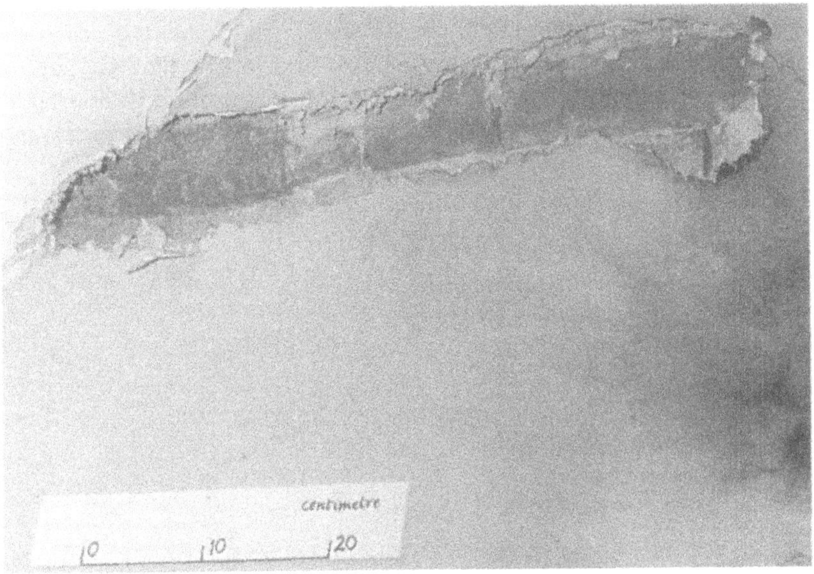

Figure 6.3. Detail of the furrow in plaster shown in Fig. 6.2. [Photograph: Dr. John Gordon.]

Figure 6.4. Damage to the outside of the house in Egham. Note broken stucco and windows. [Photograph: Dr. John Gordon.]

lightning within houses that are difficult to dismiss. As with reports of observations made within aircraft (Chapter 7), the presence of objects in the foreground and background significantly reduces the likelihood of perceptual error. The detail with which ball lightning is often described far exceeds the vague description in the preceding imaginary account.

Chapter 7

Assessment of Risk to Aircraft

Perhaps it is a result of the conditions under which observations are made that there is a greater degree of consistency in descriptions of ball lightning in aircraft than of ball lightning seen elsewhere.

7.1 Ball Lightning External to Aircraft

7.1.1 Cases where Damage Has Been Sustained

On August 12, 1956 at 12:55 p.m. local time, a Russian LI-2 propeller-driven transport plane was struck by ball lightning while flying at 3300 m in cumulonimbus clouds. The incident occurred after the airplane had flown through a stormy cold front and bad icing had occurred. Cloud peaks ascended to 5.5 to 6.0 km, the air temperature at that altitude was –2° to –4°C, and the air was very turbulent. As the aircraft passed through the storm cloud, there was severe radio interference. The bumping of the aircraft increased sharply. Immediately afterward, a dark-red (almost orange) fireball 25–30 cm in diameter rapidly approached the front of the aircraft. When it came within 30–40 cm of the nose, it veered off to the left and passed by the cockpit. As it did so, it struck the port propeller and exploded. There was a vivid white flash and a loud explosion, which was audible above the noise of the engines and reminiscent of the explosion of a torpedo in water. A fiery stream advanced along the port side of the fuselage and the aircraft was abruptly thrown upward. After the lightning discharge, the radio operator tried to disconnect the antenna and received a shock. Once the aircraft had landed, it was found that the only damage was to the trailing edge of the propeller blade. An area 40 mm in length and 5–10 mm wide had been melted 30 mm from the tip (see Fig. 7.1). Around this

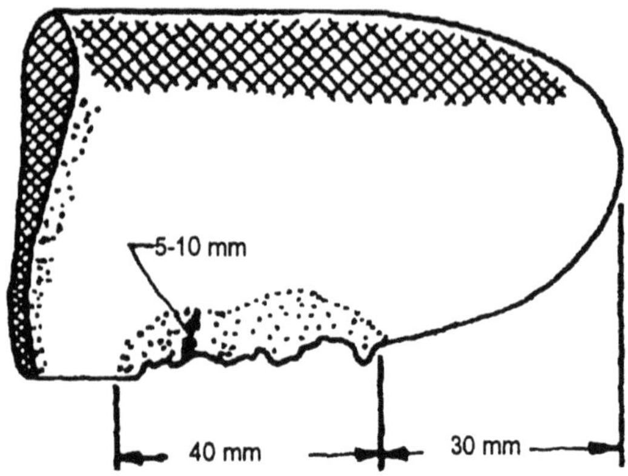

Figure 7.1. Damage to propeller blade.

region was a loose layer of soot that could easily be brushed off by hand (Ritchie 1961).

An aircraft was flying in dense, layered clouds at 400 km/h over Irkutsk, Russia in 1959. The plane was just below the center of the lower layer of cloud at 1400 m. The wing tips were not visible and there was much turbulence. At the same time as there was an ordinary flash of lightning, the plane was buffeted by a strong blow, but it remained under control. A "fist-sized" ball passed along the port side of the aircraft, accompanied by a shower of sparks. The pilot ascended to 4000 m above the clouds. The cabin was found to be strongly magnetized, as was evidenced by a 100° error in the magnetic compass. The radio was still functional, but the radio compass gave erroneous readings. After landing, the crew discovered that rivets had melted off on the port side in the front of the fuselage, but the surface itself was not destroyed (Popov 1959).

In 1965 a Russian LI-2 transport aircraft flying over the Kola Peninsula was passing through the northern part of a cyclone. There were thick nimbostratus clouds with visibility so low that the wing tips were not visible. The temperature at the altitude of the aircraft was –5°C. A sound like a rifle shot was heard when a sphere 60 to 80 cm in diameter struck the front of the aircraft. The passengers and crew observed red, streaming discharges moving along the surface of the airplane. After the collision, the radio compass rotated and the magnetic compass spun erratically for 3 to 5 min. The radio equipment was put out of action. External inspection of the aircraft after landing showed that several rivets were damaged in the forward fuselage and two holes 1.5 to 2 cm in diameter were melted in the trailing edge of the elevator (Wojkowsky 1966).

A T-33 jet trainer was flying near Moody Air Force Base in Georgia in the United States in 1952. The pilot was told to proceed to Mobile, Alabama, because of a thunderstorm. As the aircraft moved out onto a westerly heading at an altitude of 4 km, its nose was struck head-on by a "big orange ball of fire." The jolt was so severe that the pilot believed he had collided with another aircraft. The low-frequency radio compass stopped working and the crew needed radio guidance to another base. There were no marks or holes on the aircraft. The radio compass in the nose of the plane was practically melted inside and beyond repair. Once the radio compass was replaced, everything functioned normally (Altschuler 1969).

A military fighter crashed on January 7, 1948 following observations by several people of a flaming red ball that exploded in a cloudy sky (Benedicks 1954).

7.1.2 Cases where Aircraft Have Not Sustained Damage

In 1948, a TWA commercial airliner was flying from Paris to Cairo at 3400 m. There were dense clouds. A bump was felt beneath the cabin. A passenger looking through a porthole saw an orange-yellow fiery ball with a dark gray-violet layer surrounding its bright center, slightly larger than a tennis ball, rise from under the cabin. The ball was 2 to 3 cm thick, with a short tail like a rotating spiral. Its speed was equal to that of the airplane. About 30 cm from the side of the cabin, the ball exploded and ejected ahead of it a bright ray like a magnesium light almost 3 m long. The explosion of the ball was similar to a loud detonation, louder than a revolver. The ball was in sight for about 1 to 2 s (Baratoux 1952).

Ball lightning was reported near a Russian airplane in December 1956 when the plane was ascending at 2500 m in cumulonimbus clouds over European Russia. The cloud peaks were at 4500–5000 m. The ball lightning appeared in front of the airplane to the right and fragmented almost immediately with a bright flash and an audible, "sharp and muffled" explosion. The engine near to the where the ball had been seen stalled, but the pilot managed to restart it. On landing, the crew found no visible damage to the aircraft (Ritchie 1961).

In the same year, a Royal Canadian Air Force pilot was flying westward at 11 km altitude over the foothills of the Canadian Rockies near MacLeod, Alberta. There was a very severe thunderstorm—the most intense he had ever experienced in North America. Cloud pillars extended to above 12 km. He could not see the sun, which had set behind the mountains. Beneath him, the ground was in darkness. Through a gap in the clouds he saw a bright, motionless light with sharply defined edges "like a shiny silver dollar." It was buried deep within the thunderstorm, suspended over some cumulus clouds at a 4-km altitude. The diameter of the light was estimated as 15 to 30 m, and it was in sight for about 45 s (Altschuler 1969).

Other incidents have been mentioned in the literature. Jennison (1973) (see following discussion) refers to an observation of a 20-cm ball that appeared 50 cm above the trailing edge of the wing of an airplane in flight. It moved parallel to the wing at a speed of about 1 m/s before being cast off the end. The ball was not blown off despite its remarkable air speed. A similar incident was reported by Ingle (1971): "a bright ball appeared on the top surface of the wing outside the aircraft, made rapid movements to and fro for an appreciable length of time and then disappeared. It seemed quite unaffected by the air passing through it at 250 miles per hour. . . ." A further, similar observation was made on April 23, 1964 from an aircraft over Bedford, England. There was a loud bang and a whitish-blue flash of light. A ball of light the size of a football appeared on the starboard wing tip for about 2 s, then vanished (Vidler 1964).

It is interesting to note that there were numerous descriptions from pilots during World War II of luminous spheres, which they called "Foo fighters." Some of these descriptions are reminiscent of those of ball lightning (Anon. 1945a, Anon. 1945b, Chamberlin 1945). Table 7.1 summarizes the typical pattern found in these events.

7.2 Ball Lightning Seen Inside Aircraft

7.2.1 Case Histories

Mr. Tom Dewis, who when he contacted me worked on the Concorde for British Airways, was flight engineer on a VC-10 aircraft flying south en route to Nairobi, Kenya, having taken off from Rome. The aircraft was over southern Italy, about 100 miles south of Rome, at about midnight and was being held at 20,000 ft by air traffic control. The sky was very dark, with a buildup of cumulonimbus clouds, as was evident from the radar, but there was very little lightning activity. There was some static on the radio. Mr. Dewis wrote,

> Although we wished to change course to avoid the most active areas, air traffic kept telling us to "Stand by," so it was obvious that we would have to fly through the weather at this not very favorable height.
>
> There was the usual false calm that one experiences prior to the turbulence of the storm, then the bumping began. However, it was not as bad as the radar seemed to predict, and there was still not a lot of lightning around. Then the phenomenon occurred.
>
> A dim yellowish light appeared ahead of the aircraft's nose, rather like a headlamp on an old locomotive. In fact, at the time I remember thinking that we were entering a tunnel, with a train coming towards us from the opposite end. That was the impression it gave.
>
> This light expanded in circumference, again heightening the impression of a train coming towards one at high speed. By this time it was filling the windscreen with its size, and there was a corona of sorts around its outside. . . . The light was still not too bright to look at—it was like looking at the sun through a piece of smoked glass.
>
> The flight deck of a VC-10 always had a lot of air noise present in flight, both from the air conditioning and from the airflow around the

fuselage. There was very little other noise though, being so far ahead of the engines. When the light was close to the windscreen, it was possible to hear a swishing noise. At the same time, the light seemed to be rotating rapidly. . . . Unusually in that sort of event there was little or no St. Elmo's fire dancing around the windscreen.

The co-pilot, who, it transpired, had experienced something similar on a Comet, called "Duck!" and we all did what he really meant and closed our eyes. In my case, I also turned in towards my panel on the right side of the flight deck.

There was a loud report followed by a strong smell of ozone. This is usual in a lightning strike on an aircraft. But it was accompanied by, literally, a blinding flash. All I could see was a lilac-blue light, even with my eyes open. Normal vision returned very soon, however, and the instruments began to appear through the bluish light.

A very startled steward and stewardess burst onto the flight deck, saying that they thought we'd been struck by lightning. This we confirmed. They then said that they were both walking up the [darkened] aisle when this happened and they saw a ball of fire come out of the flight deck and roll down the aisle. It was about the size of a football. Both stepped smartly into empty seat spaces as the ball rolled past them [with a swishing noise]. It continued to the rear of the aircraft and disappeared.

A pity in a way that the instinct of self-preservation made them step to one side, as we may have found out what its effect was on the human body.

On arrival at our destination we examined the airframe for possible damage and found a burn mark on the radome about the size of a half-crown. . . . [The radome is the nose cone containing radar and other flight instruments.] The external effect on the aircraft was normal for a strike. . . . (T. Dewis, personal communication, 1982)

A very similar report was provided by the navigator of a U.S. Navy air transport R5D (C-54 Skymaster) aircraft flying from Morocco to London on January 6, 1952. The altitude was 8000 ft and there was light to moderate turbulence and rain turning to wet snow. St. Elmo's fire appeared on the windshield, the leading edges of the wings, and the propeller tips. Suddenly, a "ball of fire" the size of a soccer ball appeared inside the center of the windshield. This bounced off the glare shield and the control pedestal onto the flight deck, where it rolled until it struck the closed crew compartment door, whereupon it disappeared (Jennison 1992).

Professor Roger Jennison, now Emeritus Professor of Physical Electronics and former Director of the Electronics Laboratories at the University of Kent at Canterbury, England, was traveling in an Eastern Airlines all-metal aircraft over the East Coast of the United States during a thunderstorm on March 19, 1963 at 12:05 a.m.

Eastern Standard Time. He was seated near the front of the passenger cabin. There was much turbulence. The aircraft was evidently struck by lightning (he saw a bright flash of light and heard a loud bang) and some seconds later a perfectly symmetrical glowing sphere of diameter 22 $\pm$ 2 cm emerged from the pilot's cabin and traveled at constant height and speed (75 cm above the floor at $\approx$1.5 $\pm$ 0.5 m/s relative to the aircraft) and in an undeviating path down the central aisle of the aircraft approximately 50 cm from him. The blue-white sphere had no structure, and was somewhat limb-darkened and optically thick, hence appearing almost solid. It did not seem to radiate heat, and appeared to have an optical power of about 5 to 10 W. It was also seen by a terrified air stewardess as it disappeared into the toilet compartment at the rear of the aircraft.

Prof. Jennison was particularly impressed by the appearance of the ball in a totally screened environment, that the speed of the ball relative to the aircraft was comparable to the speed of ball lightning seen near the ground, and by the symmetry, which was so perfect that he could not tell whether it was rotating (Jennison, 1969; BBC television, *Horizon*, "The Day it Rained Periwinkles," February 7, 1972). In response to a later suggestion that ball lightning might be associated with strong magnetic fields, Professor Jennison (1973) remarked that he had a penknife in his trouser pocket, a steel tobacco tin in his left jacket pocket, and a steel screwdriver in his right jacket pocket, but no motion of these objects was noted as the ball passed. There was no damage to the aircraft.

Squadron Leader K. M. Wickson was the navigator on a Wellington aircraft flying in cumulonimbus clouds at 5000 ft over Tilstock, Shropshire, England on May 1, 1945. He and the other crew saw what they interpreted as ball lightning following a lightning strike to the aircraft. The opaque, blue to purple ball, about the size of a football, rushed down the fuselage along the narrow gangway and seemed to exit through the rear of the aircraft. It was bright enough to illuminate nearby objects in the darkened fuselage, and was brighter near the center than at the edges. The signaler and navigator, who were located in the fuselage and not able to see the original lightning flash, saw the ball. The navigator recorded the event in his flying log book. The lightning strike tore a landing light from the leading edge of the wing and temporarily blinded the pilot, who was unable to see for some 20 min (K. M. Wickson, personal communications, 1982, 1983, 1984).

A British Airways Boeing 737 flight from Berlin to Stuttgart, Germany had descended from cruising altitude to about 10,000 ft. There had been lightning strikes and there was turbulence, so the stewardess in charge of the flight was seated on the port side in the pantry near the front of the cabin. Suddenly, a bright, luminous ball appeared inside the pantry on the starboard side, apparently from the service door. It was bluish with a fuzzy edge, about the size of a melon (16–18 cm in diameter), and traveled quite rapidly about 70–80 cm from the floor. It swung down the passenger aisle and having traveled about 3 m down the aisle, it disappeared, seeming to pass through the port side of the aircraft. Passengers seated in the front rows saw it (Jennison 1997).

A DC6 aircraft was flying north over the Alps in heavy turbulence when a luminous ball with fuzzy edges, about 10 in. in diameter, was seen traveling rapidly along the floor of the central aisle. There was no internal damage, but on landing blackened holes were found in the nose and tail of the aircraft, 3 to 4 in. and 10 to 12 in. in diameter, respectively (Jennison 1992).

A Russian Ilyushin-18 aircraft took off from Sochi on the Black Sea coast of the Caucasus. The weather had been fairly good, but thunderclouds had been observed some 40 km from the flight path. Suddenly, at an altitude of 1200 m, a ball of fire 10 cm in diameter appeared on the fuselage in front of the cockpit. "It disappeared with a deafening noise, but re-emerged several seconds later in the passenger's lounge, after piercing in an uncanny way through the air-tight metal wall. The fireball slowly flew above the heads of the stunned passengers. In the tail section of the airliner it divided into two glowing crescents which then joined together again and left the plane almost noiselessly." The radar and several other instruments were damaged, so the aircraft returned to the airport. Two holes were found in the fuselage—one in the nose section and one in the tail plane. There were no traces inside the aircraft, and none of the passengers was hurt (TASS, January 13, 1984)[1].

Chapter 5 referred to an incident in the summer of 1938. The captain of a BOAC flying boat flying to Iraq at 2500 m in dense nimbostratus cloud saw ball lightning come in through his open cockpit window. This was shortly after the aircraft had passed the Toulouse gap. His eyebrows and some hair were singed off, and holes were made in his safety belt and a dispatch case. The ball passed through the airplane to the rear cabin, where it exploded loudly. This was seen by passenger Mr. J. Durward, formerly deputy director of the Meteorological Office of Great Britain. (Gold 1952).

In 1960, a KC-97 U.S. Air Force tanker—an aircraft used to refuel short-range aircraft in midair—was flying on instruments in cloud at 5400 m. There was light precipitation, but the temperature was above freezing. St. Elmo's fire was seen around the edges of the front windows. The pilot saw a ball about 18 in. (45 cm) in diameter, yellow-white in color, enter through the windshield and pass between him and the copilot at a fast speed. He anticipated an explosion and braced himself. The ball traveled silently down the cabin passageway past the navigator and engineer. About 3 s later, a member of the crew in the rear reported via the intercom that the ball had rolled through the rear cargo compartment and then exited over the right wing into the clouds. The aircraft was loaded with fuel, so it was fortunate that an

[1]The TASS report indicated that this event was under investigation by the Leningrad Geophysical Observatory. Professor V. D. Stepanenko, deputy director of the Main Geophysical Observatory at Leningrad, wrote to me on May 18, 1984, however, saying that this was not the case.

explosion did not occur (Uman 1968, 1986). Table 7.2 summarizes some very definite patterns evident in these descriptions.

7.3 Discussion

We now consider the effect of ordinary lightning strikes on aircraft. Experimental studies of the effect of lightning on aircraft have included simulations in high voltage laboratories, such as the Culham Laboratory Lightning Studies Unit (Oxfordshire, England), using model or real aircraft, and flying real aircraft through active storm centers (Rustan et al. 1982).

Pilots of both military and commercial aircraft normally try to give active thunderstorms a wide berth. Nonetheless, lightning strikes to aircraft are commonplace. On average, commercial airlines report about one strike per 2000 hours of operation, while military aircraft log fewer than one strike for every 10,000 hours. According to estimates, on average, each aircraft in the U.S. commercial fleet is struck slightly more than once per year.

In fact, aircraft often trigger lightning via corona discharge when flying through a heavily charged region of a cloud. Corona discharges can initiate or trigger lightning discharges because the electrical conductivity of ionized air is significantly greater than un-ionized air. In effect, the aircraft together with the electrical discharge surrounding it "short-circuit" part of the electric field beneath the cloud. This has the effect of intensifying the electric field above and below the aircraft because the potential gradient is steeper than before. In these instances, the lightning flash begins at the aircraft and extends away from it in both directions, toward the ground and the thundercloud, or toward two charge centers of opposite polarity within clouds. Aircraft also accumulate a net charge during flight by air friction (triboelectrification).

The intensity of corona discharge increases with wind speed, so aircraft flying under thunderclouds can develop particularly intense corona discharges from their extremities. These are often visible as St. Elmo's fire. There may also be radio interference in the gigahertz range (Boulay and Laroche 1982).

More than 50% of strikes to aircraft are reported to follow observations of corona and/or high-voltage streamers. Streamers are in the form of narrow filaments and branched ionization channels. The streamer discharge currents can be between 400 and 10,000 A. (The larger the aircraft, the greater the charge it can store, and therefore the larger the current.) At higher values of current, a pilot could interpret such a discharge as static discharge or as a lightning strike. Streamers may then develop into leaders, enabling electrical connection between charge centers in clouds, and currents of 100 A or more may then flow through the aircraft, feeding the leader for periods of up to hundreds of milliseconds. These are known as "continuing currents." Such currents can produce substantial magnetic fields that may undergo rapid changes, thus inducing large voltages and currents in nearby conductors.

It is interesting to note that electromagnetic resonances may be set up within the aircraft by lightning strikes (Trost and Pitts 1982). For an F-106-B aircraft, derivative-type sensors (which detect rapid changes in the magnetic field) enabled detection of periodic components in the wave forms that result from excitation of various resonances of the aircraft. These resonant frequencies are in the high-frequency range of about 7 to 23 MHz, the lowest frequency corresponding to the fuselage half-wavelength resonance and the highest corresponding to that of the wings. Intermediate frequencies probably correspond to a combination of the wings and tail. A lightning strike couples sufficient energy into the aircraft structure in this frequency range to excite these resonances. Laboratory tests with scale models confirmed these results (Trost and Turner 1982). In these experiments there was also a component at 39.8 MHz that was thought to be the second harmonic of the wing resonance. Electromagnetic resonances and standing waves are discussed in more detail in Chapter 13 in the context of ball lightning models.

Many strikes to aircraft are apparently intracloud strikes rather than strikes from cloud to ground, the latter generally being far more severe. Sometimes, however, strikes are CG encountered at high altitudes. Pilots report most strikes as "static discharges" or "triggered lightning." Triggered lightning is expected to have different properties from those of natural CG strokes.

The great majority of strikes are *not* encountered in active thunderstorms. Rather, most occur in flight through clouds with precipitation, but with little or no lightning activity noted other than the strike experienced by the aircraft. Most strikes occur at altitudes of between 1 and 4 km, so the greatest risk is during ascent or descent, especially descent (perhaps because this takes longer than ascent) (Anderson 1982). Many flight record surveys of lightning incidents show that most strikes occur to aircraft flying at the freezing level inside precipitating clouds. This is usually associated with flight altitudes of 3 to 5 km. A comparison of Soviet with other data indicated more strikes for the former at lower altitudes (Clifford 1982). (This may be because at more northern latitudes the freezing level is generally closer to the ground.) More than 40% of the ball lightning reports recorded in this chapter are from Russia.

As already noted, the altitudes recorded for three cases of ball lightning seen outside aircraft are between 2500 and 3400 m (mean value 3100 m), while the altitudes recorded for five cases of ball lightning seen within aircraft extend over a greater range, from 1200 to 6000 m (mean value 3320 m). Two cases record that the aircraft was ascending and one that the aircraft was descending at the time of the incident.

The relationship between probability of strikes and altitude is a consequence of several factors:

- Triggered lightning is more likely when the aircraft is near charge centers in clouds.
- At cloud altitudes, the likelihood of the strike to an aircraft being a CG flash is reduced because intracloud discharges are more numerous.
- At flight altitudes, a CG flash is expected to be far less severe than a similar flash near the ground.

Although passengers and crew may see a flash and hear a loud noise when lightning strikes an aircraft, nothing serious should happen because of the careful lightning protection that is an essential part of the design of all modern aircraft and their sensitive components. Most aircraft skins are made mainly of aluminum, which has a high electrical conductivity. By ensuring that there are no gaps in this conductive path, engineers can guarantee that most of the lightning current will remain on the exterior skin of the aircraft. Some modern aircraft are made of advanced composite materials, which on their own are significantly less conductive than aluminum. In this case, the composites are made with an embedded layer of conductive fibers or screens designed to carry lightning currents. These designs undergo thorough testing before they are incorporated in an aircraft.

Modern passenger aircraft have kilometers of wires and dozens of computers and other control instruments. These computers are susceptible to interference from power surges. Lightning traveling on the exterior skin of an aircraft can induce transients, called lightning "indirect effects," into wires or equipment beneath the skin. So, together with design of the exterior of the aircraft, the lightning protection engineer must also ensure that no damaging surges or transient currents can be induced in sensitive electronic equipment within the aircraft. Careful shielding, grounding, and the application of surge suppression devices when necessary avoid problems caused by indirect effects in cables and equipment. Manufacturers must confirm that every circuit and piece of equipment that is essential to the safe flight and landing of an aircraft is protected against lightning according to regulations of the aviation authorities in the country of origin.

At first lightning will attach to an extremity, such as the nose or wing tip. The light emission from the discharge, if it contacts the window of the cockpit, can cause temporary blindness of the pilots (which might cause afterimage effects). The aircraft then flies through the lightning flash, which reattaches itself to the fuselage at other locations while the aircraft is in the electric "circuit" between regions of opposite polarity. A continuing current enters at one point and exits at another, but because the aircraft is moving so rapidly (e.g., 50 m/s), the point of contact is swept along the fuselage. Current will travel through the conductive outer skin and structures of the aircraft and exit from some other extremity such as the tail. Pilots occasionally report temporary flickering of lights or short-lived interference with instruments.

The damage resulting from a continuing current may be a series of ablation pits that are roughly circular and 1 cm or so in diameter. The extent of these marks may be used together with the magnitude of the flying speed to estimate the duration of the continuing current. In some cases damage extends along the full length of the fuselage. Damage to the trailing-edge sections, to electrical and avionic equipment, and to plastic outer sections such as the radome cover is often reported (Anderson 1982). The radome presents a particular problem for lightning protection engineers. Radar cannot operate within a conductive enclosure. The radome is protected by applying lightning diverter strips along its outer surface. These may be solid metal bars or a series of closely spaced buttons of conductive material joined to a plastic strip that is bonded adhesively to the radome. In many ways, diverter strips function like a lightning rod on a building. The strips, which do not significantly impair the operation of the radar, are sized and spaced carefully according to simulated lightning attachment tests.

In some cases, lightning can attach at the leading edge of the wing and sweep back over the fuel tank region in discrete steps. Very occasionally, if the "dwell time" is sufficient, a fuel tank may be penetrated by the heating effect of the continuing current and there may be a detonation that is likely to destroy the aircraft. Such an event happened in 1978 in South Carolina. A U.S. Air Force C-130-E crashed when a wing section was blown off after fuel vapor was presumably detonated by a continuing lightning current. The fuel system, in which even a tiny spark could be disastrous, is therefore a major area of concern in aircraft design. Rigorous precautions are taken to ensure that lightning currents cannot cause sparks anywhere in the fuel system. The aircraft skin around the fuel tanks must be sufficiently thick to withstand a penetrating burn. All structural joints and fasteners must be tightly designed to prevent sparks if lightning current passes from one section to another. Access doors, fuel filler caps and vents must be designed and tested to withstand lightning. All pipes and lines that carry fuel to the engines, and the engines themselves, must be confirmed as protected against lightning. In addition, new fuels that produce less explosive vapors are now widely used.

It is often said that passengers in an all-metal aircraft are safe because the fuselage, being a hollow conductor, behaves as a "Faraday cage," and, as Gauss's law confirms, the interior is unaffected by external electric fields or currents. This is a little simplistic, because aircraft are not entirely closed metal containers—there are apertures such as windows, and radio antennae and cables pass through the fuselage. Indeed, many radio operators have received severe shocks when attempting to disconnect antennae. Furthermore, as indicated earlier, not all aircraft fuselages are constructed entirely of metal; the composites used in aircraft design have conductors embedded in them.

Windshield glasses are a sophisticated sandwich of materials. Military and commercial airliners have coated windshields composed of a bulk or surface electrical conductor, so it is very unlikely that dc fields could penetrate the

Table 7.1. Ball Lightning External to Aircraft

Ball Lightning Reports	Comments About Conventional Lightning and Thunderstorms
1. The events occur in cumulonimbus clouds. Visibility is poor.	The great majority of strikes are *not* encountered in active thunderstorms. Rather, most occur in flight through clouds with precipitation, but with little or no lightning activity noted other than the strike experienced by the aircraft.
2. There is turbulence that increases significantly just before the event (this may be described as a bump).	Turbulence is associated with active storm centers.
3. Where the aircraft altitude is stated (three cases), it is between 2500 and 3400 m.	Most strikes occur at altitudes of between 1 and 4 km, so the greatest risk is during ascent or descent, especially descent (perhaps because this takes longer than ascent). Many surveys of flight records of lightning incidents show that most strikes occur to aircraft flying at the freezing level inside precipitating clouds. This is usually associated with flight altitudes of 3 to 5 km.
4. There is radio interference or static.	This is normal in thunderstorm conditions.
5. The ball is seen approaching the aircraft from the front (bearing in mind that the aircraft is moving forward!).	Does the ball travel toward the aircraft, or does the aircraft fly into the ball?
6. The ball may strike part of the aircraft and explode with an audible sound and a bright flash. Sometimes the ball explodes without contacting the aircraft.	The explosion and bright flash may be a conventional lightning strike to the aircraft. Ball lightning may be a predischarge phenomenon similar to St. Elmo's fire.
7. A fiery stream or discharges is seen to pass along the fuselage of the aircraft from front to back, or there is a shower of sparks as the ball passes along the fuselage from front to back.	This may be the visual impression caused by conventional lightning sweeping along the fuselage.
8. Damage to the aircraft may consist of (a) melting of part of the propeller, (b) melting of rivets without damage to the skin of the aircraft, (c) damage to radio equipment, and (d) holes melted in the trailing edge of the elevator.	Conventional lightning can cause all these forms of damage.
9. Sometimes there is evidence of strong magnetic changes after the event.	Lightning currents are very high and therefore associated with strong magnetic fields, which can be intense enough to produce forces that distort metallic conductors.

windshield and produce discharge phenomena. Side windows on the fuselage are not coated, however so dc fields can penetrate the fuselage here. The electric field obtained inside the fuselage from this type of aperture is very low compared with the field outside. An aircraft fuselage thus behaves as a very effective Faraday cage (P. Laroche, personal communication, 1998; Golde 1973; Newman and Robb 1977; International Aerospace Conference on Lightning and Static Electricity 1982).

I find no evidence from the reports cited that ball lightning rather than conventional lightning has been responsible for damage to aircraft. In every instance

Table 7.2. Ball Lightning Within Aircraft

Ball Lightning Reports	Comments About Conventional Lightning and Thunderstorms
1. Weather is usually stormy, and aircraft are often flying through cumulonimbus clouds. There may be a great deal of turbulence.	See comments above.
2. The aircraft altitude at the time of observation is between 1200 and 6000 m.	See comments above.
3. Prior to the observation of ball lightning there may be a conventional lightning flash to the aircraft, which may be preceded by a build-up of electrical discharge or St. Elmo's fire near the aircraft's nose or radome.	An initial flash of lightning to the aircraft may be required to generate ball lightning.
4. Those in the cockpit of the aircraft may be dazzled.	This is normal with conventional lightning.
5. The ball may enter through the windshield of the aircraft and pass down the center of the aircraft through the passenger cabin. In some instances, ball lightning has been said to have passed from the flight deck through a metal wall into the passenger cabin.	It is interesting to note that lightning sweeps along the fuselage in the same direction, but at a much greater speed relative to the aircraft. Problems such as these are discussed in Chapters 12–13.
6. In one case, there was evidence of heat from the ball, whereas in other cases there was none.	
7. The ball may exit, silently or explosively, via the rear of the aircraft, or over a wing.	What seems to be explosive decay could be the sound of a conventional lightning strike to the aircraft.
8. Damage to the aircraft, if any, may consist of circular burn marks a few centimeters in diameter. These are found on the radome and/or the tail plane. In one case, a landing light was torn from the leading edge of a wing.	See comments above.

described here, ordinary lightning could have caused all the various types of damage. Tables 7.1 and 7.2 list the typical properties of ball lightning events associated with aircraft to see how well they can be explained in terms of accepted knowledge about lightning strikes to aircraft.

The level of consistency of description of this category of ball lightning report is quite remarkable. Ball lightning seen within an aircraft is especially interesting for several reasons. First, this high level of agreement together with the close proximity of the event to the observer, the reduced likelihood that the observer would see an external flash of lightning to the aircraft, and the availability of objects in foreground and background together render the reliability of these reports very high indeed. Second, these reports present serious challenges for many models of ball lightning, as we will discuss in later chapters. Third, the consistent movement of ball lightning inside an airplane shows that its behavior in one kind of environment, at least, seems predictable.

Chapter 8

Assessment of Risk to Trees

There have been many reports of damage to trees allegedly caused by ball lightning. Some have been discussed in earlier chapters. Additional representative examples follow.

8.1 No Clear Evidence of Contact between Ball Lightning and a Tree

In many reports, there is no clear description of contact between ball lightning and a tree, although damage is attributed to the ball rather than to conventional linear lightning. A report in Chapter 5 stated that a plum tree was allegedly shattered "to atoms" by a 6-ft (1.8-m) diameter lightning ball. There were several fatalities in the vicinity. The nearby ground was covered with several inches of large hailstones, and the surrounding garden was also left in much disarray (Anon. 1887u.). In a similarly vague report discussed in Chapter 5, during a violent thunderstorm, several fireballs, one of which was seen to have come out of an excavation full of stagnant water, were observed moving close to the ground in different directions, making a "crackling" sort of noise. A man and several animals were killed, and there was much damage to trees and houses nearby, for which the fireballs were blamed (Flammarion 1905). The splintering of a tree in an incident at Shreveport, Louisiana in 1866 was attributed to a meteor, but is more reminiscent of ball lightning (Corliss 1982). Chapter 4 records an incident dated September 1, 1971, in Minneapolis, Minnesota (Anderson and Freier 1972) in which a ball was not observed directly, but several neighbors described a persistent red-orange glow seen near an oak tree following thunder that almost coincided with a lightning flash. There was a singed path in grass leading from the tree to a house, where electrical installation was damaged. The present author interprets this as a side flash from the tree to the electrical supply.

Crossley (1924) gave an account of an observation at Telemarken, Norway on August 21, 1900. During a thunderstorm with an impressive lightning display at 8 p.m. local time, a yellow streak 1 in. (2.5 cm) wide came gradually downward, straight for the observer. As it reached the window, it changed to a brilliant yellow ball of fire the size of a cricket ball (~10 cm diameter). The ball fragmented with a loud explosion and shot out violet flames, which spread in all directions. The following day, the observer found a track or furrow in the ground beginning 20 m from the window. The furrow was 44 m long and of variable width (7 to 24 cm) and depth (4 to 12 cm), and in some places it disappeared below ground. It had circumvented a large boulder of granite, from which two fragments had been broken. The surrounding area was covered with small pieces of earth and tree branches. Two or three small trees had been uprooted (Cade 1969, Singer 1971).

On July 9, 1947, in Kirkcudbrightshire, Scotland, a fireball was reported to have run along an electrical wire and blasted a large oak tree to pieces (Corliss 1982).

8.2 Reports of Contact between Ball Lightning and a Tree

Some reports are more specific than others in stating that ball lightning made contact with a tree, which was damaged. One such incident from Conwy, North Wales was described in Chapter 5. In 1992 ball lightning was reported to have struck an oak tree about halfway up the trunk, then rolled down the tree, and disappeared with a very loud crack and explosion, dispersing in flashes. The witness described "waves of lightning" passing from the ball into the ground and sparks radiating out of it in all directions. The bark and trunk were split, showering splinters of wood over a distance of about 45 m (Stenhoff 1992a).

Chapter 4 describes the death of a 100-ft (30-m) gum tree within 48 h of an incident in which ball lightning was reported (A. McEwin, personal communication, 1983). A gum tree was also a casualty in a similar incident in Australia (W. Hill, personal communication, 1974, D. Moseley, personal communication, 1983). This occurred during a violent thunderstorm in February or March in the early 1960s at Warriewood, New South Wales. These three people saw a round, orange-yellow, opaque ball about 15 ft (4.6 m) in diameter fall rapidly into a clearing in the bush that served as a firebreak. The ball rolled down the hill as a stone would and came close to a house, hitting a 60-ft (18-m) angophra gum tree, some 300 years old, about 15 ft (4.6 m) above the ground. It then exploded, having been in sight for about 15 s. The tree, which had been in blossom, was killed instantly. Red gum dripped from its trunk for about a fortnight. All living foliage in the path of the ball was "cooked." There was evidently a side flash to the electric mains in the house that caused extensive damage.

Three men were sitting at the bar of a golf club at Selsdon, Surrey, England, on April 14, 1980 at about 7 p.m. There was a violent thunderstorm with much lightning, and the atmosphere was very humid, so the patio doors were open. One of the witnesses wrote:

> I noticed about 40 yards [37 m] away a pale, orange, nebulous ball which floated into the top of a Lebanon Cedar tree (reputed to have been planted by Elizabeth I). The size of the ball was approximately . . . the apparent size of a full moon. . . . After about two seconds, no more, there was an extremely loud clap of thunder which apparently originated from this ball of light, the result of which was that the glass containing a large gin and tonic which I was holding disintegrated. At that stage my friends retreated from the open patio doors. The tree, which is a listed tree, had to have concrete poured into its bole and some of the upper branches removed for safety reasons. (C. G. Palmer, personal communication, 1982)

The ball was opaque and of uniform brightness across its surface. Although pale, it was bright enough to be clearly visible in daylight. Its appearance did not change significantly throughout the observation. At its closest, it was about 20 m away. The estimated distance and subtended diameter suggest a true diameter of 35 to 65 cm.

Cade and Davis (1969) describe an observation at Ketama in the Rif Mountains in Morocco. A witness on a veranda was watching lightning when he saw "a glowing orange ball, about the size of a football" emitting a hissing sound. It was traveling roughly horizontally above the road at fast walking speed. About 25 yd (23 m) away from the house, the ball struck a tree. There was a loud explosion like a roar, and the tree trunk was split in two.

Flammarion, in a chapter discussing the effects of (ordinary) lightning on trees, described damage to a pine tree near Jare, Landes, France.[1] A "fireball fell" onto the tree,

> which it shivered into myriad slender strips, about 2 metres long, many of which were caught on the branches of pines within a distance if 15 metres. Only a stump, 2 1/2 metres in height, remained standing. At the same time three other pines, which stood 18 and 25 metres away from the first, were destroyed. The bark had been stripped off each, but only as far as the incision made for extracting the resin. (Flammarion 1905, p. 167)

Thus the extent of damage reported varies. In 1956, ball lightning of a diameter 15 to 20 cm, which appeared bluish-white while distant from the observer but red when nearby, was reportedly seen rolling down a tree. There was no ordinary

[1]Cade and Davis (1969) give the year as 1904, and Flammarion gives the date as June 25.

Figure 8.1. Tree destroyed by ordinary lightning. [Photograph: Dr. Eric Wooding.]

lightning at the time. The ball then bounced on the wet ground up to a height of 4 m, moved slowly along a barn, and finally circled a tree and struck a fence post, exploding loudly. A very weak trace was marked on the tree, apparently unlike that produced by ordinary lightning. After the ball disappeared, ordinary lightning struck the barn and started a fire (Pflegel 1956).

8.3 Conventional Linear Lightning and Trees

Every day, thousands of trees worldwide are struck by ordinary lightning. There is sometimes no damage, or the effects range from superficial damage to virtually total destruction (Fig. 8.1). Superficial flakes of outer bark or, frequently, the entire bark may be removed; wood and bark strips may be ejected; or the tree may be reduced to slabs and slivers. In this latter case, the slivers and fragments may be projected over considerable distances and may themselves represent a serious hazard.

After lightning strikes to coniferous trees and to deciduous trees with rough bark such as oaks, it is frequent to find a furrow 5 to 25 cm wide that spirals along the grain of the wood along part or all of the trunk, and exposes the outer layers of sapwood. Trees with smoother, thin bark such as birches may have large, irregular patches or sheets removed (Taylor 1977), although such trees rarely suffer structural damage, probably because most of the lightning is discharged over a large surface area of the trunk (Golde 1973). Taylor (1977) lists other microscale symptoms of ordinary lightning flashes to trees. Most trees struck are not killed; most recover, although they may be weakened, and some are subsequently attacked by bark beetles. Sometimes, however, a single lightning flash can kill a group of trees (Uman 1986).

Between 10,000 and 15,000 forest fires are started by lightning each year in the United States (Taylor 1969). Where these are caused by the combustion of a tree rather than, as is more usual, by burning of materials or a tree stump on the ground, it seems that most are the result of flashes of "hot" lightning—lightning with a long continuing current.

When the outer surface of a tree's bark is wet, it still offers a much higher resistance than that of the cambium (the inner bark and first few rings of live wood) and the sapwood. Experiments show that when lightning strikes the top of a tree, it is first discharged through the trunk, but as the current progresses down the trunk, flashover will occur on the surface (Golde 1973). The resistance per unit length of a tree is not constant (partly because of variations of cross-sectional area of the trunk), and its resistance has a significantly negative thermal coefficient. By combining measured resistance values with a progressively increasing lightning current, Golde (1973) calculated the variation with time of the potential gradient along the trunk $\Delta V/\Delta x$ and compared it with the impulse breakdown field of air.

This calculation showed that if the top of a tree is struck, the current at first passes through the trunk. However, as $\Delta V/\Delta x$ increases, it eventually reaches a value where surface flashover will occur.

This is probably followed by a shock wave caused by the return-stroke current, which is responsible for showering fragments and slivers of wood over the surrounding area. The spiral pattern described above could be caused by variations in resistivity in the cambium together with electromagnetic forces (Golde 1973).

Trees have a resistance of several kilo-ohms per meter, so there is a serious risk of a side flash (see Chapter 2) if a building, a person, or a grounded metal conductor is nearby. Lightning safety codes recommend cutting branches that approach a building to within 1 to 2.5 m. If a tree is taller than a nearby structure, the codes state that the clearance between the tree and a structure should be at least one-third to one-half of the height of the structure. In circumstances where a tree may have been struck by conventional linear lightning, or allegedly by ball lightning, the possibility of a side flash should be considered, especially if these codes have not been observed.

8.4 Does Ball Lightning Damage Trees?

The cases in Section 8.1 offer no convincing evidence that ball lightning was the cause of damage to the trees since it is not clear that the reported ball made contact with the tree. Where damage is described, it is entirely consistent with the effect of an ordinary lightning strike. We have seen in other chapters that ball lightning reports have a tendency to attribute all observed effects to ball lightning rather than to conventional linear lightning.

While the reports cited in Section 8.2 offer much clearer evidence of contact between the ball and a tree, the damage reported in each case is also completely consistent with a scenario in which conventional linear lightning has struck a tree. The Conwy, 1992 report appears in fact to describe a flashover.

However, we note that a large proportion of reports of tree damage describe a manifestation of ball lightning that precedes tree damage that we have attributed to conventional linear lightning. It therefore seems that ball lightning may be a precursor of a lightning strike to a tree. This sequence of events is thus inconsistent with the idea that ball lightning is an afterimage or combustion resulting from an ordinary lightning strike to a tree.

Chapter 9

Photographs and Videotapes

9.1 General Considerations

Since the earliest days of photography, investigators of ball lightning have hoped for a definitive image of this elusive phenomenon. Skeptics have used the absence of compelling photographic evidence to support the argument that ball lightning does not exist. About 65 still photographs alleged to be of ball lightning have been published, and these have mostly been disappointing and easily explained. In recent years, moving films and videotapes have offered the opportunity to acquire more convincing evidence with considerably greater information density. Nonetheless, there is still the possibility of erroneously identifying traces on such media as ball lightning.

The small number of published photographs allegedly of ball lightning is perhaps unsurprising given its evidently unpredictable and transitory appearance. The probability of having a camera available during such an event is fairly low, and even if one were available, the chances of capturing a satisfactory image on film are small. There are many rare atmospheric and meteorological phenomena of which there are few or no photographic records. [Minnaert (1993) gives several examples.]

Barry (1980) carried out a comprehensive survey of alleged ball lightning photographs. Since his book was published, several further photographs and a videotape have been presented as evidence. The results of Barry's survey are summarized and subsequent photographic evidence is discussed in this chapter.

9.2 Causes of Misidentification

In evaluating still photographs alleged to be of ball lightning, investigators should look for evidence of camera movement or double exposure. Photographs of stationary or moving single or multiple light sources can, with camera movement, show traces with the following characteristics:

- A second image, perhaps less intense, of a building or other object.
- Spacing of any of the alleged ball lightning traces or other bright images that is equal to the spacing of the first and second images of background structures such as buildings.
- Alleged ball lightning traces that are parallel.
- Regularly modulated traces suggest camera movement while photographing a source fed by alternating current mains, for example, a streetlight. These have a frequency of 60 Hz in the United States and 50 Hz in Great Britain. The modulation in brightness has twice the frequency of alternation of the power supply. However, photographs of several streetlamps with deliberate camera motion to reproduce this effect may (Bauer 1938) or may not (Poulter 1935, 1954) show such modulation (Fig. 9.1). Much will depend on ambient lighting and properties of the film. Corresponding

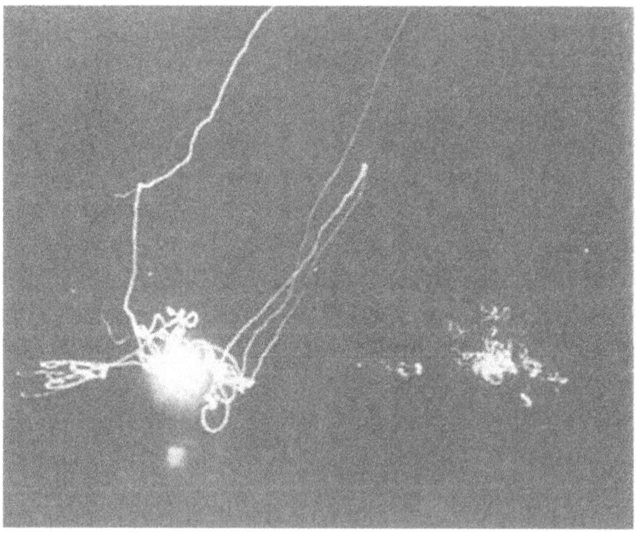

Figure 9.1. Still photograph of several streetlamps with deliberate camera motion. Like some other photographs in this chapter, there are multiple traces characteristic of photographs produced in this way. [Reproduced from Poulter (1935) with the permission of the Controller of Her Majesty's Stationery Office.]

multiple images of other objects may not be evident if the background is poorly illuminated. In a color photograph, the color may correspond to the color of streetlights, advertising signs, etc.

- In photographs where there has been a double exposure, images of objects that are actually opaque may appear transparent or translucent.

In cases (of which there are many) when an image is observed on film, but the photographer did not directly observe ball lightning, investigators should establish whether optical effects within the camera could be the cause or whether a spurious image has been generated by camera motion. Internal reflections within the camera or lens flares can also produce spurious images. Lens flares are produced by intense light shining directly into the lens. These can either project the shape of the aperture opening or create rings of light. Such photographs are characterized by symmetry about a line connecting the flare to a bright source in the photograph. Similar effects may be observed in movie or videocameras. Symmetrical motion of a primary and secondary luminous region about the optical axis of the lens may provide evidence of spurious images caused by artifacts of the camera. Autofocus features may cause images to be defocused.

Defects in photographic emulsion of the negative (or positive), such as blemishes, foreign particles, or scratches, can also produce spurious images. It is essential that the negative be subjected to microscopic analysis. Unfortunately, very few negatives of photographs of alleged ball lightning are available for scrutiny.

Some alleged ball lightning photographs are photographs of familiar phenomena taken under unusual circumstances. The circumstances under which a photograph or film has been taken and the credibility of the photographer are critical to its value.

9.3 Still Photographs

A number of photographs of alleged ball lightning are of exceptionally poor quality and have insufficient detail for evaluation or to yield useful data (Fig. 9.2) (e.g., Dmitriev 1971a,b).

Several photographs show regularly modulated traces. Jennings (1962) (Fig. 9.6) and Abbott (1934) (Fig. 9.3) published photographs with this description. Petersen (1952) (Fig. 9.4) and Davidov (1958) (Fig. 9.5) published photographs in which part of the trace shows this feature. As indicated earlier, photographing a device such as a streetlamp fed by ac mains while moving the camera can produce similar traces. In the Petersen photograph, background images are duplicated, indicating camera motion. Campbell (1987, 1992) has argued convincingly that camera motion and stationary light sources were the cause of the Davidov photograph.

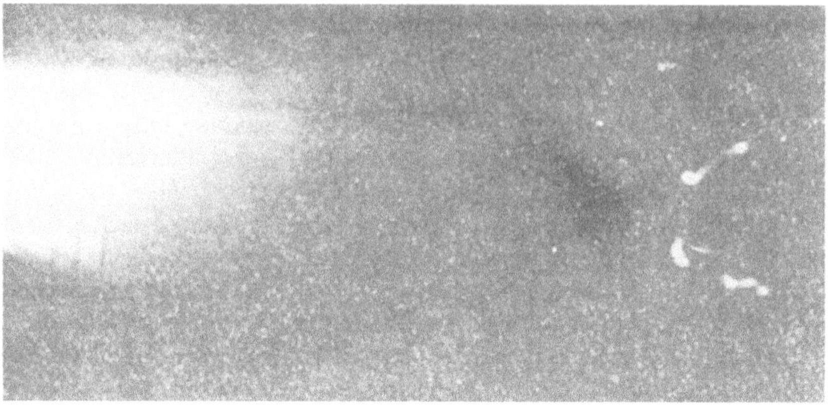

Figure 9.2. Still photograph made during a storm showing in the right of the frame an intensity-modulated image that was identified as that of ball lightning. [Reproduced with permission of *Priroda*, from Dmitriev (1971a,b).]

The Jennings photograph (Fig. 9.6) was taken during a storm at Castleford, Yorkshire, England, one morning in August 1961 at 2 a.m. The photographer, Mr. R. C. Jennings, heard a loud explosion and the window in his room shook. According to him, the streetlights were not illuminated (Lane 1965). The ball was

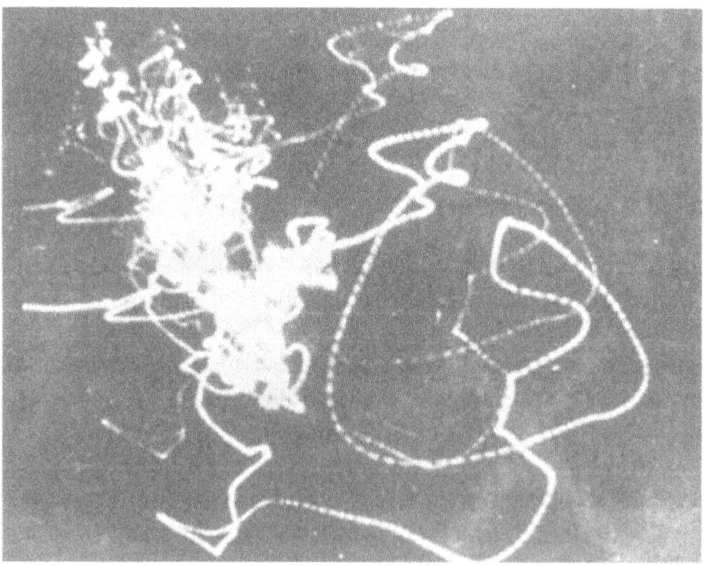

Figure 9.3. Still photograph taken during a thunderstorm in 1908 by R. J. Spickerman. Modulation of the traces suggests camera motion and ac mains lighting sources. [Reproduced from Abbot (1934).]

Figure 9.4. Still photograph taken by R. J. Bird. The luminous trace was identified as that of ball lightning. Poulter (1954) noted the obvious camera motion and questioned the validity of the photograph. [Reproduced with permission of *Geophysica*, from Petersen (1954a,b,c).]

accidentally photographed during the thunderstorm and no direct observation was reported. The modulated image undergoes a slight color change along its path from red through yellow to white. Jennings said that the slight blurring of the background was caused by camera motion. There is some evidence of reflection in the windows of buildings in the background (Barry 1980). Davies and Standler (1972) identified the trace as a 140-W sodium vapor streetlamp. The Electricity Board stated that the lamp was, in fact, illuminated at the time of the event. This and other features of the investigation suggest that the image was of the lamp. Campbell (1981) supported this interpretation with an excellent, detailed analysis.

Several photographs (Figs. 9.7–9.10) show duplicate or multiple traces at different places on the plate. (Abbot 1934; Manteuffel 1938; Wolf 1956; Shagin 1960; Müller-Hillebrand 1963, 1966). This is also probably indicative of camera motion with multiple light sources, and thus tends to invalidate interpretation of the traces as being due to ball lightning. Barry also attributes this cause to the trace on the photograph published by Young (1934) and Dixon (1955) (Fig. 9.11). There was a second, identical trace on the Davidov (1958) photograph.

Figure 9.5. Still photograph taken by B. V. Davidov during a thunderstorm at Kharkov on August 27, 1957 showing a trace subsequently identified as that of ball lightning. A sootlike residue was reportedly found at the window where the trace apparently ended. Campbell (1987) concluded that cause was camera motion and a stationary lamp. [Reproduced with permission of *Priroda* from Davidov (1958).]

A photograph of an erratic luminous trace that is neither duplicated in the frame nor shows signs of modulation was sent to the present author by Mr. W. T. Cowhig (Fig. 9.12). In 1937 he was attempting to photograph ordinary lightning from his bedroom window on a stormy summer's night at Bown Road, Rugby, England. He did not see a ball image at the time, but he thought he may have been preoccupied with the camera because he was attempting to take photographs by opening and closing the shutter roughly every 5 s. When the negative was developed, it showed horizontal luminous light traces that appeared to culminate in a tear-shaped ball. In his garden he had intertwined a cable along the fence to take power to a hut at the bottom of the garden, about 9 yd from the house. One of the dominant lines on the trace was parallel to this cable. On seeing his photograph, a neighbor remarked that the area was notorious for unusual lightning during thunderstorms.

Stakhanova (1997) published a time-exposure photograph showing two vertical, meandering traces without the degree of tortuosity ordinarily associated with ordinary lightning (Fig. 9.13). This was taken in Sverdlovsk, Russia, with a camera fixed to a windowsill during a thunderstorm in late 1979 by a student of the Riga

Figure 9.6. Still photograph taken by R. C. Jennings during a thunderstorm showing a modulated trace. This was identified with ball lightning although ball lightning was not seen at the time. Davies and Standler (1972) and Campbell (1981) concluded that the image was of a streetlamp with camera motion. [Reproduced from R. C. Jennings (1962).]

Institute of Aviation Engineers. One of the traces on the photograph appeared to pass behind a tree that was 5 to 7 m from the window, and points of light apparently from part of the trace are visible through the branches, which is evidently inconsistent with camera motion while viewing a stationary light source. Stakhanova interpreted the trace as consistent with a ball falling to the ground in front of a tree 30 to 50 m from the camera, bouncing on the ground, and then extinguishing. The observer noticed neither the ball nor any thunder, and the trace on the slide was a surprise to him. There is some modulation of the traces, but not in a way that suggests light sources from alternating-current mains.

Rutgers (1958) published a photograph (Fig. 9.14) that similarly shows an erratic, single, unmodulated trace. This was taken from a window in Arnhem, the Netherlands, in 1958. Westphal (1958) and Wolf (1958) discussed this photograph and concluded that it was not of ball lightning.

Holtzer, Workman, and Snoddy (1938) published photographs very similar (Fig. 9.15) to those just discussed. Rapid and random motion of the dart leader of

Figure 9.7. Still photograph showing a modulated trace identified as that of ball lightning. The modulation and the presence of a faint duplicate trace suggest camera motion and a stationary light source. [Reproduced from Zoege-von-Manteuffel (1938) by permission of *Umschau Zeitschriftenverlag*.]

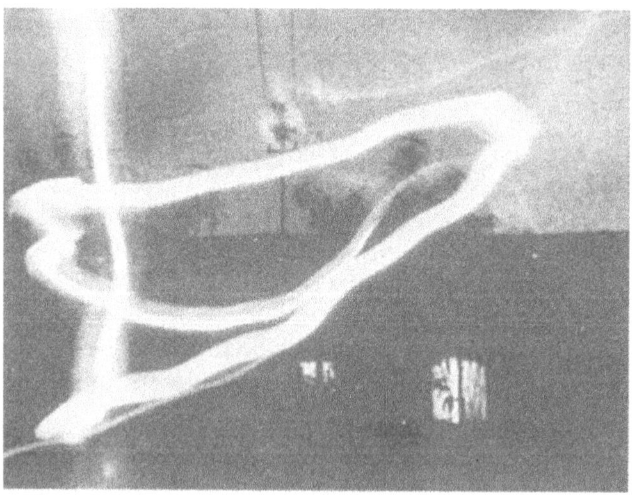

Figure 9.8. Still photograph taken by H. Jack during a thunderstorm in 1955 showing a trace identified as that of ball lightning. The presence of a faint, duplicate trace suggests camera motion and a stationary light source as the cause. [From F. Wolf, "Interessante Aufnahme eines Kugelblitzes," *Naturwissenschaften* **43** 415, 1956. © Springer-Verlag. Reproduced with permission.]

Figure 9.9. Still photograph showing traces identified as due to ball lightning. The presence of multiple traces suggests camera motion and multiple, stationary light sources as the cause. [Reproduced from Shagin, 1960.]

natural lightning produced the erratic trace. Rapid motion in a localized region could give the impression of a small region of luminosity.

In some alleged ball lightning photographs, the ball image is related to an image or images of ordinary lightning discharges. Such photographs have been used to support the idea that ball lightning is formed from a lightning channel. Bauman

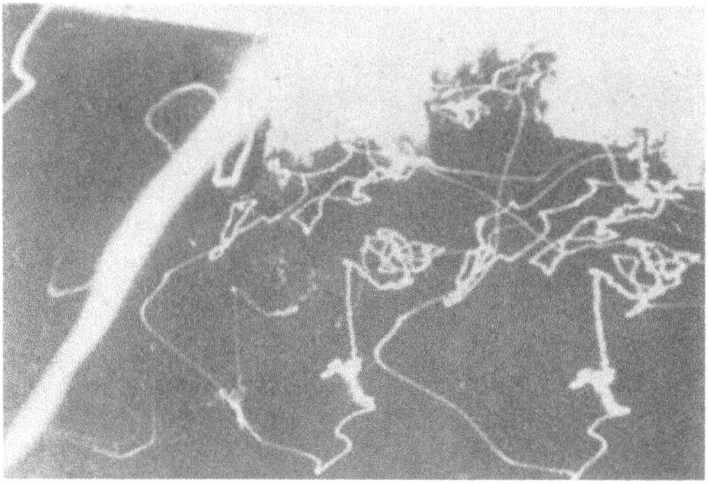

Figure 9.10. Still photograph showing traces identified as due to ball lightning. The presence of multiple traces suggests camera motion and multiple, stationary light sources as the cause. [Reproduced from Müller-Hillebrand (1963).]

Figure 9.11. Still photograph showing trace identified as ball lightning, but subsequently explained by camera motion. [Reproduced from Dixon (1955) with permission of *Weather* and the Royal Meteorological Society.]

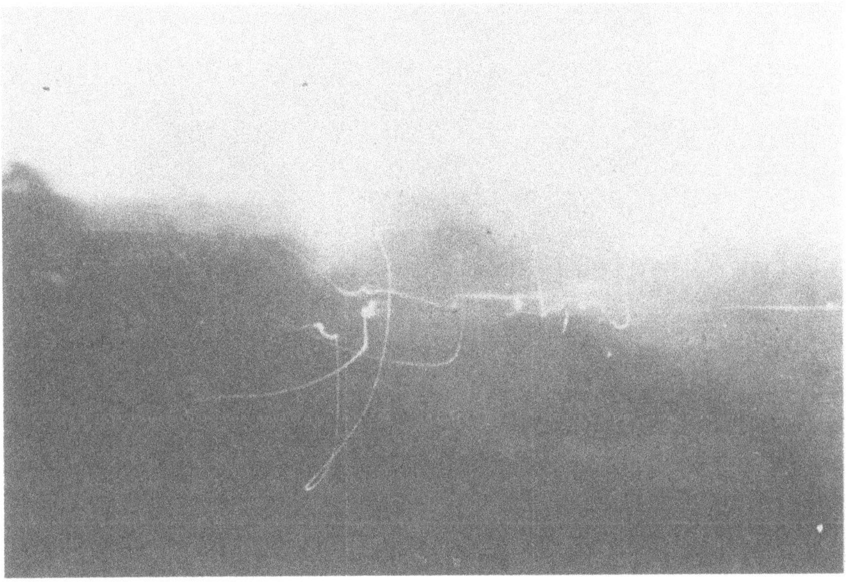

Figure 9.12. Still photograph taken by W. T. Cowhig in Rugby in 1937 during a thunderstorm showing trace identified as ball lightning, although nothing was seen at the time.

(1937) (Fig. 9.16) and Norinder (1939) (Fig. 9.17) published photographs of discharges from clouds that were alleged to have terminated above the ground and produced ball lightning from the point of termination. Merhaut (1944) (Fig. 9.18) published a photograph of ordinary CG lightning with a bright, independent region some distance to the left of it on the plate, level with the lower tips of the ordinary lightning channel; it was claimed that ball lightning had developed from this bright region. The trace in the photograph published by Norinder was unusually broad, but independent ball lightning is not visible. In the Bauman and Merhaut photographs, it is impossible to determine whether the bright region at the point of termination is actually above or on the ground, and the bright regions are of amorphous rather than spheroidal shape. They bear some resemblance to the videoframes by Hermant (personal communication, 1997) showing lightning striking a pine tree (Figs. 9.19a and b). It is suggested that a similar explanation might apply to the Bauman and Merhaut photographs.

B. P. Everett, a principal engineer working at Telecom Australia, took a color photograph in 1979 (Fig. 9.20) that shows a broad trace apparently emanating from a conventional lightning discharge. Power or telephone lines are visible in the foreground, although the broad trace does not appear to relate to them. It is difficult to explain because there are no signs of camera movement or double exposure.

Figure 9.13. Time exposure photograph taken by A. E. Ermolaev in June 1979 in Sverdlovsk, Russia during a thunderstorm. The trace was interpreted as ball lightning, although no ball was seen at the time. Stakhanova (1997) noted that the tree obscures parts of the trace, suggesting that the trace was a real image behind the tree; also noted was the absence of blurring. [Reproduced from Stakhanova (1997) with kind permission.]

Unfortunately, more detailed information about this photograph is unavailable (Larsson 1998).

Similar traces were obtained by Tompkins and Rodney (1975, 1976, 1977) and by Eriksson (1977) under more strictly controlled conditions. These are discussed later.

Some images allegedly of ball lightning show clearly defined spheroidal structures with defined boundaries bearing more direct resemblance to descriptions of visual observations. Norinder (1939, 1965) published a photograph of an ellipsoidal image (Fig. 9.21) recorded during a thunderstorm by engineer H. Schneidermann in 1935. Schneidermann was in the habit of photographing ordinary lightning from a balcony of a building in Berlin at night. The ball, of a diameter of about 30 cm and at a distance of 200 m, was reported to have moved past a building in the vicinity. It was in sight for about 10 s. A second ball was also reported but not photographed. Independent reports of ball lightning observed in Berlin the same night appeared in the newspapers the following day. Norinder drew attention to the similarity of the image to photographs of luminous globes reportedly produced after an electrical short-circuit of the terminals of a generator near a waterfall (Brand 1923).

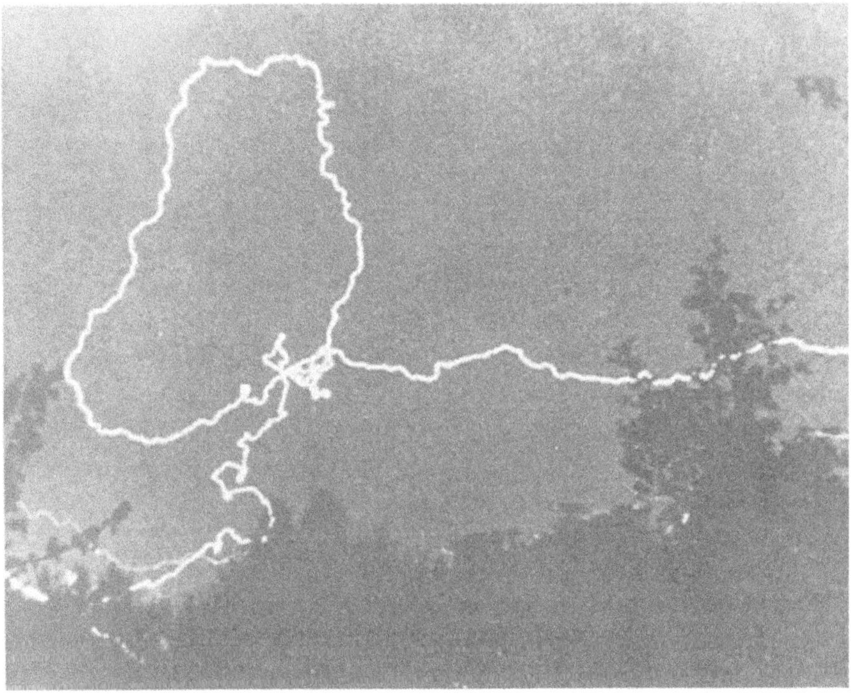

Figure 9.14. Still photograph taken in 1958 at Arnhem, the Netherlands, by J. Veenstra. Westphal (1958), Wolf (1958), and Barry (1980) concluded that it was not ball lightning. [Reproduced from Rutgers (1958) with permission of *Phys. Blatt.*]

Prochnow (1930) published a photograph (Fig. 9.22) that included a circular, luminous image with houses silhouetted in the background, but other spurious images on the film led Walter (1929) to suspect the cause as a double exposure with camera motion.

Charman (1976) published a photograph (Fig. 9.23) of a fairly clearly defined circular image on a photograph by Mr. M. R. Lyons taken in the hills of Derbyshire, England, during early summer 1972 while the weather was dry but thundery (although it rained less than an hour later). The photographer, who was short-sighted, did not see anything unusual in the viewfinder when he took the photograph. On processing his own film, he found one frame in a sequence of 35-mm exposures taken at intervals of about 15 s that showed what appeared to be a luminous ball less than 8 cm in diameter with hazy edges. If the image was of a real object, either its lifetime must have been less than 30 s or it must have been moving faster than 10 cm s^{-1}. Charman remarks that the image does not seem consistent with photographic artifacts such as lens flares, static marks, underagitation, or air bubbles during processing.

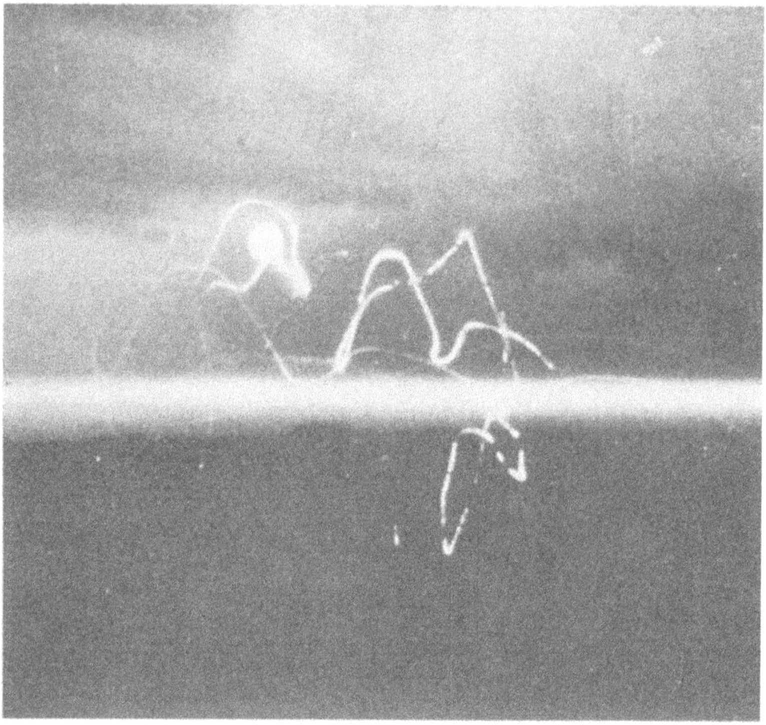

Figure 9.15. Fast-moving film photograph taken during a thunderstorm. The rapid and random motion of a dart leader in a linear lightning discharge produced the trace. This photograph should be compared with many of those attributed to ball lightning. [Reproduced from Holtzer, Workman, and Snoddy, 1938.]

Mr. G. A. Jones, the deputy chief executive of Ilford, Ltd. wrote to Charman and offered to have an expert in photographic fault analysis examine the photograph. Unfortunately, Mr. Lyons had moved and attempts to contact him to secure the negative were unsuccessful. Once a print had been examined, Mr. Jones wrote:

> It seems fairly clear from the print that what we are seeing is a true image on the film at the time of exposure and not some subsequent artifact. There are, however, several puzzling features: (1) The image appears as a sphere surrounded (at an angle to the lens axis) by several diffuse rings. I thought at first that the effect might be caused by rapid movement of the object during the duration of the exposure. However, closer examination of the image makes this explanation most unlikely. (2) The record could also be interpreted as a shell with a bright interior. It is difficult to see how such an unsubstantial object could be brighter inside than outside and, in any

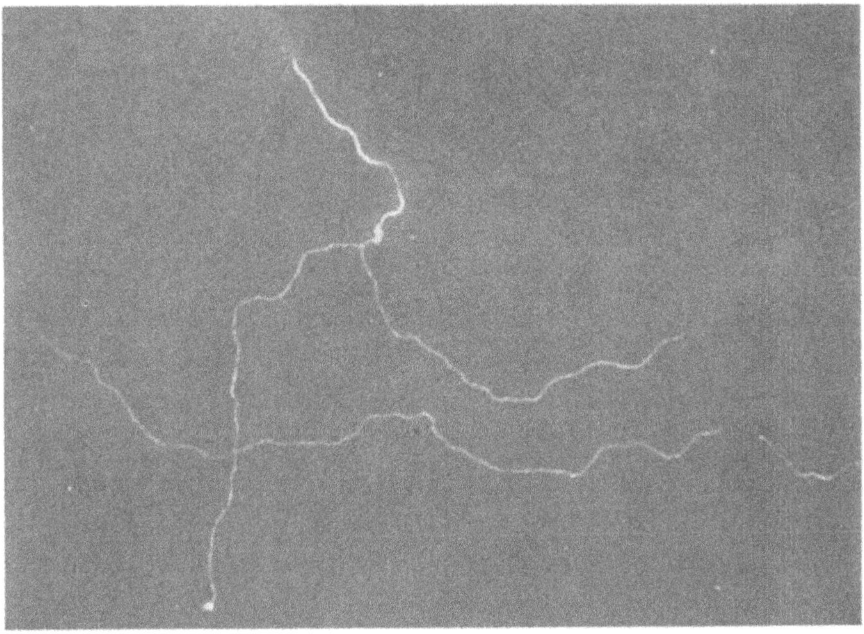

Figure 9.16. Still photograph of a lightning discharge that apparently terminated above ground (see trace at left of plate). The point of termination was reported to have developed into ball lightning. [Reproduced from Bauman (1937).]

case, close observation of the top right portion shows that the ivy leaf growing from the wall is either quenching the glow or casting a shadow on it! This could not be so if the object were luminous. (3) The surface luminosity appears quite low. In fact, the "shadow" on the lower left side of the sphere is exactly that which would appear on an opaque sphere lit from the same direction as the leaves. The tones in the diffuse rings suggest an even lower brightness. Although photography is not a good method of measuring luminosity, even with a standard in the picture for comparison, examination of the density in the negative might give more information than we can gain from examination of the print.

To summarize, the picture does apparently record a white ball, but one which appears not to be highly luminous. The only explanation I can think of is that the object is a translucent ball of light-coloured material surrounded by a flattened ring or rings of some form of emission which is either luminous, and quenched by the ivy leaf, or is recorded by the light incident on the leaf which thus throws a shadow on it. (G. A. Jones, personal communication to Prof. W. N. Charman, 1976)

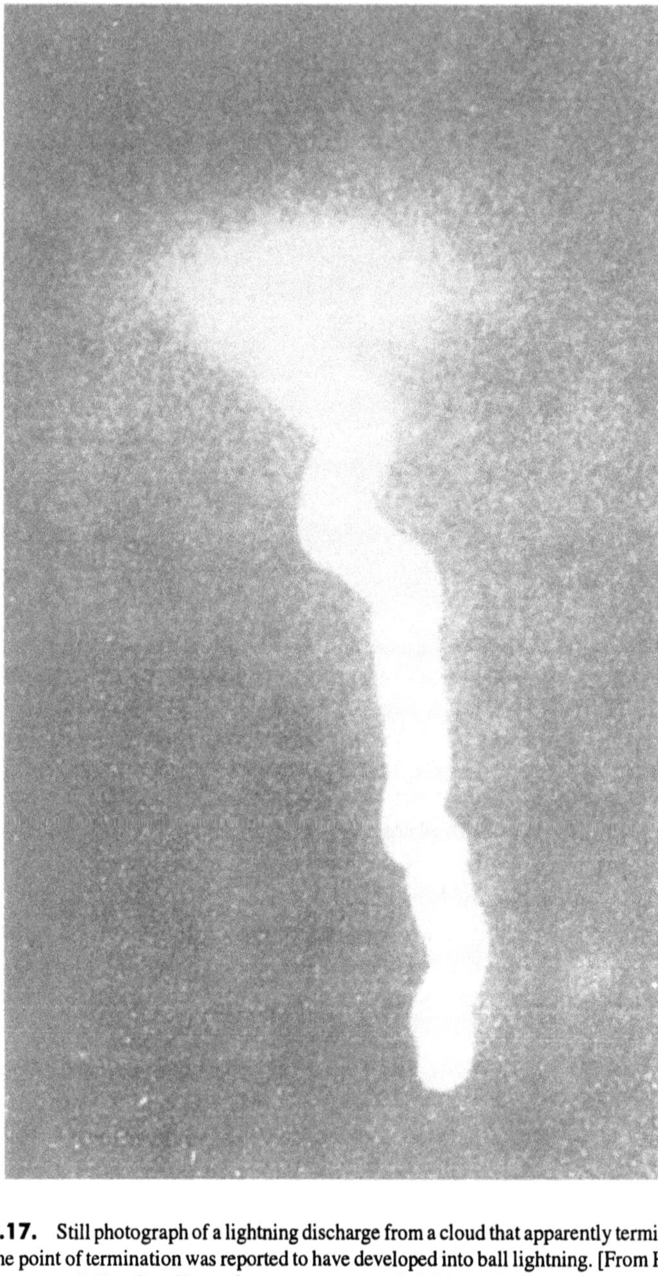

Figure 9.17. Still photograph of a lightning discharge from a cloud that apparently terminated above ground. The point of termination was reported to have developed into ball lightning. [From H. Norinder, "Om blixtens natur," *Kungliga Ventesskaposocietetens Arsbok* **94**, 39, the Royal Swedish Academy of Sciences. Reproduced with permission.]

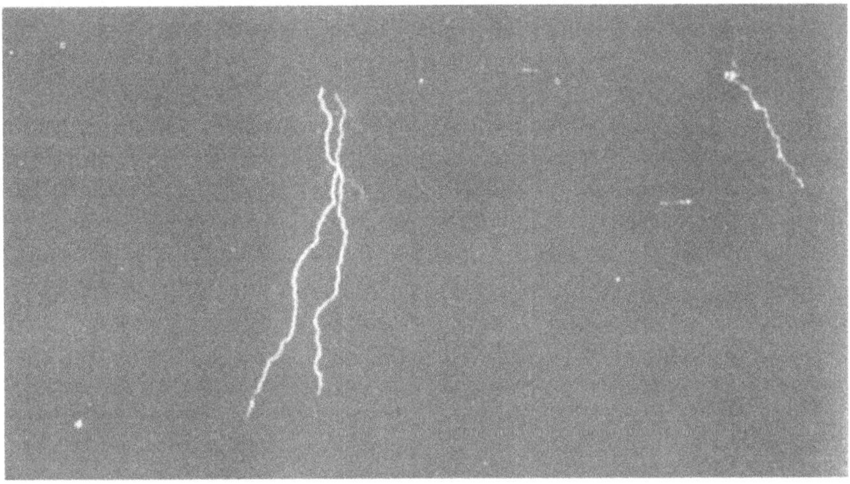

Figure 9.18. Still photograph taken during a thunderstorm. The image in the bottom left-hand corner was reported to have developed into ball lightning. [From O. Merhaut, "Eine bemerkensweithe Blitzaufnahme," *Naturwissenschaften* **32** 212, 1944. © Springer-Verlag. Reproduced with permission.]

Figure 9.19. A still from a video sequence showing linear lightning striking a pine tree. [Photograph: A. Hermant. Reproduced with his kind permission.]

Figure 9.19. Continued.

A circular, luminous trace was obtained on two consecutive photographs (Figs. 9.24a and b) taken 3 s apart and published by Stakhanova (1997). The photograph was taken in September 1985. Two people were on a fishing trip to Undugun Lake, 100 km from Chita in Russia. One morning, 30 min before sunrise, one of them observed a luminous ball with a bright point at its center that appeared from behind the hill on the other side of the lake. He woke his companion, who watched the ball with him. It moved with the wind toward them, gradually descending. As it arrived at the bank of the lake about 250 m from the observers, the ball stopped for a short while and then continued on along the bank. At this time, the witness took two photographs of the ball with an exposure time of 0.5 to 1 s. He steadied himself on the roof of his car to do so. Photometry of the film yielded estimates of the parameters of the ball. Its diameter was about 30 cm, its speed was 2 to 4 m/s, and its brightness was that of a 100-W lamp.

A professional weather observer and amateur photographer, Christian Witz, was taking time exposures during a severe storm on the late evening of July 4, 1989 at Senning, Lower Austria. There were many nearby CG flashes. A few moments after a CG flash about 200 m away, Witz saw something form about 200 to 300 m from the ground flash, as if "in a spiral, rotating movement," about 2 to 3 m above the ground. This was a very bright, almost blinding white, luminous ball with fuzzy outlines that appeared almost as large as the full moon. It provided a steady

Figure 9.20. Still photograph, originally in color, taken during a thunderstorm in 1979 by B. P. Everett in Australia. The broad trace apparently emanating from a conventional lightning discharge appears unrelated to the power or telephone lines in the foreground. There are no signs of camera movement or double exposure (Larsson 1998). [Photograph kindly supplied by Dr. A. Larsson.]

luminosity. It very slowly descended and then extinguished "like a candle flame in carbon dioxide" about 7 s after it had first appeared. When the ball appeared, Witz was already taking a time exposure and he released the shutter just after the light went out (Keul 1994). Explanatory hypotheses included reflection of a bright object in the wing mirror of the car, or flashover from lightning. Campbell suggested an error in the time of the event, photography of Saturn in a clear sky, not in clouds, and mirage effects producing apparent downward motion. Keul found the reflection/flashover hypothesis more plausible (Keul 1996).

Other images resemble pyrotechnics. Some of the best-documented and most famous photographs alleged to be of ball lightning were taken and published by J. C. Jensen (1933a,b, 1934), a scientist engaged in research on thunderstorm electricity. Between 1929 and 1931 he carried out extensive studies in Nebraska in the United States involving simultaneous electric field and pressure measurements and photography of lightning discharges at night. During a severe storm at about 9:40 p.m. on August 30, 1930, two cameras in a fourth-story window above the trees

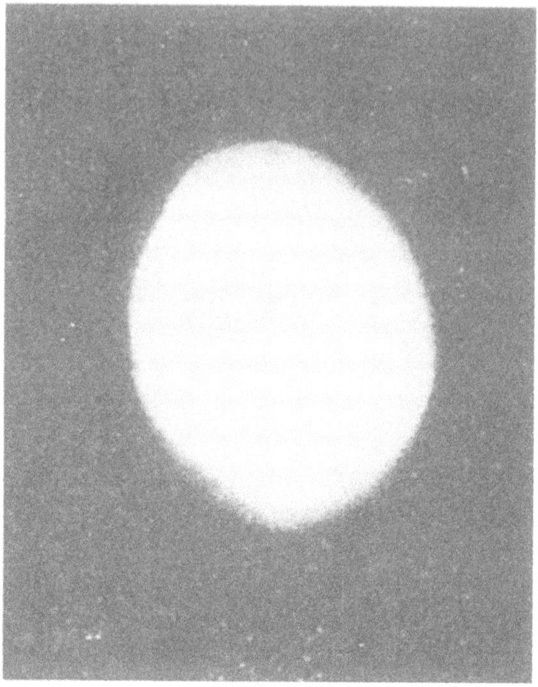

Figure 9.21. Still photograph taken during a thunderstorm in 1935 by H. Schneidermann. [From H. Norinder, "Om blixtens natur," *Kungliga Ventesskaposocietetens Arsbok* **94**, 39, the Royal Swedish Academy of Sciences. Reproduced with permission.]

were being used to record discharges. One was a 5 × 7 Graflex with an f4.5 lens and the other an Eastman 1A Kodak with an f6.3 lens. There was a great deal of dust in the air near the thundercloud. In the wake of one of the brilliant lightning flashes, Prof. Jensen saw a shapeless mass of lavender or rose color and pyrotechnic appearance that seemed to float slowly downward, apparently becoming brighter as it did so. Two or three of the globular structures appeared to roll for about 100 ft along a pair of 2300-V power lines, then bounced onto the ground, disappearing explosively. Both cameras took five pictures of the phenomenon over about 3 min (Figs. 9.25). Prof. Jensen estimated the diameter of the balls as 28 ft and 42 ft, respectively, at a height above the ground of 92 ft. This would have placed them above the power lines (Singer 1971).

It was subsequently suggested that students who had ignited fireworks during the thunderstorm had hoaxed Prof. Jensen (Wolf 1956). Lane (1966, 1968) pointed out, however, that the event occurred during the college vacation and that the reported duration of some of the balls—at least 3 min—seemed excessive for fireworks. He also suggested that it would have been impossible for students to

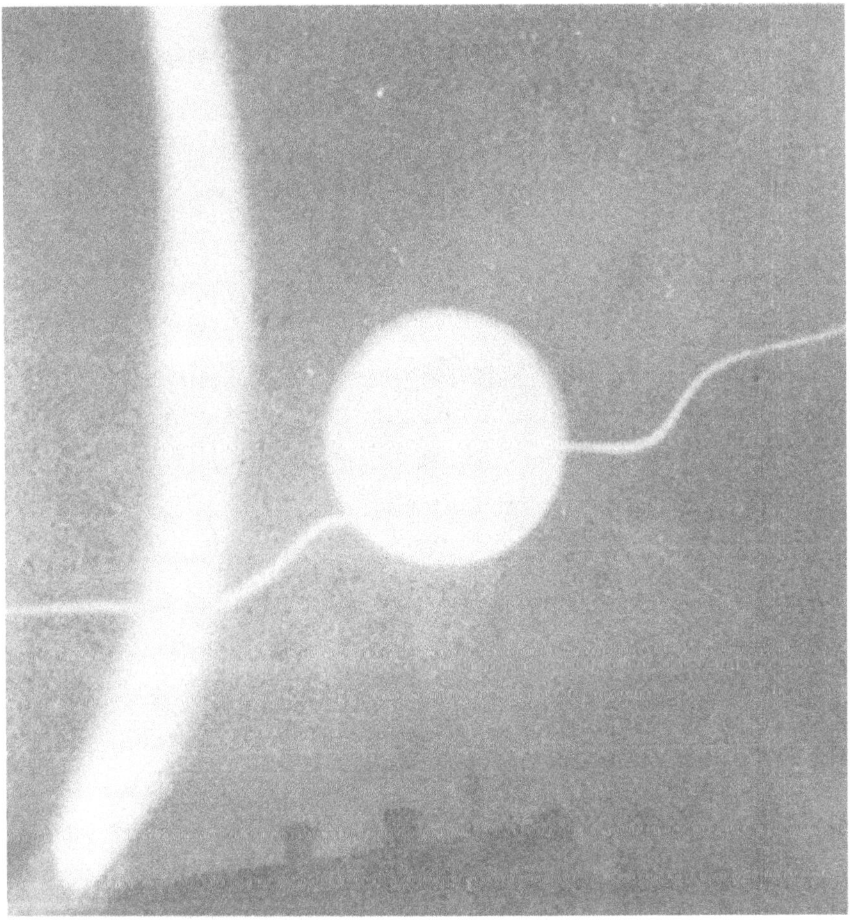

Figure 9.22. Still photograph taken during a thunderstorm. It was alleged to show the image of at least one ball lightning at different distances from the photographer. Walter (1929) suggested that the photograph was a result of double exposure and camera motion. [Reproduced from Prochnow, 1930a,b.]

anticipate the storm or to know that Jensen would be photographing it. L. E. Salanave of the University of Arizona, Tucson, studied the negatives of Jensen's photographs and expressed the view that they were not fireworks, having attempted unsuccessfully to duplicate the effects by photographing fireworks at a similar distance (Salanave 1965; Lane 1966, 1968).

Kuhn (1951) published a photograph (Fig. 9.26) alleged to be of ball lightning of pyrotechnic appearance. The ball was described as having risen behind a structure while emitting residual material in the form of streamers. Berger (1973) dismissed the photograph on the basis of a field investigation by Prof. H. F.

Figure 9.23. Still photograph by Mr. M. R. Lyons taken in Derbyshire, England, during summer 1972. Charman stated that the image does not seem consistent with lens flares, static marks, underagitation, or air bubbles during processing. [Reproduced from Charman (1976) with kind permission of Prof. W. N. Charman.]

Schwenkhagen, who had concluded that lightning to a mast supporting 10-kV power lines had caused a short-circuit discharge on the lines. He proposed that the ball consisted of molten metallic drops spraying from the power line. Barry (1980) points out that downward motion might be more likely in such circumstances and that lightning strikes to the mast had not been mentioned by the observer. He likened the photograph to a comparable account and very similar illustration by von Haidinger (1868a). [Campbell (1992) dismissed this latter observation as an astronomical mirage of the star Antares.]

A photograph (Fig. 9.27) of a trace superficially of pyrotechnic appearance was taken by Werner Burger one night in the summer of 1978 in Sankt Gallenkirch,

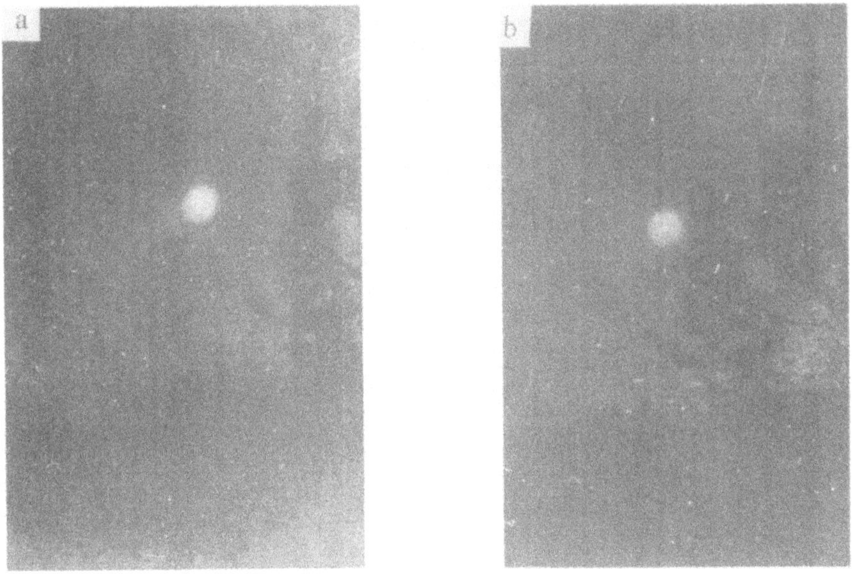

Figure 9.24. A circular, luminous trace was obtained on these two consecutive photographs taken 3 s apart in September 1985 at Undugun Lake, 100 km from Chita in Russia. Two observers saw a luminous ball with a bright point at its center. [Reproduced from Stakhanova (1997) with kind permission.]

a mountainous region in Montafon, Vorarlberg, Austria (Keul 1992). Keul, a psychologist, carried out a field investigation 12 years after the event, and supported this by consulting experts in pyrotechnics and lightning protection. Burger was taking color time-exposure photographs using a camera mounted on a simple tripod as he watched an approaching squall line. Just as he was taking a time exposure, he heard a sudden, strange sound "like a Christmas sparkler or a wire-brush moved over an edge (intermittently)." Immediately, in front of him, a "fireball fell down." In his surprise, he released the wire controlling the camera just before the disappearance of the object. He estimated that the fireball was in sight for 2 s. Keul suggested that this might have been an overestimate.

A pyrotechnics expert from a fireworks company rejected the suggestion that this was a firework rocket. A meteor or meteor-related phenomenon was suggested (Crew 1992a,b), but was rejected by two astronomers. A lightning protection expert suggested that it was an upward lightning stroke. Other lightning specialists disputed this (Keul 1996). Keul pointed out that this was inconsistent with the description given by the witness. Other suggestions that were made included a combustion phenomenon, a military object, or a hoax (White 1993). Keul found a number of experts in relevant fields who disputed each of these hypotheses; as a psychologist, he firmly rejected suggestions of a hoax (Keul 1996).

Figure 9.25. Two of five still photographs taken with two cameras by physicist Prof. J. C. Jensen, who was recording a severe storm in August 1930. It was suggested that students who had ignited fireworks during the thunderstorm had hoaxed Prof. Jensen. This interpretation has been disputed. [Reproduced from Jensen (1933b).]

Figure 9.26. Still photograph alleged to be of ball lightning. The ball rose from behind a structure, appearing to emit a luminous residue. [From E. Kuhn, "Ein Kugelblitz auf einer Moment-Aufnahme?" *Naturwissenschaften* **38**, 518, A51. © Springer-Verlag. Reproduced with permission.]

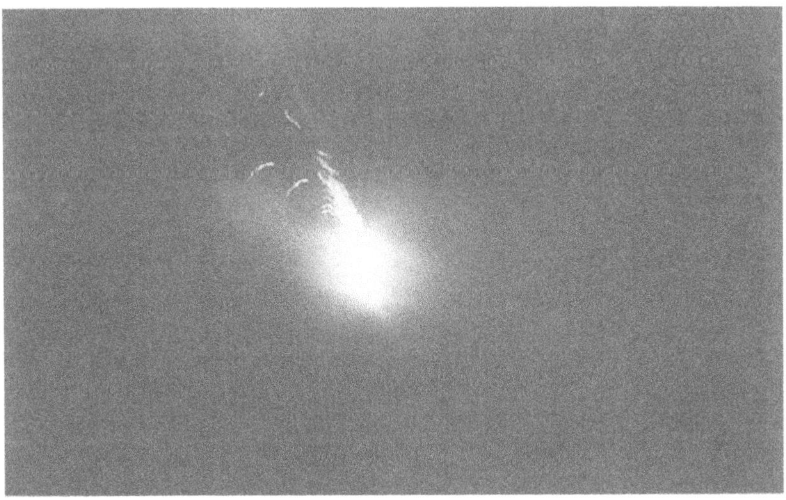

Figure 9.27. Still night-time photograph of a trace of pyrotechnic appearance taken by W. Burger in summer 1978 in Sankt Gallenkirch in Austria (Keul 1992). Suggested explanations such as firework rockets, meteor or meteor-related phenomenon, and an upward lightning stroke were rejected by experts in relevant fields (Keul 1996). [Reproduced by permission of the photographer, Herr W. Burger, © 1978.]

9.4 Films and Videotapes

Just after 9 a.m. on January 11, 1973, a cold, slightly misty morning with an overcast sky, Mr. Peter Day was driving east near Cuddington toward Aylesbury, Buckinghamshire, England, when he noticed an orange ball of light low on the northern horizon traveling in the same direction as his car. When he was eventually able to stop, he filmed the ball through the open window of his car for about 23 s using a Super 8 motion-picture camera that he habitually carried with him, until the ball suddenly vanished. He heard no sound. He obtained 380 frames of the object (Fig. 9.28). He interpreted the phenomenon as a UFO. The ball lightning "explanation" was suggested when the U.K. branch of the Committee for the Scientific Investigation of Claims of the Paranormal arranged a screening of this and another film at a branch of Kodak in London in September 1978. Scientists present at the screening, including the present author, considered ball lightning very unlikely. An investigation by Campbell suggested that the ball was fuel being dumped by a U.S. Air Force F-111 from Upper Heyford, which subsequently crashed near North Crawley some 30 km away from Day's location. Day resisted this interpretation (Campbell 1991).

On September 10, 1989 at 12:30 a.m. Greenwich mean time (GMT), Mr. Ray Cahill used a hand-held Sony CCD-V8AF-E camera to record 10 min of a thunderstorm from the first-floor bedroom of his house at Willesborough, Ashford,

Figure 9.28. Still from motion-picture film of an orange ball of light traveling eastward near Cuddington, Buckhinghamshire, England filmed by Mr. Peter Day on January 11, 1973. Several scientists considered ball lightning very unlikely. Day rejected the suggestion that the ball was fuel being dumped by military plane. [Photograph © Peter Day. Photographic credit: Steuart Campbell.]

Kent, England. The film was made through a double-glazed window overlooking some unlit houses about 50 m away. There was heavy rainfall. To the left of his field of view was a roundabout with streetlights, which appeared orange-white on the film.

When he played back the tape, he found that several flashes of lightning had been recorded. Apart from the streetlights, these appeared to be the only sources of illumination and offered a momentary clear view of the neighboring houses and gardens, which otherwise were not visible on the recording. One very close flash (with almost simultaneous lightning and thunder) provided a very diffuse, unfocused image of the houses. Some 4 s later, a red-orange, mottled, elliptical image with irregular edges, occupying about one-sixth of the frame, was seen to pass from the lower left edge of the frame to the upper right, disappearing from the edge of the field of view (Fig. 9.29). The image was not seen in the monochrome viewfinder. The image was more oblate near the edges of the field. It took 1.3 s to pass across the frame. The frame frequency was 25 Hz, so just over 30 frames recorded the image. There appeared to be certain geometrical features that remained constant throughout the 1.3 s. Within the bright image was a dark spot perhaps 20% of the external diameter of the image, as dark as the background. It passed from the bottom left of the image when near the corresponding corner of the frame, through the center when the image was closest to the optical axis of the camera, to the top right of the image when the image was in the top right of the frame. The annulus formed by the bright part of the image appeared to have modulated brightness around its circumference so that there were eight brighter regions, fairly regularly spaced. Movement of these brighter regions suggested rotation of the ball. No foreground or background was visible while the image was in frame, so evidently, if a luminous object had produced the image, it did not illuminate its surroundings. The intensity of the image was considerably less than that of the streetlights, although its color was rather similar, if more red.

Immediately after the image disappeared from the frame, the camera was turned toward the roundabout, where streetlights could be seen in focus. The view yielded by the preceding lightning flash suggested that the camera was pointed roughly toward the ground floor of the neighboring houses when the image was recorded, but the subsequent view of the roundabout implies that there must have been camera movement during the recording of the image (Meaden 1990, Stenhoff 1990).

Prof. R. C. Jennison of the University of Kent examined the videotape (Jennison, Lobeck, and Cahill 1990). Controlled experiments with the same camera from the same viewpoint implied that the image was of an object at a distance of 10 m with a diameter of 20 cm and a speed of 1 m s^{-1}. The uncertainty factor in these estimates was 2. The estimated total optical power was 50 mW in 4π steradians within the range detected by the camera. The conclusion of the authors

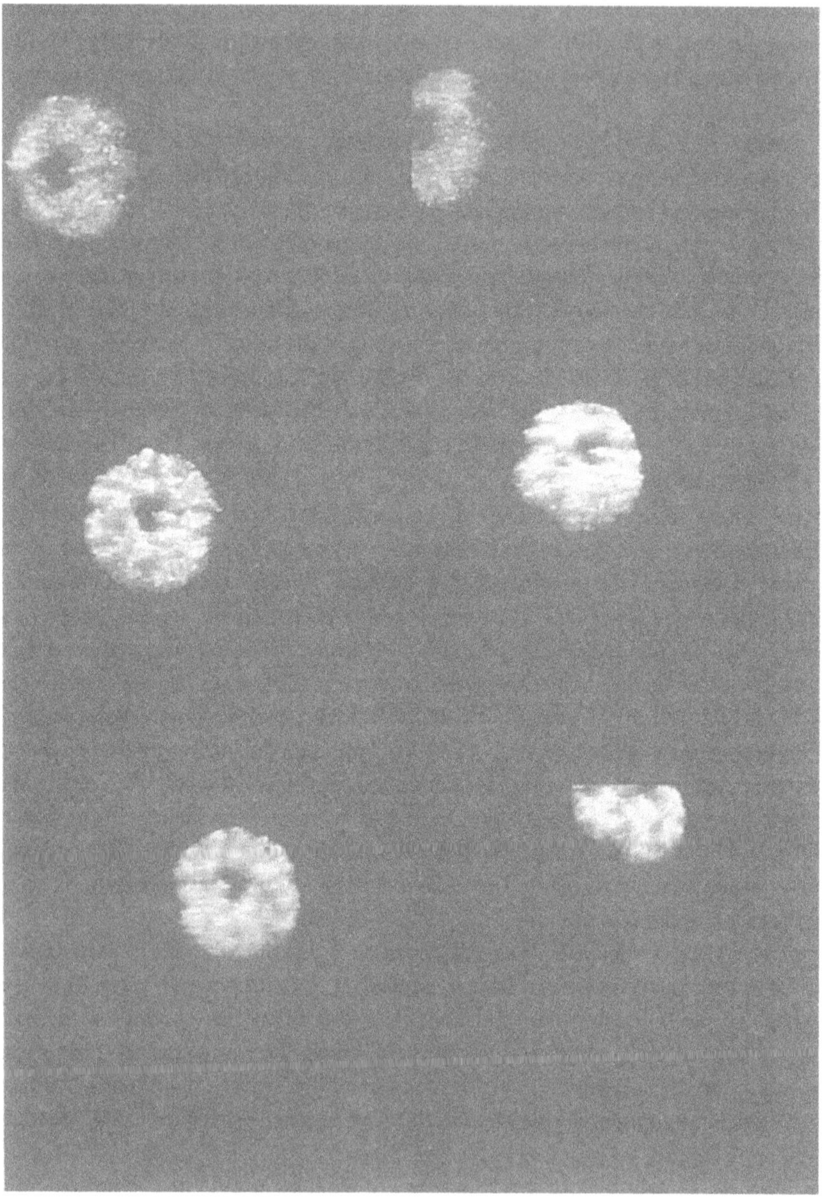

Figure 9-29. Frames from a videotape made on a camcorder during a thunderstorm on September 10, 1989 by Mr. R. Cahill at Willesborough, Ashford, Kent, England. Analysis by Bergstrom and Campbell (1991) showed that the object was most probably an image of the stop plane of the recorder illuminated by a streetlamp. [Photograph © R. Cahill and R. Lobeck. Reproduced with their kind permission.]

was that a luminous, toroidal object with a node–antinode circumferential structure had produced the image.

Stenhoff (1990) examined copies of the videotape at the invitation of the British Broadcasting Corporation (BBC), which had broadcast an extract in a number of news programs. He subsequently saw the film frame by frame at TVS, Maidstone, the television company to which Mr. Cahill had sent his tape, and from which the BBC had obtained their copy. Stenhoff suggested that a raindrop refracting light from the streetlamps might have produced the image. Meaden (1990) considered this explanation and also the possibility of a lens flare.

Analysis of the tape and comparisons by Bergstrom and Campbell (1991), however, showed that the object was most probably an image of the stop plane of the recorder illuminated by a streetlamp. Detailed analysis demonstrated how the videocamera could have caught the lamp, thus becoming out of focus (owing to the autofocus mechanism), and showed how the recorder could produce the image. Certainly the apparent motion of the image in relation to the optical axis of the camera and its appearance support this argument. Jennison (1992) accepted an explanation of this nature.

9.5 Instrumented Observations

Tompkins and Rodney (1975, 1976, 1977, 1980) reported two probable ball lightning events recorded by the Prairie Meteorite Network in the United States, one of which was close enough to be analyzed in detail (Fig. 9.30). The network consisted of 16 stations that photographed the night sky in seven Midwestern states every night over a 10-year period. Tompkins and Rodney searched about 12,000 photographic records of the network for ball and bead lightning. The cameras used a $13\frac{1}{3}$-Hz chopping shutter, closed for approximately two-thirds of the cycle. Thus both ball and bead lightning would be expected to appear as a sequence of images connected to a lightning flash. The regular spacing of the ball images was identified as ball lightning events because this spacing was determined by the chopping shutter speed and by the nontangential way in which the image sequence joined the lightning channel. In the cases found, the ball appeared to have left the lightning channel well above the tip of the stroke at angles from the vertical of 20 and 30°. In the photograph studied in detail, the series of images produced was consistent with a ball of a diameter of 2 to 4 m falling at about 62 m s^{-1} and drifting with the wind as it fell. There were also many other photographs consistent with conventional linear lightning terminating above ground and a ball being emitted, but the image of the ball would only be clearly visible when emitted from the side of the channel. The number of probable ball lightning photographs obtained enabled the frequency of ball lightning in this region to be estimated as one per 4800 km^2 night-years.

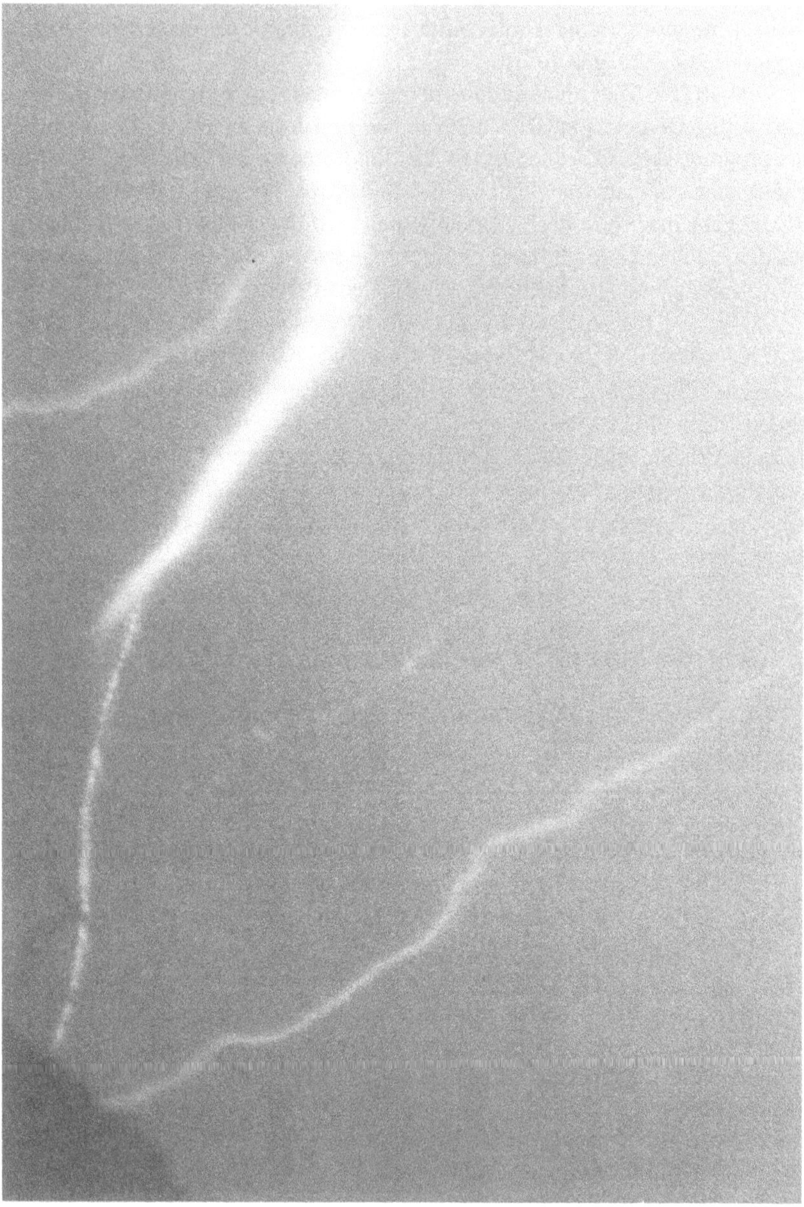

Figure 9.30. One of two probable ball lightning events recorded by the Prairie Meteorite Network. In both cases, the ball appeared to have left the lightning channel well above the tip of the stroke at angles from the vertical of 20 and 30°. In the photograph studied in detail, images were consistent with a ball of a diameter of 2 to 4 m falling at about 62 m s^{-1} and drifting with the wind as it fell (Tompkins and Rodney (1975, 1977)). [Photograph reproduced by kind permission of Dr. D. R. Tompkins, Jr.]

During a thunderstorm on the evening of November 12, 1976 in the Transvaal Highveld region of South Africa, an image was recorded by an automatic videotape recording system used to study conventional linear lightning. The system consisted of a television camera with a silicon diode array target tube (Eriksson 1977a) (Fig. 9.31). The image appeared detached and round and was seen in the vicinity of a branch in the main channel of a CG lightning flash that consisted of a series of four return strokes, the decay of the last of which was unusually tortuous. The round image appeared to become brighter in the intervals between strokes. Its altitude was estimated as 300 m and its diameter as 5 m, although the latter estimate was rendered uncertain by image blooming.

Firing rockets with grounded wires attached into thunderclouds and thus initiating lightning return strokes has been the basis of some very important lightning research. Such "triggered" lightning experiments at the St. Privat d'Allier lightning research station in the Massif Central region of France yielded several observations of phenomena resembling bead and ball lightning (Fig. 9.32). These suggested there might be a relationship between the two phenomena. The decay of triggered lightning was found to be beaded. The beads had an initial diameter of about 40 cm and decayed gradually, with a total lifetime of up to 0.3 s. During

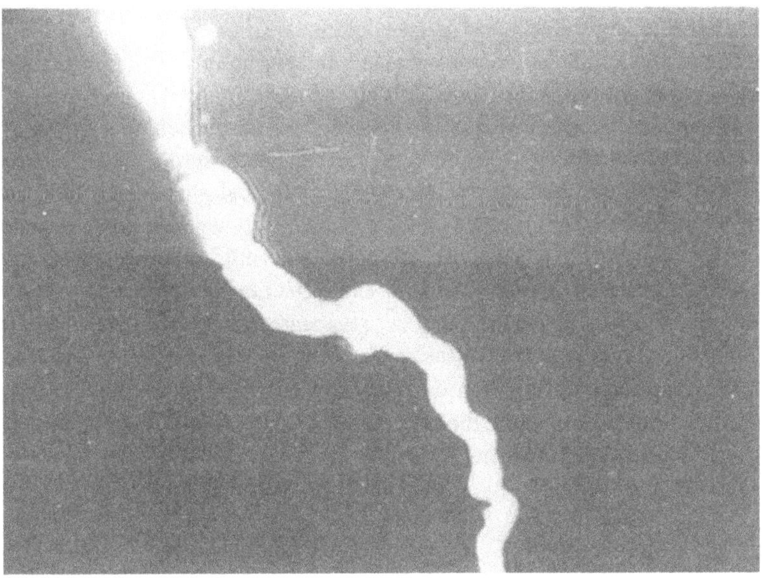

Figure 9.31. Frame from a videotape sequence showing a luminous ball image that appeared near a lightning flash about 300 m above the ground. The diameter of the ball was less than 5 m and its duration was about 20 to 40 s. [Photograph reproduced from Eriksson (1977a,b).]

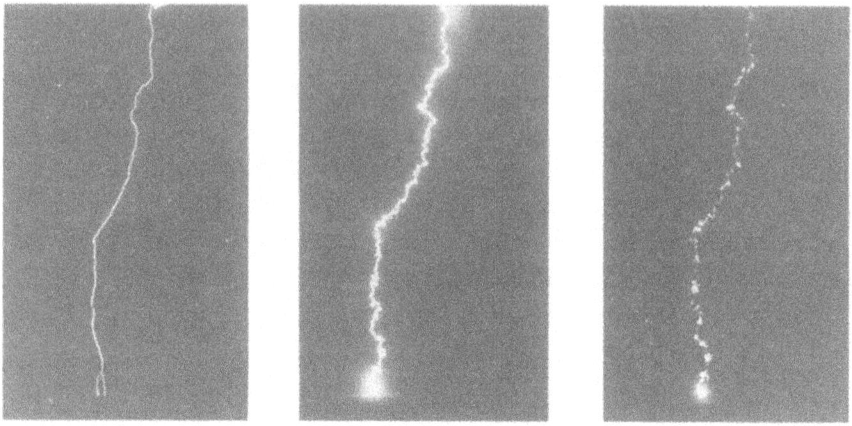

Figure 9.32. Frames from movie films of the persistent illumination that sometimes remained near the ground following triggered cloud-to-ground lightning flashes even after the main channel luminosity had decayed. The ground luminosity would rise with a velocity of 1 to 2 m s^{-1}. It was speculated that the luminosity could be produced by gases from the soil, heated by the flow of current. [Reproduced from Hubert (1975a) by kind permission of Dr. P. Hubert.]

strokes of long duration, the channel, which was initially straight, became progressively more tortuous, and the largest beads were found in regions of maximum tortuosity. The larger beads had longer durations, so that at the end of the decay only one or two luminous spheres remained; these usually moved upward, perhaps by convection, with speeds of 1 or 2 m s^{-1}. Continuing currents could not easily be used to explain the persistence of the spheres because the behavior of one sphere was apparently unaffected by a subsequent stroke that occurred during its lifetime.

Other luminous phenomena of roughly spherical shape and an estimated diameter of 25 cm were observed on the ground next to wooden posts surrounding the rocket launchers. Their appearance coincided with lightning flashes to soot-coated pipes placed above the grounding cable that ran between the launchers. The luminosity disappeared periodically, but reappeared after each new strike. It was thought by the investigators to be the result of currents circulating in the soil around the grounding cable during each flash that caused outgassing and a local electric discharge at the points where the posts punctured the upper layer of soil (Fieux, Gary, and Hubert 1975, Hubert 1976).

9.6 Conclusions

We conclude that the evidence provided by still photographs alleged to be of ball lightning is very questionable. Still photographs taken by chance will always be a matter of controversy. The likelihood of obtaining probative photographic evidence of ball lightning through chance observation is small. Videotapes (or films) have the potential to yield more useful data, but there is still the possibility of error. The results of instrumented and experimental observations are much more interesting.

Wooding (1992) discussed the viability of a systematic attempt to record ball lightning using instrumented monitoring systems and concluded that such an exercise would not be practically or economically feasible. Nonetheless, experiments used to monitor other atmospheric or astronomical phenomena may occasionally reveal images of ball lightning. Ball lightning investigators might usefully capitalize on the opportunities thus provided. Discussions with scientists involved in such experiments concerning their viability for observing luminous phenomena of the brightness and duration of ball lightning may be fruitful.

Chapter 10

The Existence of Ball Lightning

10.1 The Problem of Random, Transient Phenomena

Unpredictable, short-lived phenomena such as ball lightning present a particular problem for physical science. Because such phenomena are normally reported by casual, untrained observers who are unfamiliar with many effects and processes that occur in the atmosphere, there is plenty of scope for misidentification both of commonplace and rare but understood events, or for inaccurate reporting. The language observers use to describe what they have seen is likely to be imprecise, subjective, and unscientific, and their interpretation tends to be instinctive. A further frustration for a would-be investigator is the inevitably low yield of information, especially quantitative data, from a description of a visual observation of an event of very short duration.

As a consequence, many phenomena reported in this way have traditionally been relegated to mythology, or to what has become known as *folk science*. This may be particularly true of phenomena whose earliest reports predate the scientific era.[1] Ravetz uses the term "folk-science" to mean "a body of accepted knowledge whose function is not to provide the basis for further advance, but to offer comfort and reassurance to some body of believers" (Ravetz 1971).

[1]The scientific era here is taken to have begun in the seventeenth century. Until the late seventeenth century, science was a relatively unorganized activity. Practitioners of science included those of independent means, those of other professions (such as clergy and medics), and those in receipt of grants from royal patrons. The name "scientist" was not invented until the nineteenth century. The beginnings of science as a coordinated activity were with the foundation of the learned societies—the Royal Society of London in 1662, the French Académie des Sciences in 1666, and the Berlin Academy of Sciences in 1700. By 1790 there were approximately 220 such organizations (Rose and Rose 1969).

Bookshops abound with literature concerning topics on or, indeed, beyond the fringes of science, such as unidentified flying objects, psychic phenomena, the Loch Ness monster, and the Bermuda Triangle. Many scientists describe the study of such phenomena as *pseudo science*, a more pejorative term than folk science. Their skepticism is directed both at the phenomena and at the way in which they are discussed (Gardner 1952, 1957; Evans 1973; Rothman 1988; Schick and Vaughn 1995). No doubt many, or perhaps all of these phenomena are spurious. However, there is perhaps a danger that scientists could occasionally dismiss phenomena of interest simply because *discussion* of them has been nonscientific (or even antiscientific).

The complexity of perceptual processes and memory was discussed in Chapter 3. A scientist skeptical of phenomena such as ball lightning may express little faith in eyewitness testimony and emphasize the significance of distortions of perception and memory. Such an attitude gains support from inconsistencies in eyewitness descriptions of even relatively familiar events such as crimes and accidents.

Matters are not entirely hopeless, however, in circumstances where a reported phenomenon leaves behind traces or damage. Just as in criminal investigations where forensic science may be critically important in determining what really happened, an analysis of traces or damage may yield considerably more useful information than the most detailed eyewitness account. Chapters 4 to 9 discussed such important reports and their interpretation. In the case of ball lightning, however, very little evidence has been found of "strong interaction" with the environment because most damage attributed to ball lightning appears to be caused by ordinary lightning. This, together with the weakness of photographic evidence, means that physical evidence for the existence of ball lightning is scarce. This, however, may result from the properties of the phenomenon rather than implying that ball lightning does not exist.[2]

[2]Because of the difficulties presented by transient phenomena, the Smithsonian Institution in Washington, DC set up the Center for Short-Lived Phenomena in 1968 to operate a global environmental alert system for the rapid communication of scientific information on short-lived natural events. The center was established to improve opportunities for research by quickly disseminating information on short-lived events while environmental changes were still occurring and to improve the exchange of scientific information through the development of an effective global communications system. This included a network of more than 3000 scientists and scientific field stations in 148 countries on every continent and every ocean of the world. The center communicated data and information on significant changes in biological and ecological systems, as well as volcanic eruptions, the birth of new islands, major fireball events and meteor falls, and environmental pollution events. Rapid receipt of information about events permitted research teams, with their instruments and equipment, to enter event areas in as short a time as possible to collect important data that might otherwise have been irretrievably lost to science. (The center has since narrowed its areas of investigation and is now called the Global Volcanism Program.)

10.2 Ockham's Razor and Other Philosophical Questions

William of Ockham, born about AD 1290, was a theologian, not a scientist. In his writings, the "razor" appears as *Frustra fit per plura quod potest fieri per pauciora*, which translates as "It is vain to do with more what can be done with fewer" (Russell 1945). A modern interpretation of Ockham's razor is that "if two theories equally fit all the observed facts, the one requiring fewer or simpler assumptions is to be accepted as more nearly valid" (Rothman 1988). Rothman points out that it is important to recognize that Ockham's razor is not an empirical natural law, but, "rather, a heuristic principle that leads scientists along the path of the most efficient investigations." Rothman writes, "On close examination, this interpretation of Ockham's razor appears to be more a convenient prejudice than a useful rule. If two theories equally fit all the observed facts, then it makes no difference which of the two theories you choose. One of them is no more valid than the other. It is true that you would probably prefer to use the theory with the fewer or simpler assumptions, or with the simpler mathematical equations, but that has nothing to do with validity." Another rewording is "The simplest explanation for an observation is the most likely to be the correct one" (Feynman, Leighton, and Sands 1965). Rothman criticizes this interpretation because "the simplest explanation means different things to different people."

In his discussion, Rothman is particularly considering claims of paranormal events. He suggests that when we look at claims of unusual events:

1. We must assure ourselves that the event actually took place before trying to explain it.
2. If we are sure that the event happened, we should try to explain it by normal means. This is a useful way of applying Ockham's razor even though it is not amenable to scientific proof.

What he says is quite correct. However, when we apply this method to casual observations by untrained observers, we may run into a difficulty. The level of confidence we attribute to the testimony of eyewitnesses in general may determine the outcome of the process. A rigorous application of Ockham's razor, given the present level of knowledge about ball lightning, will always place the "ball lightning interpretation" low on the rank order for simplicity. Adoption of the ball lightning interpretation implies the following assumptions: (1) the eyewitness description is essentially accurate and (2) ball lightning exists notwithstanding the persistent difficulties it has presented for both theoretical and experimental scientists.

In particular, in applying Ockham's razor there is a danger of weighing the probability of eyewitness error against the apparent improbability of available theories. If an investigator is disposed to reject reports of ball lightning, he merely need adopt a low level of confidence in the competence of the reporters. Of course this is illogical since in general terms the quality of empirical data is independent of the plausibility of theories. Campbell (1992a,b) writes, "We then observe that no theory exists which can explain all the reported characteristics of ball lightning and that no-one has been able to create ball lightning in laboratory conditions which simulate those in the open. These facts can be explained most simply by proposing that ball lightning does not exist."

This problem, applied to UFOs, was discussed by Walker (1968). He wrote, "One recourse, or course, is to deal only with 'hard data' and to simply refuse to deal in any way with eye witness reports, contending that such observations are unlikely because they are too bizarre or have previously been reported only by 'crazy people.' This kind of reaction reflects scientific closed-mindedness. . . . On the other hand, the opposite position of complete, unquestioning faith in observer reports is no better." Instead, Walker recommends an initial attitude of benevolent skepticism in investigating eyewitness reports of such phenomena.

Phenomena reported in this way form a kind of bridge between the physical sciences and psychology. There is a danger that physicists will resort to psychological or perceptual explanations to avoid the need to provide a physical explanation. As Keul (1993a) has rightly pointed out, ball lightning research must be an interdisciplinary process involving psychologists as well as physicists. Walker (1968) recommends that eyewitnesses are subjected to a general medical evaluation, including a neuro-opthalmologic examination in order to determine physical or psychological factors that may affect the accuracy of reports.

Until and unless the existence of ball lightning is established beyond reasonable doubt, the best we can do, however, is, with some caution, to apply Ockham's razor in attempting to explain ball lightning reports. Wherever possible, we must explain ball lightning reports in terms of those phenomena that we already understand, while giving eyewitnesses a certain measure of credit for their integrity and observational competence in the absence of evidence that we should do otherwise. The objective at first is to establish whether the phenomenon exists and determine its observed properties. Once those questions are resolved, other, more borderline cases can be revisited if this seems appropriate.

10.3 Skeptical Views

Many scientists, such as Lodge (1905), have been skeptical of ball lightning. Many have favored the afterimage interpretation—these include Kelvin (1888), Argyle (1971), and Berger (1973) (see Section 3.3.5). The survey by Humphreys (1936), with skeptical conclusions, was discussed in Chapter 1. Humphreys thought an afterimage could be one possible interpretation.

Michael Faraday was not skeptical of the existence of ball lightning, but doubted that it had "anything to do with the discharge of ordinary electricity," or was "at all related to lightning or atmospheric electricity" (Faraday 1839).

Schönland (1950) wrote, "It appears that no professional observers of the weather, such as meteorologists, have ever seen a fire-ball." However, by this date reports had been published from scientists and meteorologists (see later discussion).

Berger, after 30 years of experimental research into lightning phenomena on Mount San Salvatore near Lugano, Switzerland, which included the study of thousands of photographs and over 1000 oscillograms, reported that he had never found evidence for ball lightning. Based on his own experience and a study of the literature on the subject, Berger stressed the need to distinguish between subjective and objective observations. He claimed that all published photographs of ball lightning had been otherwise explained. He recommended research on afterimages produced by flashes on the human retina, and that new reports of ball lightning should be followed up by field investigations by high-voltage engineers and physicists (Berger 1973).

Campbell (1992a,b) defined ball lightning "(if it exists) as an electrical discharge phenomenon." Having, it seems, once considered ball lightning to be a phenomenon of some significance, he came to question "the existence of corona discharges in mid-air, but not necessarily the existence of other luminous phenomena." He claims that "there is no photograph, film, or video recording which can be accepted unreservedly as showing ball lightning." He concludes by explaining that he does not claim that ball lightning does not exist, but "merely propose[s] the null hypothesis." He correctly indicates that the onus of proof is on those who advocate the existence of ball lightning.

He presents a number of examples of ball lightning reports that he has very successfully explained, but the explanations he offers for some others strain credibility and seem at odds with Ockham's razor. For example, he suggests astronomical mirages as a cause of some ball lightning reports, where necessary introducing observer error to explain discrepancies in times, altazimuth values, etc. (Keul 1996).

Hubert (1996) writes, "It is possible to quote a Russian professor, an eminent plasma specialist passing through Paris, who declared publicly, 'In my Institute, when at the end of a seminar people begin to talk about ball lightning, I consider that it is time to end the seminar!'"

However, some experienced lightning specialists disagree. Uman (1987, p. 23) refers to "the relative wealth of similar ball lightning observations over a period of centuries; reports that leave little doubt as to its reality." Dr. Earle Williams is of the opinion that the consistency of ball lightning reports is fairly persuasive, and that reports of ball lightning inside aircraft are among the most puzzling (E. Williams, personal communication, 1998).

10.4 Reliability of Reports

10.4.1 Ball Lightning Reports from Scientists

Scientists have often reported ball lightning. Tuck (1971) and Davies (1987) refer to ball lightning reports from Niels Bohr and Victor Weisskopf, and Davies also refers to another from Sir Martin Ryle. Barry (1980a) cites observations by Loeb (1966), Marsh (1895, 1896a,b, 1899), Rotch (1903a,c), and Caballero (1890b). In Chapter 7 we gave Professor Roger Jennison's impressive report of ball lightning seen at close range inside an aircraft.

Some reports of ball lightning from scientists seem to lend themselves to more conventional interpretation—for example, perhaps, the reports by Lennox, Froome, Smith, and Swithenbank in Chapter 3. In many cases, scientists have endeavored to explain their own observations of ball lightning in familiar terms. While some reports from scientists may indeed be explained by other phenomena, reports written by scientists often have high information content, with detailed quantitative estimates. Most scientists are, no doubt, cautious in recounting personal experiences of a controversial nature, so their reports are usually very well thought out.

This is a report written only 4 days after an observation by Dr. Eric Dunford, until 1998 director of space science at the Rutherford-Appleton Laboratory, Oxfordshire, United Kingdom.

> *Ball lightning* seen on Thursday 18th April 1968.
>
> *Climatic conditions:* The ball was seen at about 1915–1930 hours when an intense thunderstorm had been in progress for about an hour. At the time torrential rain was falling and there was a high wind. The sky was heavily overcast.
>
> *Description:* There had been spectacular lightning flashes very close to the house. A flash occurred almost due north at a distance I would estimate to be between half a mile and a mile. Immediately following the flash an intense spherical ball of light appeared, its angular diameter I estimate to be about half the sun's diameter. Its brightness was sufficient to light up the surrounding fields. My own impression was that its color was very similar to that of the lightning itself but my wife and mother-in-law said it had a red or yellowish tinge. I believe there was a tail of light (or smoke) attached to the ball. The whole event lasted about five seconds (long enough for my wife to get out of her chair and come to the window) and during that time the ball did not move very much. If its distance was half a mile then I estimate its height to be 60–100 ft (above some distant trees).

> There was no explosion but a fairly slow disappearance with a hazy or smoky appearance.
>
> *Surrounding area:* In the direction of the occurrence of the ball open fields slope downwards to the Thames Valley. The nearest house in that direction is three quarters of a mile away. Maidenhead is about four miles in the same direction.

In 1983, Dr. Dunford completed a questionnaire in which he gave further details of the event. The three witnesses were in the ground floor of a house and viewed the event through window glass. The ball appeared late in the storm and seemed to follow a cloud-to-ground lightning flash, which may have struck a tree or the earth. The ball was formed in midair and remained stationary throughout its lifetime. Its appeared to subtend between 0.25° and 0.5°. If it was half a mile away, its diameter would be about 4 to 8 m. The ball was spherical and white in color, and it was not possible to see through it. It appeared to be about as bright as an ordinary lightning stroke. The brightness was uniform across its surface, and both brightness and diameter remained fairly constant (Dunford 1968, 1983; personal communications).

During a violent thunderstorm that occurred in a monsoon in the Murree Hills, India, at the foot of the Himalayas, at 7 p.m. one evening in 1938, Dr. J. C. Bass, now director of research at the Plessey Company, plc, saw a green ball of 25 cm diameter enter a room through an open window, fall to the floor (with which it remained in contact), and travel slowly along the floor and up the leg of a wooden table. On top of the table, it came into contact with a water-filled tin bath where it disappeared, having been in sight for about 5 s or more. The appearance of the ball did not change throughout the time of observation (Bass, J. C. 1983; personal communication).

A number of scientists recall ball lightning experiences from childhood. The late Dr. Jack Katzenstein of the Atomic Energy Organization of Iran gave the following account (Katzenstein, J. 1976; personal communication):

> I was 12–13 years old at the time of the occurrence. I was sitting on the floor of my parents' bedroom at my boyhood home in Shreveport, Louisiana, USA, reading a newspaper. The time was between 2 and 3 p.m. . . . and a thunderstorm was in progress. There was a tremendous crash and flash of light. Subsequently, I learned that a tree in the yard had been struck by lightning. I looked up from the newspaper and saw a luminous ball, predominantly red in color, but with a bright white center, float in through an open window which was covered with a copper screen. The ball was about 8 inches [20 cm] in diameter, less than a foot [30 cm] but larger than a ball for croquet or bowls (6 in. diameter [15 cm]). The ball floated across

> the room about 10 ft [3 m] and contacted a metal floor lamp. At the instant of contact, the ball exploded with the approximate force (report and shock wave strength) of a large firecracker. I felt the sting of the shock wave hitting the exposed parts of my body. After the explosion the ball disappeared. There were no signs of any burns or any after-effects on the wire screen of the window, the floor lamp or any other place in the room. As I said, a tree some 20 ft [6 m] from the window had been struck by lightning and a 4 inch [10 cm] strip of bark removed, but the tree itself remained standing.

Dr. L. Harrison Matthews, FRS, reported a ball lightning event that occurred during a summer evening in June or July of 1916 or 1917. He was sitting at a desk in Clifton, Bristol, England, and looking out of a window during a tremendous thunderstorm.

> There was a telephone post on the other side of the road and while the rain was pelting down in bucketsful a glowing pink ball that appeared to be about the size of a football appeared at the top of the pole and quickly slid down it to the ground where it appeared to bounce two or three times as it ran along the ground and then disappeared without any loud explosion. There was no crash of thunder but a hissing noise. The pole was about 30 feet [9 m] high and the top [was] level with the roofs of the houses. Although it was so long ago I remember it very vividly. There was no damage. (Matthews, L. H. 1982; personal communication)

The fact that the luminosity was first seen near the top of the pole would suggest St. Elmo's fire, but the subsequent reported motion seems inconsistent with this because the electric field would be expected to be most intense at the tip of the pole.

F. J. Hiorns, a Fellow of the Institute of Physics, reported a ball lightning event

> in 1939 or 1940 when I was around 14. We were living in a house at Kingsthorpe Grove, Northampton. On the day in question there was a violent thunderstorm with very heavy rain, and my mother and I were alone in the house. . . . It was then raining heavily and as we looked out of the rear window a lightning ball moved across and below the window. . . . When it had arrived within a few inches of the sloping roof it disappeared. So far as I am able to recall, there was instantaneously a very loud bang. We subsequently discovered that the main chimney stack was split down the middle, and I was afterwards told by my father that it had been struck by lightning.
>
> From my personal observation (1) the ball was well defined, probably about 8 inches [20 cm] across, spherical and of uniform appearance. (2)

> It traveled a distance of about 6 feet [1.8 m] in 3–4 seconds at a uniform speed. (3) It moved from left to right—i.e., away from the chimney which was damaged. (4) The noise occurred when the ball disappeared, and I think that the chimney damage must have occurred then. (F. J. Hiorns 1983; personal communication)

The statement that the ball preceded a lightning flash to the chimney is contrary to any suggestion that it was a positive afterimage.

Dr. James L. Guthrie reported "an excellent example" of ball lightning:

> in 1944 when I was 13 years old. . . . We lived about 100 ft [30 m] from a steel water tower which was struck frequently by lightning. During one storm in the evening, my mother, two younger brothers and I saw a ball descend from a chandelier in our living room and settle onto the floor. It was pale yellow, about the size of a football or volleyball, perhaps smaller, and threw off small sparks. It behaved like a balloon, seeming to have little mass, and compressing slightly on the bottom when it touched the floor. We wanted to play with it, but my mother told us to keep away from it. It was not at all frightening.
>
> The ball rolled slowly across the floor, to a wall, and up the side of an upright piano, then across the top of the piano to where my father and I kept our violins. I was hoping to give up the violin at that time, and remember wishing that the ball would set the violins on fire. But instead of that, it disappeared. I cannot remember exactly how it vanished, but I think that it "popped" like a balloon but with only a slight sound. It may have accelerated its pace or have begun to spin just before it popped.
>
> I ran to look at the violin cases but there was no sign of burning or heat. I do not remember any odor. The whole episode lasted 10–15 seconds.
>
> The surface of the ball was active—perhaps effervescent is the right description. It looked fuzzy, but the "sparks" may not have left the surface. I think there was a faint hissing sound during the entire lifetime of the ball, but I am not certain of that point. The strangest thing about the ball was its slow pace and apparent weightlessness. I can't believe that it contained much energy (J. L. Guthrie 1983; personal communication)

Dr. A. T. Donaldson, a physicist, also recalled an experience from his school days. This was one late afternoon in 1963 or 1964 and occurred at Presentation College, Reading, Berkshire, England. There was a severe storm with much lightning and torrential rain.

> Following a flash of lightning a bright ball of light rather like a descending military flare tracked slowly down close outside our classroom. The ball appeared to stop descending at about 20 feet [6 m] above ground level. It then moved horizontally, again quite slowly, and although very bright gave the impression of changing color. Throughout this entire sequence there was a loud buzzing sound. There followed a tremendous crash and the ball had disappeared. If my memory serves me, one or more windows in a class further along the building were shattered. I would guess that dozens of pupils saw the event, along with several teachers. (A. T. Donaldson 1982; personal communication):

The ball was about 50 to 100 yards (45 to 90 m) from the observers. The diameter of the ball was between 1 and 3 ft (30 cm to 90 cm). During its lifetime of about 10 s, the ball changed color from red to green to blue to white. It was bright enough to be clearly visible in daylight. In its final moments, it may have accelerated. It was about 150 yards (135 m) away when it disappeared.

Professor Sir Brian Pippard, F. R. S., reported a particularly well-attested case of ball lightning observed at the Cavendish Laboratory, University of Cambridge. The reported event occurred soon after 4 p.m. on August 3, 1982, while an unusually violent thunderstorm was in progress. During the storm, the laboratory and its surroundings were struck several times. There was no structural damage. Immediately after one flash near the Bragg building, several observers reportedly saw at least one luminous ball.

A physicist on the ground floor of the Mott building, who was seated with his back to the window, saw his room momentarily illuminated as if by a very bright object moving rapidly toward the west between the Bragg and Mott buildings. Another observer on the first floor saw the space between the buildings "filled with a luminous haze" at least to first-floor level, and he interpreted this as sheet lightning. On looking to the west, he observed a blue-white light, about the apparent size of the moon, which he thought was motionless at an elevation of 10 to 15°, and which was visible for about 3 to 4 s. Another observer in the same room may have seen the same phenomenon momentarily earlier, as she had the impression that it was receding, perhaps expanding as it went. It had originally been about the size of a grapefruit. The estimates of actual diameter and of angle subtended imply an approximate distance from the observers of 12 m.

A further report came from three people who saw a ball moving over the ground to the west. They agreed that it subtended about the same angle as the moon, and was very bright, blue-white in color, and visible for about 4 to 5 s before it abruptly vanished.

An administrative assistant in a duplicating room on the ground floor was closing a small window when she was startled by a noise that suggested the window had been knocked in. A bright, spinning, sparkling object of pyrotechnic appearance entered past her head, rebounded from a copying machine, and departed as it had arrived. There was no damage to the window. Another assistant in the same room was also convinced that something had entered the room (Pippard 1982; *The Times* August 23, 1982).

10.4.2 Reports from Other Professional People

Schönland (1950) was in error when he remarked that no professional meteorologist had ever seen ball lightning. Chapter 7 included a report from Mr. J. Durward, a former deputy director of the British Meteorological Office. Mr. Durward had also reported ball lightning on a previous other occasion, while driving near Loch Tummel, near Pitlochry, Scotland, with his son. A thunderstorm was in progress. He stopped the car to enable his son to open an iron gate across the road, when he saw ball lightning about 12 in (30 cm) in diameter float out from some pine trees. The ball struck the gate at the far end from the boy and he received a severe electric shock that paralyzed his arm for several hours (Gold 1952).

An interesting ball lightning report of an observation made by a coastguard, Mr. Anthony Dalton on June 8, 1977, was sent by telex to the Meteorological Office in England. The report read:

> At 2:27 a.m. . . . very large (estimated bus-size) brilliant yellow-green transparent ball with fuzzy outline descended from base of towering cumulus over Garn Fawr Mountain [near Strumble Head, Fishguard, Pembrokeshire, Wales] and appeared to "float" down the hillside. Intense light emitted for about three seconds before flickering out. Severe static on radio. Object slowly rotated about a horizontal axis and seemed to "bounce" off projections on the ground.
>
> Weather at the time: wind northwest three or four, mainly clear sky with local cumulus and cirrus and cirro-stratus and cu[mulo]-nimbus in the North and Northeast (distant). Good visibility. Cattle and seabirds in immediate vicinity became disturbed. (A. Dalton, personal communication to Meteorological Office, 1977)

Further investigations revealed the following information. Before the ball descended from the cloud base, there was a glow visible in the cloud. The estimated size of the ball was 20 ft (6 m) based on a scale provided by farm buildings and a house in front of the ball. Its distance from the observer was about $1\frac{1}{2}$ miles (2.4 km). The size remained constant. The ball was globular. It was possible to see through the ball. The internal structure appeared like fibers. Initially, the light was similar to that of a neon advertising sign, but then it became much more intense until it cast enough light for objects to be perceived with clarity. It then flickered out. The ball was in sight for about 7 s. The vertical (downward) speed was estimated as about 5 miles per hour (2.2 m s^{-1}). Figure 10.1 is a sketch showing the path of the ball. If Mr. Dalton's observation was of ball lightning, this is one of the largest balls reported.

Mr. P. M. Bagnall, network director of the British Meteor Society, wrote the following detailed ball lightning report on the same day as his experience, which took place at Wallsend, Tyne and Wear, England:

> 1974 June 8: At 1:36 a.m. this morning I awoke to find the room bathed in a dim orange light. Whether it was the light that woke me or the violent thunderstorm I cannot decide, but at first I thought someone had come into my bedroom. It was only when I rolled onto my back that I noticed an orange sphere floating some 60 cm from the window and about 1.6 m off the floor near the foot of my bed. I wasn't sure if I was awake, asleep, dreaming or what—but I started counting. The object must have "floated" about 1 m before I realized that it was probably ball lightning. I had never seen "kugelblitz" before but from the few bits and pieces I had picked up it seemed to have all the trade marks.
>
> After I had gathered my senses I climbed out of bed and glanced at my watch—I could just make out the time. At about 2.5 m from the window the ball stopped and appeared to just hang in empty space. Some 20 s had elapsed. I timed it at 1:36 a.m. and 47 seconds.
>
> I tried to gather some information on its appearance but I wasn't really sure what! I noticed it was revolving about once every 10 s; it was limb-darkened with a central intensity of, say, 25–30 W. By this time I had stopped counting—nothing seemed real any more. There was no sound

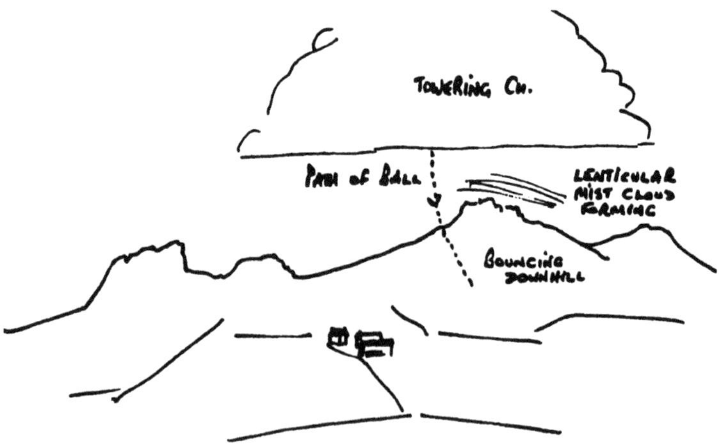

Figure 10.1. Sketch of path of ball lightning reported by Dalton (1977; personal communication).

> coming from the ball and no heat—or at least I thought so until I reached out towards it and found a slightly warm feeling at a distance of about 10 cm from its ill-defined limb. It was a funny sort of heat—not convection but a sort of still heat and dry, unlike, say, the heat from a gas burner which feels moist. I tried clapping my hands to see if sound had any effect on it, but to no avail. I really had no idea what to do. I estimated its diameter at about 15 cm but that was all I could think of doing. Finally the ball moved upwards towards the ceiling, at about the same velocity as it had crossed the room, and passed through it like a Hollywood Ghost. I managed to time the disappearance at about 1:37 a.m. and 18 seconds having looked at my watch about 3 s before the ball passed through the ceiling.
>
> I turned the light on but there were no signs of burning or any other damage. I hope to God I never see another—at least not under those circumstances! (P. M. Bagnall 1974, 1983; personal communications)

10.4.3 Instrumented Observations

Chapter 9 described at least three occasions on which phenomena resembling ball lightning have been detected by instruments in the course of scientific work.

10.5 Scientists and Skepticism

The probability or otherwise of ball lightning as an empirical phenomenon cannot be assessed except by an evaluation of ball lightning reports; certainly, a prejudice about the improbability of mechanisms that have been proposed to explain it should not be a determining factor. Scientists are not immune to prejudice. Indeed, Beveridge has argued that extensive prior knowledge may present its own barriers to research:

> When a mind loaded with a wealth of information contemplates a problem, the relevant information comes to the focal point of thinking, and if that information is sufficient for the particular problem, a solution may be obtained. But if that information is not sufficient—and this is usually so in research—then that mass of information makes it more difficult for the mind to conjure up original ideas. . . . Further, some of that information may be actually false, in which case it presents an even more serious barrier to new and productive ideas. (Beveridge 1950)

Philosopher of science John Ziman has written that "science is the search for a consensus of rational opinion among all competent researchers." He goes on to say:

> The scientific enterprise is corporate. . . . It is never one individual that goes through all the steps in the logico-deductive chain; it is a group of individuals, dividing their labour but continuously and jealously checking each other's contributions. The cliché of scientific prose betrays itself: "Hence we arrive at the conclusion that . . ." The audience to which the scientific publications are addressed is not passive; by its cheering or booing, its bouquets and brickbats, it actively controls the substance of the communications that it receives.
>
> In other words, scientific research is a social activity. (Ziman 1968)

One therefore might not be surprised to see a certain amount of pressure on scientists to conform and not to be seen to be too eager to embrace ideas that are not a part of the scientific consensus. Some scientists may fear the consequences for their professional standing if they study phenomena with a poor reputation in scientific circles, or may think that the likely yield of such studies will not justify the risk of damage to their reputation. Medawar (1969) wrote: "Good scientists study the most important problems they think they can solve. It is, after all, their professional business to solve problems, not merely to grapple with them. The spectacle of a scientist locked in combat with the forces of ignorance is not an inspiring one if, in the outcome, the scientist is routed."

The history of science shows that persistence in studying such phenomena, often in the face of skepticism, can provide information of great value to science. The French Académie des Sciences was formed in 1666 to promote original work in the mathematical sciences, such as geometry and astronomy; and in the physical sciences, such as chemistry, botany, and anatomy. Toward the end of the eighteenth century, the Académie exemplified a view that was becoming commonplace among scientists: that they were coming close to a full understanding of the mechanisms of the universe. This period has thus been described as the Age of Reason, or as the Enlightenment.

Against this context of neat rationality, there was an accumulation of phenomena that did not fit so tidily into the scheme of things. In the preceding three centuries a body of reports had developed, mostly from ordinary people, of stones that fell from the skies. These are now known as meteorites. Such was the skepticism toward these reports that most scientists of the day rejected the observations because they believed that such phenomena were impossible. Lane (1945) quotes Paneth [from his 1940 Halley Lecture of the Royal Astronomical Society, London, entitled: The Origin of Meteorites]: "By the eighteenth century learned men became too enlightened to believe the stories of fiery bodies coming down from a cloudless sky and burying themselves, after loud explosions, in the ground." Despite vast numbers of eyewitness reports and the recovery of fragments, the academy continued to deny that stones had fallen from the sky.

Scientist Ernest Chladni carried out a systematic study of meteorites in about 1794. Despite this, the French academy continued to go on record as denying that meteorites could have an origin outside the Earth's atmosphere (Olivier 1965). The physical existence and nature of the phenomenon were eventually established by a substantial fall of meteorites on April 26, 1803 at l'Aigle, France, which was investigated by Jean Baptiste Biot.

A meteorite also fell in Weston, Connecticut on December 26, 1807. Yale geologists recovered fragments and identified it as an extraterrestrial object (Olivier 1965). Even after this, President Thomas Jefferson stated "I could more easily believe that two Yankee professors would lie than that stones would fall from heaven" (Lane 1945, p. 97). Lane (1945) quotes Olivier (1925): "In the face of all this evidence we have an example of stupidity and bigotry, exhibited by the foremost body of scientists of the day—men who doubtless considered themselves, and were so considered by others, the most advanced and 'modern' of their time—which for all ages should stand as a warning to any man who feels that he can give a final verdict upon a matter outside his immediate experience."

There was similar controversy within the Académie in 1890 concerning the existence of ball lightning. Following a report of a large number of luminous spheres, whose behavior was very reminiscent of ball lightning, and which were seen during a tornado, there appears to have been a lively debate. During it, a member of the Académie suggested a cautious approach to ball lightning because reports probably resulted from optical illusions. In the ensuing discussion, the evidence of ignorant peasants was discounted. However, the former Emperor of Brazil, a foreign member of the Académie, remarked, perhaps somewhat indignantly, that he had seen ball lightning.

Thus scientists are not immune from prejudice and dogma. Sir Richard Woolley, the Astronomer Royal, is supposed to have said: "Space travel is utter bilge." This was in 1956, one year before the launch of the first Sputnik [C. H. Townes, in Moskovits (1995)]. Similarly, Cromer (1993, p. 159) writes of the discovery of parity violation in nuclear physics: "The story of the entire scientific community's giving up a basic principle on the basis of hard evidence would be more glorious if it were not for the fact that parity violation had been observed thirty years earlier . . . [but] these investigators discounted their own observations . . . because they violated an accepted principle."

Gregg (1941), on the other hand, speculates thus: "One wonders whether the rare ability to be completely attentive to, and to profit by, Nature's slight deviation from the conduct expected of her is not the secret of the best research minds and one that explains why some men turn to most remarkably good advantage seemingly trivial accidents. Behind such attention lies an unremitting sensitivity."

I am not advocating an uncritical approach to ball lightning reports. Cromer (1993) writes: "Science must be skeptical, not open-minded. . . . The burden of proof must be on the believer. If the evidence is convincing enough the skeptics will in time accept almost anything, even that the continents are drifting about the face of the earth, but until the evidence is there the only sane course is to reject all claims that are unverified and inconsistent with current knowledge."

Ball lightning is not (usually) considered to be a paranormal phenomenon, but in common with some paranormal phenomena, it is a topic on which there is no clear scientific consensus of the sort described by Ziman. Cromer (1993) writes of paranormal phenomena: "Probably nothing disturbs non-scientists about science more than its seemingly dogmatic rejection of claims of the paranormal. But science isn't rejecting the claims themselves so much as the evidence used to support them. Scientific evidence, by our definition, must be strong enough to win a consensus."

10.6 Conclusions

The argument that ball lightning does not exist is supported by the large number of already well-understood phenomena that resemble its reported characteristics, and by the unreliability of eyewitnesses in describing transient events, especially after long intervals of time. The persistent difficulty of scientists in evolving a model of ball lightning to explain its reported characteristics has also been mentioned in support of this argument. However, there are examples of phenomena, now entirely accepted by science, for which there was once no viable model, or for which the only evidence was descriptions from untrained observers.

The counterargument that ball lightning exists receives support from the significant number of reports from scientists and other trained observers, and from the basic consistency of many aspects of reports, notwithstanding that ball lightning is a concept unfamiliar to most of the people who report it. Especially consistent are the reports of ball lightning inside airplanes (see Chapter 7). Although there is a shortage of physical evidence for ball lightning in the form of traces that may not be explained by other means and an absence of compelling photographic evidence, this may be a consequence of the properties of the phenomenon.

A rational approach to the ball lightning problem must avoid prejudice generated by the nonconformity of this phenomenon to the "standard" scientific method, by the difficulty of proposing a viable model for ball lightning, or by the association of the phenomenon with "folk science." Reports of ball lightning must be judged individually, collectively, and comparatively on their own merits.

Chapter 11

Ball Lightning Theories and Experiments

11.1 Present Status of Ball Lightning Theory

Even after 150 years of scientific discussion, ball lightning research remains an immature field of study as defined by Ravetz (1971). Uman (1987) remarks, "Despite the relative wealth of similar ball lightning observations over a period of centuries, reports that leave little doubt as to its reality, there is still no consensus as to the physical mechanism or mechanisms responsible for ball lightning. Unfortunately, a significant fraction of the theoretical literature on ball lightning could best be described as rubbish, so the uninitiated reader should read the literature with more than the usual level of skepticism." Ten years later, after the first decade of international conferences devoted to ball lightning, Singer (1997) indicated that matters remained unresolved: "this natural phenomenon . . . has thus far eluded satisfactory (or perhaps only conclusive) scientific explanation. . . . The continuing lack of a convincing explanation has led to consideration of many advanced theories as they appeared in other areas of science, including attempts to apply such concepts as antimatter, monopoles, and micro black holes."

Until very recently, ball lightning theory developed in a very unsystematic fashion and many irreconcilable strands of argument diverged in all directions. It is only in the past decade that there have been conferences and symposia dedicated to the discussion of ball lightning. Before then, theoretical proposals were initiated by publication of quite brief and tentative papers stimulating an occasional letter by way of debate. Some of these papers could better be described as light-hearted musings rather than disciplined contributions to discussion. Even in the recent era of freer and more systematic debate, the heterogeneous nature and variable quality of ball lightning theory is still evident and consensus is a long way off.

There are several fundamental weaknesses in most theories. The most serious, as Jennison (1997) has pointed out, is that many theories do not correlate with observational evidence. Some theories predict overt behavior by ball lightning that is simply not reported by eyewitnesses. Equally, other theories fail to predict frequently reported observational characteristics of the natural phenomenon. Some are devoted to explaining the large destructive energy manifested in damage which, as we have seen, can often be explained as due to ordinary lightning. Some theories focus on specific ball lightning reports and neglect the larger body of observational evidence.

11.2 Aims of Ball Lightning Theory

The epistemological problem arising from the vague definition of ball lightning has already been discussed (Chapter 1). However, the mundane physical explanations offered in Chapter 3 cannot explain the formation in the atmosphere of persistent, luminous, spherically symmetric, independently mobile regions that rarely exhibit upward motion suggesting convection. It is a matter of opinion whether perceptual or psychological explanations are sufficient to explain ball lightning reports, but some scientists eminently qualified in these areas believe that they are not (Charman 1971a,b, 1979).

The goal of every ball lightning model is to explain as many of the diverse properties of this phenomenon as possible. The most successful models explain or predict the following:

1. *The relationship between ball lightning, thunderstorms, and linear lightning*. Most reports of ball lightning describe thunderstorm conditions. Many reports describe the formation of ball lightning following a nearby CG flash, and many others describe disappearance of the ball coincident with a nearby CG flash.
2. *The luminosity of ball lightning, which persists for several seconds, with a fairly constant luminous power output.* The luminous power of ball lightning in the optical region is described as typically constant at perhaps 40 W. The light from ball lightning may result from incandescence or luminescence. Incandescence is the emission of light by a substance as a result of raising it to a high temperature. Luminescence is the emission of light by a substance for any reason other than a rise in its temperature. An example of luminescence is the emission of light from electronically excited gases in neon lamps. Atoms of substances emit photons of electromagnetic energy when, having been in an excited state, they return to the ground state. There are various causes of excitation. In incandescence, excitation is by thermal energy. If the cause of excitation is a photon, the

process is called *photoluminescence*; if it is an electron, it is called *electroluminescence*. *Chemiluminescence* is luminescence resulting from a chemical reaction. Such radiation is most often generated by oxidation, where excitation of a molecule that is itself not undergoing oxidation is provided by the transfer to it of the oxidation energy. *Bioluminescence* is the luminescence produced by a living organism. Luminescence that persists significantly (more than ~10 ns) after the exciting cause is removed is called *phosphorescence*; if it does not, it is called *fluorescence*.

Powell and Finkelstein (1970) indicated that if the radiation from ball lightning is thermal in origin, an input of several kilowatts is required to explain the reported brightness. They therefore suggested that the ball glows because of electroluminescence, which they defined as an unusually efficient conversion of electrical energy to light. They describe experiments showing anomalously high-efficiency glows from oxygen and nitrogen, with unusually persistent afterglows (e.g., 0.5 s). They calculated that a power input of several hundred watts would produce effects that might conform to descriptions of ball lightning (see Chapter 13).

3. *The shape and size of ball lightning: roughly spherical with a modal diameter of 20–50 cm, again, both remaining fairly constant.* In most internal energy models, there is a problem of explaining containment, i.e., the balance between outward (pressure) and inward radial forces. In some models, such as that of Kapitsa (1955), there is a negative feedback mechanism that maintains the constant size of the ball (see Chapter 13).
4. *The independent motion of ball lightning through the air, mostly in a horizontal direction.* For internal energy models there is difficulty in explaining the absence of convection if the model ball is less dense than atmospheric air. Some models involving charge suggest that thunderstorm electric fields maintain the ball in equilibrium against convection, although the balance of forces would easily be disturbed by a loss of charge or by a change in the external field. In external energy models where heating occurs, the external electric or electromagnetic field must provide confinement against thermal motion.
5. *Observers report the formation of ball lightning inside closed rooms and inside aircraft.* This presents a difficulty for models that require ball lightning to be formed at the point of impact of a lightning flash or from the lightning channel itself, or for models that require the ball to have a constant, external source of energy.
6. *The silent and explosive modes of decay of ball lightning.* While ball lightning exhibits remarkable stability over a period of several seconds, some balls decay explosively, apparently spontaneously or on making contact with an object such as a grounded conductor. However, caution is required here. In circumstances where ball lightning appears to behave as

a precursor of a CG flash, the flash may be interpreted as explosive decay of the ball lightning.

Several authors (e.g., Hill 1960, Hubert 1996, Sinkevich 1997f) have pointed out that there may be several independent, nonprosaic mechanisms responsible for the generation of ball lightning reports. The persistent intractability of this problem may support the view that more than one mechanism is required to explain all or a significant proportion of the reports. Only recently has there been an attempt to correlate reported characteristics with particular models (Hubert 1996).

11.3 Classification of Models

Finkelstein and Rubenstein (1964) provide a scheme of classification of ball lightning models. They identify *chemical* (gas burner) and *electrical* models (involving charges, electric currents, and/or significant electric or magnetic fields). Like Uman (1969, 1984, 1986), they distinguish between *self-powered* models that run on internal energy resources until they are consumed, and *externally powered* models that derive their energy from external resources. They further suggest that the model may be categorized according to whether the matter within the ball is *stationary, oscillatory*, or *turbulent*, or, in the case of electrical models, whether the electromagnetic quantities are *dc, ac,* or *noisy*. In the chapters that follow, we group models according to whether they are self-powered (internal energy source) or externally powered. Within these broad categories, we have grouped models by similarity.

11.4 Plasma Models

In some work on ionized gases, it is convenient to define a fourth state of matter known as *plasma*. In general terms, we usually think of plasma as hot, ionized gas, and in elementary accounts it is sometimes said that plasma is the highest temperature state of matter (Morse 1991). Lightning is the only natural example of plasma found within the Earth's troposphere. However, in the universe as a whole, plasma is considered the most abundant state of matter, probably accounting for more than 99% of its total mass. Many ball lightning models are plasma models.

One general, qualitative definition of plasma is that it describes "any state of matter which contains enough free, charged particles for its dynamical behavior to be dominated by electromagnetic forces" (Boyd and Sanderson 1969). This definition encompasses the solid state, since it applies to electrons in metals and semiconductors, and therefore contradicts the generalized statement in the previous paragraph. However, the name "plasma" is more generally applied to ionized gases where very low degrees of ionization are sufficient for the gas to exhibit electromagnetic properties and significant electrical conductivity. Other definitions are "a large collection of approximately equal numbers of positively and negatively charged particles" (Shohet 1971), or "a quasineutral gas of charged and neutral particles which exhibits collective behavior" (Chen 1974). This collective behavior results from the nearest-neighbor interactions being dominated by long-range coulomb forces exerted by many distant particles. This contrasts with a neutral gas, where the forces between particles are short range.

Both natural and artificial plasmas are of interest to physicists. Dijkhuis (1979, 1980), Roth (1990, 1993, 1995, 1997) and others have suggested that plasma physicists may gain much from a study of ball lightning. One major area of potential technological application is in the possible creation of a viable and significant energy source using thermonuclear fusion. It is believed that fusion reactions can only be sustained in the plasma state. For two positively charged nuclei to fuse together to make a larger nucleus with a significant yield of energy, it is necessary for the nuclei to come close enough together for the short-range strong nuclear force to come into effect. To do so, they must overcome their electrostatic repulsion. The temperature must be sufficiently high that positive ions have enough kinetic energy to overcome their electrostatic repulsion. Therefore both temperature and density must be very high for thermonuclear reactions to occur. In main-sequence stars, these conditions are achieved because of their very large mass and the consequent magnitude of the inward, radial gravitational force which, under equilibrium conditions, is balanced with the outward radial component of the pressure force. Gravitational forces cannot contain terrestrial plasmas, which are of a much smaller scale. In research directed toward achieving controlled thermonuclear fusion, plasma confinement schemes involving magnetic fields have been most popular. Lasers have been used to attempt inertial confinement of plasmas. Confinement can also be provided by electromagnetic waves.

When plasma is in a magnetic field, the charged particles spiral around the magnetic field lines—negative charges in one direction, positive charges in the other. This process is called *Larmor gyration*. The radius of the spiral, the *Larmor radius* or *gyro radius*, given by $\rho_L = mv_\perp/qB$, depends on the velocity $v_\perp$ and mass m of the particles as well as the magnetic flux density B of the magnetic field. Ions are significantly heavier than electrons, and particles of all kinds have a range of speeds. Thus, collisions will occur between particles that cause them to move from orbiting around one field line to orbiting around another. This process, called *plasma diffusion*, allows some particles to move across the field lines.

A more intense magnetic field causes a smaller radius of spiral. If the field across an orbit is nonuniform, the variable spiral radius will cause the charged particle to move across the field lines—negative charges in one direction, positive charges in the opposite direction. This separation of charge sets up an electric field. The combined magnetic (B) and electric (E) fields cause a net force on the plasma in the direction $E \times B$, away from the stronger magnetic field region and toward a region of lower flux density. Consequently, plasma placed in a nonuniform magnetic field will tend to move *away* from regions of higher flux density. Magnetic confinement utilizes this fact that plasmas are *diamagnetic*, that is to say, the magnetization due to the plasma is in the direction opposite to that of the applied field.

A simple confinement scheme called a magnetic "bottle," which essentially consists of a vessel containing plasma with a coil at each end, is unsuccessful because plasma "leaks" from the ends. This is because the component of force on particles in the direction of the magnetic field is zero. Most confinement schemes are therefore based on closed vessels with no "ends," the most common of which has the form of a torus (or Tokamak).

Even in these circumstances and with ingenious magnetic field configurations (Gill 1981, Dendy 1993), there are technological difficulties because plasmas are subject to a wide range of instabilities. Instabilities, among other things, can cause the plasma to make contact with the walls of its vessel, which has a deleterious effect on the desired processes. Instabilities feature in some of the formation mechanisms proposed in ball lightning models, as described later.

All current-carrying plasma columns, whether of cylindrical or toroidal shape, are subject to instabilities. The designs of plasma machines such as the Tokamak yield important information about plasma confinement. Consider a simple toroidal configuration, with exclusively poloidal current and exclusively toroidal magnetic field. In a simple torus, the magnitude of the B field is inversely proportional to r, the major radius. Owing to the diamagnetic behavior of plasma, the plasma will undergo an $E \times B$ drift away from the major axis as described earlier. To avoid this drift, toroidal plasma machines introduce a twist, or rotational transform (sometimes called a *rotational correction*) in the magnetic field. This effectively provides a short circuit to prevent formation of the electric field responsible for the drift (Chen 1974). However, there are also many other instabilities, which are very difficult to control.

It can be seen that the magnetic confinement of plasmas, even under carefully controlled conditions using external magnetic fields, is a nontrivial problem. Under atmospheric conditions, with no such control, a region of plasma would be expected to have a very short lifetime, as the short duration of a lightning flash confirms. Many different processes contribute to the de-ionization of the plasma. These include recombination of ions and electrons, various forms of diffusion, and the formation of negative ions (Nasser 1971). Oxygen readily forms negative ions and so the lifetime of air plasma is significantly shorter than one composed of some other gases. The characteristic times associated with this process are probably somewhere between $\sim 10^{-8}$ s (Smirnov 1975a,b) and up to about 1 ms (Hill 1960).

Various mechanisms to slow the rate of recombination have been proposed. For example, Maiorov, Tkachev, and Yakovlenko (1991a,b 1992a,b, 1993a,b, 1994a,b) attempted, using many-body dynamics and thermodynamics, to explain the absence (or anomalous delay) of recombination. A plasmoid consisting of a mixture of metastable, supercooled plasma with a perfect (ideal) gas, might, they argued, conform to the properties of ball lightning. Protasevich (1993, 1995a,b) studied and described the properties of gas-discharge plasma and the possible influence of the concentration of water vapor, with the vapor acting as a medium for insulation. He showed that the process decreasing the rate of plasma decay occurs most efficiently when air humidity is 95–98%. Under these conditions, he argued, cold nonequilibrium plasma would be formed, with a lifetime from tens of milliseconds to seconds.

11.5 How Is Ball Lightning Formed?

11.5.1 Lightning Impact Models

The models discussed in this section offer relatively mundane explanations for ball lightning, and it is only the fact that the mechanisms are not well established that determine that they should be discussed here rather than in Chapter 3. These models are applicable only to circumstances where ball lightning is formed at the point of impact of ordinary lightning.

In Section 4.2.2, mechanisms were suggested by which lightning currents could heat materials on the ground and initiate chemical reactions or combustion. This may be supported by triggered lightning experiments with rockets attached to grounded wires that produce roughly spherical, luminous phenomena with an estimated diameter of 25 cm on the ground near wooden posts surrounding the rocket launchers (see Chapter 9). The investigators attributed the spheres to currents circulating in the soil around the grounded cable during each flash and causing outgassing and a local electric discharge at the points where the posts penetrated the upper layer of soil (Fieux, Gary, and Hubert 1975, Hubert 1976). The absence of convection was not explained (Charman 1979).

A similar heating effect might occur locally when lightning strikes a non-metallic object of moderate resistance, such as a tree, and produces a glowing mass (Singer 1971). Combustion or chemical reaction might produce persistent luminosity. Some ball lightning observations support this (e.g., Séguier 1852, Stenhoff 1992a).

The suggestion by Harris (1843) that ball lightning might consist of globules of molten metal at red heat or metal vapor generated by electrical discharges was supported by experimental studies (van Marum 1800, Jones 1910). Jones (1910) reported the accidental production of a glowing sphere following a short circuit in a copper wire. The sphere rolled across a table, leaving a line of scorch marks that ended at a crack in the table 1 or 2 mm wide. The ball disappeared. A small copper sphere about 1 mm in diameter was found under the crack.

Harris's suggestion is also supported by some ball lightning reports (e.g., Potts 1910) in which appearances of ball lightning were evidently associated with the melting of lengths of telephone wire or antennae by ordinary lightning. De Jans (1910, 1911a–d, 1912a–e) maintained that such phenomena were not the same as natural ball lightning. Charman (1979) has suggested that the beaded decay of lightning triggered by firing into thunderclouds rockets with grounded wires attached (see Chapter 9) might be explained in this way.

This idea is also supported by experimental work involving high-current short circuits that produced luminous phenomena reminiscent of ball lightning. (Jones 1910, Brand 1923). Brand described the generation of luminous phenomena reminiscent of ball lightning by a short circuit in an electrical power system. In one event, a ball 5 cm in diameter was generated which moved with the wind over a distance of about 50 m.

Silberg (1962, 1965) reported the accidental short-circuiting of a circuit breaker on a submarine, causing erosion of silver electrodes. This produced green fireballs 10–15 cm in diameter with a duration of about 1 s. The green color was attributed to the copper onto which the silver electrodes were attached. Silberg (1965) first used these experiments to support a ring model (see Chapter 12) and later (1977) carried out an analysis based on the nonlinearity of the voltage-current characteristics of a high-current arc to support an electromagnetic standing-wave model (see Chapter 13). Model (c) of Lowke, Uman, and Liebermann (1969) offers a more straightforward interpretation of these observations; this is described in Chapter 12.

Golka (1991) claimed that it was possible routinely to produce fireballs similar to those produced in U.S. submarines and in aluminum-skinned aircraft during flight. At the Third International Ball Lightning Symposium at the University of California, Los Angeles, he played a videotaped demonstration of 60-Hz short-circuit experiments. The videotape showed luminous spheres of molten metal floating on water some seconds after the current between submerged electrodes was interrupted. Other experiments based on electrical short circuits were carried out by Dijkhuis and his colleagues, and are described briefly in Chapter 12.

Singer (1963, 1971) reasoned that the luminosities in the events and experiments described here, if caused by ionization, could not be explained simply in terms of cooling by thermal conductivity. This would cause the ball to cease to be visible over a much shorter period than that reported. He suggested other processes, such as relatively low-energy, long-duration electronic deexcitation mechanisms in gases or active nitrogen (Bayes and Kistiakowsky 1960), with energy transferred from the gas to the metal atoms. He further remarked (1971) that the persistence of geometrical form in these experiments offered evidence of the formation of a stable geometrical structure such as a vortex ring (see Chapter 12).

Wooding (1972) suggested that a process analogous to the ablation of a solid surface by a high powered laser pulse might produce ball lightning. He envisaged that coherent radiation might be produced from a thundercloud with a low efficiency of conversion of electrical energy to radiation, and would escape at the cloud or ground, or from any part of the channel where there was a change of direction. He suggested that this mechanism could explain ball lightning formed near a solid surface near a lightning stroke, or the passage of energy through windows or aircraft portholes. Coherent radiation is featured in other models.

Yakovlenko (1992a,b,c) proposed an experiment to form a nonideal plasma bunch by ionizing a vapor with a laser having a photon energy close to the ionization energy, claiming on theoretical grounds that such a plasma should recombine anomalously slowly. Ignatovich (1992) proposed a model of ball lightning produced by a "frozen" shock wave from a point explosion that is blocked by laser radiation along its front. The postulated structure thus created was in the form of a spherical capacitor. Handel formulated a maser theory based on the creation of a population inversion in some rotational energy levels of the water vapor present in a volume of air (see Chapter 13).

Andrianov and Sinitsyn (1977a,b) carried out experimental investigations to try to simulate the behavior of erosion products that they supposed would be released when lightning strikes and melts sand to form fulgurites. A fulgurite is a long, hollow tube with corrugated glassy walls. Its diameter is up to 5 cm and its length may be 20 m (Uman 1986). They suggested that such erosion products might form ball lightning. To model the fulgurite, they used a chamber at reduced pressure, sealed with a Plexiglas diaphragm. Various materials were used for the walls of the chamber. An electrical discharge produced a rapid increase of pressure in the chamber to several hundred atmospheres, so that the diaphragm ruptured, allowing the gas to escape through the aperture into the atmosphere. Several milliseconds afterward, various luminous forms considered to be vortex rings were produced. These were about 4 cm in diameter, with speeds of 10–20 m s^{-1} and with lifetimes of up to 0.1 s (and occasionally up to 0.3 or 0.4 s). While this duration falls far short of that reported for ball lightning, it is considerably longer than the characteristic recombination time (~1 ms) for this volume of fully ionized gas within the atmosphere. Similar experiments were reported by Bychkov et al. (1997).

Avramenko et al. (1990a,b, 1992a,b) studied plasmoids obtained using an erosion discharge in a cylindrical channel with dielectric walls. They found, for certain geometries of the discharge unit and discharge conditions, that it was possible to obtain a plasma with unusual properties reminiscent of ball lightning. They argued that the plasmoids obtained are capable of autonomous existence in a still atmosphere and in a gas flow, and that they are accelerated in the latter to a velocity of approximately 200 m/s.

The models mentioned in this section fail to explain ball lightning formed in an enclosed space or remote from a lightning channel. They mostly predict that ball lightning is hotter than its surroundings, but fail to explain the absence of convection.

11.5.2 Formation from a Lightning Channel

The suggestion has repeatedly been made that ball lightning could be formed from linear lightning, perhaps by some detachment process, and like the lightning channel, is composed of plasma (Singer 1971, Charman 1979). However, the very short duration of ordinary lightning compared with the reported duration of ball lightning is a consequence of the very rapid deionization processes that occur in atmospheric conditions. Like other plasma models, these models must necessarily therefore propose a mechanism by which the lifetime of the plasma may be extended. Furthermore, these models encounter serious difficulties in accounting for ball lightning formed within rooms or aircraft, unless in the former case the ball lightning is formed from a plasma arc within the room.

Planté (1875a,b) suggested that ball lightning is formed from an incomplete discharge of ordinary lightning that terminates above the ground. Carlheim-Gylensköld (1905a,b) thought that ball lightning was formed from an ordinary lightning channel as a rotating spherical vortex composed of ionized air.

Kozlov (1975, 1976, 1978c) proposed that low-frequency relaxation pulses might be generated by the discharge track of a lightning channel emerging from a cloud, but which terminates in the air. The pulses would be reflected to and fro along the channel at about 3–30 Hz (which is comparable with the frequency of multiple strokes along a single lightning channel) gradually damping, but then being followed by further pulses. Although the entire channel would emit light, the greatest intensity would be found at the tip of the track, where the electric field would be most intense. In some circumstances, only the head of the channel would be visible and this would be seen as ball lightning. Kozlov estimated a ball lightning diameter of 6 cm to 3 m, a duration between 0.5 s and 9 min, a temperature of 300 to 500 K, and a power output of 50 kW.

Brovetto, Maxia, and Busetti (1976) proposed that a negative step leader of ordinary lightning approaching the ground would, by electrostatic induction, cause a grounded prominence to emit a positive charge. This model would presumably only apply where such a prominence was available. Rather than the prominence emitting an upward-moving positive streamer, it was suggested that the electrification process might be inhibited so that instead a sputtering process would occur, causing positive charges to be swept away together with the air and creating a vacuum bubble. This bubble could detach from the prominence, bearing on its surface the positive charge. Luminosity would be provided by corona discharge. An equilibrium radius would be achieved when atmospheric and electrostatic pressures were balanced. For a ball 20 cm in diameter, the energy associated with these pressures would be about 1.7 kJ. Although the ball would be denser than atmospheric air, it could be supported by the upward force due to the upwardly directed electric field of the storm and would move up or down as the magnitude of this field varied.

Recently (1996) Hubert has suggested that some ball lightning may be a form of lightning leader that is considerably slowed down near the ground, and with its length reduced to approximately the size of its diameter. Ohnishi and Suganuma (1997) carried out a numerical simulation of ball lightning thus formed. The photograph published by Norinder (1939) seems to show such a phenomenon. White (1994) comments that some reported ball lightnings behave more like rocket lightning leaders than discrete masses of air plasma, with vertical velocities too high for masses near air density. They often travel against the wind and even accelerate. This behavior is more reminiscent of lightning leaders, except these phenomena are slower than normal lightning leaders.

Singer (1971) remarked that no explanation has been offered for the conditions responsible for rounding of the advancing end and for the extended lifetime of ball lightning if it is composed of the same substance as ordinary lightning.

Current-carrying plasma columns, such as the lightning return stroke, are subject to a compressive magnetic force that acts radially inward on the column. This is known as the "pinch effect." The pinch effect has been suggested as a mechanism for the formation of bead or ball lightning from the lightning channel (Uman 1962, 1967, 1968b). The effect has been extensively studied in plasma machines that apply the effect to confine the plasma. It can also be a source of instability in the plasma. The effect has been considered relevant in estimates of lightning charge and channel diameters (Hill 1963, Jones 1968).

A simple, hydromagnetic model of the pinch effect, known as the Bennett pinch, yields the unstable equilibrium condition

$$\mu_0 I^2 = 8\pi NkT, \tag{12.1}$$

where μ_0 is the permeability of free space, I is the axial plasma current, N is the total number of particles per unit length of the column, k is Boltzmann's constant, and T is the absolute temperature (Rose and Clark 1961). This model neglects inertial and inductive effects, and so is applicable only to systems in which the pinch current remains fairly constant during the time taken for the plasma particles to achieve thermodynamic equilibrium. This is not so in the lightning channel.

Uman (1969) has used the Bennett relation to show that the pinch effect is unlikely to be important in the reduction of the channel radius in the transition from stepped leader to return stroke. It may be important while the return stroke core heats from, say, 3000 to 30,000 K, causing an increase in pressure to 10 atm. Equation 12.1 shows that a balancing magnetic pressure could be achieved if the current exceeds 80 kA for a 1-cm channel, or 8 kA for a 0.1-cm channel.

A more sophisticated analysis of the dynamic pinch, called the "snowplow" model, assumes infinite plasma conductivity and yields the following equation of motion:

$$\frac{\mathrm{d}}{\mathrm{d}t}\left[\pi\rho_m(r_0^2 - r^2)\frac{\mathrm{d}r}{\mathrm{d}t}\right] = -\frac{\mu_0 I^2}{4\pi r}, \tag{12.2}$$

which can be solved numerically if an expression for the pinch current can be obtained (Uman 1964, Rose and Clark 1961).

Instabilities in the plasma pinch are a considerable source of nuisance in experiments whose ultimate goal is controlled thermonuclear fusion. However, they are central to the mechanisms proposed for the formation of bead and ball lightning from the lightning channel. Several modes of instability are possible with the linear pinch. Many describe fluting of the column in a longitudinal direction. Those of more interest in models of ball and bead lightning are the sausage and the kink instability (Boyd and Sanderson 1969).

The *sausage* instability is an azimuthally symmetric instability where the column tends to constrict at one or more axial positions. The axial current is the same at all points, so induction will be greatest at the narrowest part and produce the strongest magnetic field here, so the diamagnetic plasma continues to narrow and the instability grows. The rapid motion of the B field at the constriction induces an electromotive force that accelerates ions across the gap between the adjacent fat portions of the plasma.

Uman (1962) and Johnson (1965) published ball and bead lightning models based on the sausage instability. Uman concluded that the regions of a lightning pinch that would be most luminous would be the constrictions because of the greater current density there. This would produce the visual effect known as bead lightning. The luminosity would persist after the current had ended because of the enhanced electron and ion densities in this region, and afterglow (see later discussion) might be responsible for further extension of the lifetime of the beads. More recently Uman (1983, 1998, personal communications) has doubted that the pinch effect is significant in the formation of bead lightning, although he thinks that magnetic forces will have some effect on the overall mechanical and thermodynamic behavior of the return stroke channel.

Carpenter (1962, 1963) proposed that ball lightning was formed when a segment of the lightning channel, presumably formed because of a sausage instability, formed a plasmoid that had the structure of the Hill spherical vortex. The cylindrical segment was considered to have a strong trapped magnetic field without an electric field. The magnetic field caused rotation of ions and electrons about the axis, while ions and electrons would migrate in opposite directions along the rod, thus forming a dipole. The electric field of the dipole and the magnetic field together would cause the ions and electrons to form an umbrella shape at each end. These two "umbrellas" would then extend toward each other until a closed conduction path was formed on the spherical surface and passed through the center of the rod. Currents were on the order of 10 MA. Electron velocities were approximately ⅓ c

and would give rise to X-radiation and scattered high-energy electrons. This would imply that ball lightning would be the most intense radiation source occurring naturally in the atmosphere. Please refer to Section 12.7 for comments concerning ball lightning and radioactivity.

Johnson (1965), in contrast, suggested that when the sausage instability caused the isolation of a segment of the lightning channel, the collapsing azimuthal B-field would induce an electric field that would cause poloidal circulation of the charged particles in a toroidal vortex plasmoid.

The *kink* instability appears as a kinked column with a radial displacement of the pinch column by an amount varying sinusoidally along its length. Because of the diamagnetic property of plasma, it is driven away from the stronger B-field on the inside of a kink and the instability tends to grow. The tortuosity of natural lightning channels has sometimes been attributed to the kink instability, but there are doubtless many other mechanisms that contribute to tortuosity and which may be more important.

Ritchie (1961), Bruce (1963b), and Wooding (1963) developed ball lightning models based on the kink instability. These theories and several that followed predicted the formation of various kinds of vortex plasmoids. Wooding suggested that the vorticity of ball lightning could be produced impulsively by asymmetric expansion of a lightning stroke, especially if this took place near a solid object containing an aperture, for example, a building containing a chimney or doorway. Ritchie suggested that the kink instability could cause a loop segment to become detached from the main channel, and that a plasma toroid with a toroidal current might thus be formed.

Bruce, however, suggested that with a kink in the lightning channel, the stronger field on the inside of the bend and the weaker field on the outside would cause the diamagnetic plasma to flow through the weakened region. As the plasma escaped through the B field, he suggested, it would roll up to form a vortex containing activated states of nitrogen and oxygen. These would undergo slow decay, providing steady luminosity unless an explosion occurred because of an increased rate of decay. This might occur because of a disturbance such as the introduction of fresh oxygen. He estimated the volume of escaping gas from the jet velocity as several cubic decimeters, although while obtaining approximate agreement, Barry (1966) pointed out that this estimate was based on an excessive value of the high-current period of the return stroke.

Bruce estimated the total energy content of the ball as 10 kJ. This was stored by activated states of nitrogen atoms and molecules between 2 and 12 eV. The Rayleigh afterglow of "active" nitrogen, which is light produced by recombination of atomic nitrogen (Bayes and Kistiakowsky 1960), was used to account for the extended luminosity of the spheres. The presence of "active" nitrogen could also explain the explosive mode of decay of some ball lightning. However, Powell and

Finkelstein (1970) concluded that the intensity of afterglow radiation was too low to be the source of light in ball lightning.

In possible support of these theories, Alexeff and Rader (1995) describe studies of several hundred photographs of ultrahigh voltage discharges that show closed current loops. They suggest that these closed current loops may be precursors of ball lightning. There is a distinct threshold in voltage and/or current below which the closed loops do not occur. This threshold current fits other experimental data, but is well above the currents usually observed in natural lightning. (Such phenomena might, therefore, be associated with positive giants or superbolts. See Chapter 2.)

11.5.3 Focusing of Cosmic Rays

Arabadji (1956), Chalmers (1957, 1967), and Leonov (1965c) considered the possibility that thunderstorm electric fields might focus heavier reactive particles such as cosmic rays into a small region that would become ball lightning and would be sustained by nuclear reactions. Leonov proposed that this model would predict an increased incidence of ball lightning during solar flares when the cosmic ray flux has also increased; furthermore, ball lightning would occur mostly at high altitudes. Markson (1978) and Markson and Muir (1980) proposed that solar variability modulates the Earth's electric field. They found that galactic cosmic radiation is inversely correlated with solar wind velocity. Cosmic radiation is the primary source of atmospheric ionization and hence could influence the electrification of clouds. Lethbridge (1981) confirmed a maximum in thunderstorm frequency 3 days after high cosmic ray counts.

Singer (1971) remarks that such focusing of otherwise diffuse cosmic rays would require strong fields extending over large geometrical regions.

11.6 Energy Content of Ball Lightning

Kapitsa (1955) reasoned that an upper limit may be set on the radiation time of ball lightning with an internal energy source by considering the radiation time of a fireball formed by a nuclear explosion, which would be expected to contain the largest possible store of energy. A fireball of a diameter of 150 m would radiate for ~ 10 s. Energy content is proportional to volume and hence to the cube of the diameter, while light emitted is proportional to the surface area and hence to the square of the diameter. Thus radiation time should be directly proportional to diameter, so a sphere with typical ball lightning dimensions (~10 cm) would glow for a maximum of 10 ms. He therefore concluded that ball lightning was fed by an external source of energy (see Chapter 13). If it were not, the decay processes must be retarded in some way (see Chapter 12).

Barry (1980a,b) completed the most extensive study of the energy density of ball lightning, derived from reports of observations and from experiments and models. The six observations he discussed were:

Event reported by	Features	Comment (MS)
1. Morris (1936), Goodlet (1937)	"Tub of water" event.	See text.
2. Covington (1970) and Zimmerman (1970)	Ball lightning descended into a wharf piling and shattered it into splinters.	The ball lightning may have been a precursor of a CG flash and the damage was caused by the latter.
3. Stenhoff (1976) and Wooding (1976)	A woman was struck by ball lightning and her clothing damaged.	Estimates of energy for the ball lightning were between 440 J (Barry 1980a,b) and 3 kJ (Wooding 1976). Energy density estimates were between 0.8 and 105 J cm^{-3} (Barry 1980a,b).
4. Dmitriev (1967a, 1969a,b)	Ball lightning was seen following CG lightning. Samples of gases were collected and analyzed.	The ball energy was estimated as 530 J and its energy density as 0.4 J cm^{-3} (Barry 1980a,b).
5. Wittmann (1971)	A ball descended vertically onto a tree, where it disintegrated into several smaller spheres. Some of these landed on asphalt and melted it.	The smaller spheres that melted the asphalt may have been burning material from the tree.
6. Anderson and Freier (1972)	Ball lightning was not observed, but rather a persistent glow following a lightning flash. There was an electrical power failure. A crooked trail of scorched grass was found connecting a tree to a metal rainspout.	CG lightning may have struck the tree and flashover through the soil to the grounding system of the house may have produced the trail.

The Dorstone case (Morris 1936, Goodlet 1937) is the most often cited in arguing that the energy content of ball lightning exceeds 1 MJ. Unfortunately details of the case are scarce and there is no record of a field investigation of the report. Originally published as a letter in a daily newspaper (see Chapter 1), this report came to the attention of the scientific community as a result of comments made by the eminent physicist and lightning experimentalist Sir Charles Vernon Boys. This was during a discussion at the Institution of Electrical Engineers in London on January 7, 1937 following the presentation of a review paper by Prof. B. J. Goodlet entitled "Lightning."

Boys said, "The importance of this observation is that for the first time, so far as I know, a measure of the energy in the red hot ball has been roughly found. I have ascertained that the ball appeared to be of the size of a large orange, and after 20 minutes the water was too hot for Mr. Morris to put his hands into it. The amount of water was about 4 gallons." (At the time of the event, Boys was an octogenarian, so it seems most likely that the extra data had been acquired through correspondence with Mr. Morris rather than by a field visit or interview.)

In response to the discussion, Professor Goodlet estimated the energy content of the ball using these data. He estimated that Mr. Morris's description of the heating of the water suggested a temperature rise to 60°C; thus if the initial temperature was 20°C, $\Delta\theta = 40°C$. Four Imperial gallons of water have a mass m_1 of 40 lb = 18 kg. The specific heat capacity c of water is 4200 J kg^{-1} K^{-1}, thus $\Delta Q = m_1 c \Delta\theta = 3.1$ MJ. However, he then carried out a calculation based on the assumption that all the water was raised to 100°C and that a mass m_2 of 4 lb of water (1.8 kg) was vaporized. However, this was not supported by Mr. Morris's description of the event or by Sir Charles Boys's subsequent inquiries. The latent heat of vaporization of water L is 2.26 MJ kg^{-1}, hence $\Delta Q = m_1 c \Delta\theta + m_2 L = 10$ MJ. These large estimates of energy were significant in shaping the development of ball lightning theory in subsequent years. Assuming the ball had a diameter of 10 cm, and that the energy was 3.1 MJ, the energy density of the ball can be estimated as 6 kJ cm^{-3}. Ritchie (1963) suggested an upper limit for energy density of 0.17 kJ cm^{-3} for a fully ionized plasma, and Balyberdin (1966) and Dmitriev (1967a) of 2 kJ cm^{-3} for the explosive energy of TNT.

In Chapter 2 we saw that the energy available from an average lightning flash is on the order of 200 MJ. In principle, then, a lightning flash could easily provide sufficient energy to heat or vaporize this quantity of water. The present author measured the resistivity of water collected in a rain barrel as 12.1 kΩ m. For a barrel with a diameter of 0.45 m and a water depth of 0.75 m, the resistance of the water contained in the barrel would be about 60 kΩ. If the rise in temperature of the water were due to ohmic heating, a typical lightning current of 55 kA could provide 10 MJ of energy in about 60 ns.

Uman (personal communication, 1997) has suggested that it is possible that lightning caused an arc from nearby electrical power lines, continuing current then being supplied by the electrical mains (see also Uman 1986, p. 46). If the electrical mains supplied 30 A and this alone was responsible for heating the water, 10 MJ of energy could be supplied in less than 0.2 s. These calculations neglect consideration of rather complex heat transfer processes within the water (Holman 1981), or electrolytic processes, but they demonstrate that either of the above scenarios could be energetically viable.

Many models discussed in the next two chapters have been designed to account for the supposedly very high energy content of ball lightning. It appears that upper estimates of energy can generally be explained without attributing that energy to the internal energy of ball lightning. The survival of those who have encountered ball lightning at close proximity offers support to this view. Ball lightning is therefore, in my view, probably a relatively low-energy phenomenon, perhaps with energy of up to 3 kJ. Nonetheless, as explained later, its presence is not entirely benign because it may warn of a severe hazard from ordinary lightning flashes.

Support for this view is provided by Stepanov (1997), who divided reports that allowed energy estimates into two groups: (1) those where an external discharge through the ball was improbable, such as those seen indoors; and (2) those that occurred outdoors. He found average values for group (1) of about 100 J and for group (2) of about 100 kJ. He therefore concluded that the large energies in the second group resulted from the flow of an external current through the ball and the release of external field energy from a large volume. He concluded that the internal energy of ball lightning is only about 100 J. Stakhanova (1997) similarly concluded that the large energy associated with some cases is from electric currents that have nothing to do with the internal energy of ball lightning.

Amirov et al. (1997) argued that certain surveys showed that the most energetic ball lightning events occurred in weather without thunderstorms. Apart from noting the unreliability of surveys with an unknown signal-to-noise ratio, it is appropriate to comment that the type of damage described in the relevant surveys is generally consistent with ordinary lightning damage. Uman (1987) refers to "bolts from the blue" in which lightning occurs from a clear blue sky or from distant thunderstorms beyond the view of the observers. He also refers to observations of an "anomalous" multiple-stroke flash with 32 current pulses in excess of 1 kA. For ten of these strokes, a field mill network determined their charge source to have been located in clear air at the top of the photographed channels. We also note that positive ground flashes, which are associated with larger peak currents and charge transfer than negative ground flashes, are relatively common in winter thunderstorms, which produce fewer flashes in total. Some of the above events could be explained in this way.

If we take the diameter of ball lightning to be typically between 0.2 and 0.5 m, which is consistent with modal values from the surveys discussed in Chapter 1, then its volume is between 4.2×10^{-3} and 65×10^{-3} m^3. If its energy is 3 kJ, its energy density is about 0.046 to 0.72 J cm^{-3}. Uncertainties in estimates are very high because the energy density is inversely proportional to the cube of the estimated diameter. If the ball contains 3 kJ of energy and persists for 2 to 5 s with constant luminosity and size, its average total power output is between 600 W and 1500 W. Observers report that most balls are bright enough to be clearly visible in daylight. This is probably comparable to a 150-W filament lamp, which has a luminous efficiency of about 20% and thus emits about 30 W of power in the visible part of the spectrum. Thus the luminous efficiency of the process responsible for ball lightning may be about 2 to 5%.

Wien's displacement law, $\lambda_{max}T = 2.9 \times 10^{-3}$ m K, may be used to define a *blackbody temperature*, that is, the temperature of a black body that emits radiation of the same description as the body under consideration. For a yellow-orange ball, for example, with $\lambda_{max} = 588$ nm, the blackbody temperature would be about

4900 K. Red ball lightning would be cooler and blue ball lightning hotter. However, a black body is a theoretical ideal so these estimates are unrealistic, hence these inconsistencies. Furthermore, Barry (1980a) points out, as discussed in Section 11.2, that the source of luminous energy in ball lightning may perhaps not be thermal.

As we have seen, overestimates of energy and energy density usually arise from confusion about the cause of damage because of the close association between CG lightning and ball lightning (Stenhoff 1985, 1988b, 1992b, Hubert 1996). Ball lightning often precedes a CG flash to the immediate vicinity, which coincides with the apparent decay of the ball. This lends support to the idea that some types of ball lightning are a predischarge phenomenon similar to but not identical to St. Elmo's fire. However, while ball lightning may not in itself represent a hazard, like St. Elmo's fire it may be a precursor of an impending lightning flash. The most energetic damage following such events may be attributed to the effect of an ordinary CG flash. The scale and nature of such damage often suggests that it was caused by large electric currents, perhaps of hundreds of amperes or greater. It seems quite implausible that ball lightning could contain within its small volume sufficient charge per unit volume or sufficient electrical potential energy to deliver such large currents. It is much more credible that an ordinary CG flash was responsible. With these considerations in mind, we now embark on a survey of models proposed to explain ball lightning.

Chapter 12

Models Based on an Internal Energy Source

12.1 Heated Sphere of Air

The simplest model for ball lightning is a heated sphere of air. This idea was first discussed by Uman (1968b), who reported on his work with Lowke. They considered a starting central temperature of 10,000 K for spheres with radius of 5, 10, and 20 cm and found that cooling by thermal radiation and conduction was surprisingly slow. For a 10-cm radius, which is typical of ball lightning, a temperature drop from 3000 to 2000 K takes about 4 s, and during this time the sphere would maintain a fairly steady radius. Since a typical duration for ball lightning is 2–3 s, this suggests that ball lightning can remain hot throughout its lifetime, even without an external input of energy. However, a hot sphere of air would undergo rapid upward convection in the absence of any restraining forces, and this is not often observed with most ball lightning. Furthermore, a sphere of air of 10 cm radius and 5000 K would be about as bright as a 1-kW incandescent lightbulb (Uman and Helstrom 1966), but for every 1000 K fall in temperature, the luminous intensity would be expected to fall by at least an order of magnitude. However, those who report ball lightning fairly consistently describe its constant brightness throughout its duration (Hubert 1996).

In a later paper, Lowke, Uman and Liebermann (1969) considered three models based on the same idea: (a) cooling spheres of air, (b) cooling spheres of air containing small traces of sodium vapor, and (c) cooling spheres of mixtures whose composition by mass was 87.5% carbon vapor and 12.5% air, or 75% copper vapor and 25% air. They predicted upward convective behavior in models (a) and (b). In model (c), although the mass density could approximate that of air, there would be insufficient emission of light. They considered the possibility in model (c) that chemical reactions at the boundary with the air might provide relatively steady luminosity. Model (a) might explain reports of upward-moving luminous spheres formed from a lightning channel or at the point of impact of lightning whose brightness rapidly decreases, but few reports support this description. Model (c), with postulated chemical reactions at the ball–air interface, might explain reports of ball lightning formed following a lightning strike to a metal surface. This might include flashover within a building. Neither of these models could explain ball lightning formed within a metal aircraft.

12.2 Plasmoid and Vortex Plasma Ring Models

Bostick (1956, 1957) experimentally discovered plasma equilibrium configurations, which he called *plasmoids*. He defined a plasmoid as a geometric structure of limited extent composed of plasma whose form and stability are determined by the magnetic field it carries along with itself. The lifetimes of such plasmoids were a few microseconds. A magnetic field is trapped within the plasmoid as it is formed and there is a continuing toroidal current (Lindberg, Witalis, and Jacobsen 1960).

Shafranov (1957a,b) investigated the equilibrium conditions for bounded systems of a conducting gas in a magnetic field. He recognized that a theoretical investigation of structures such as those studied experimentally by Bostick could be of relevance to the theory of ball lightning. Shafranov thought that ball lightning might be a closed current produced at the instant of the thunder discharge. Most ball lightning models based on plasmoids have considered whether some configuration of closed-loop currents within a plasmoid could produce magnetic fields that caused self-containment. Interest in these problems was motivated by attempts to explain ball lightning, the existence of magnetic fields in cosmic space, the magnetism of stars and of the Earth, and by experimental investigations of high-current gas discharges.

Shafranov obtained the equilibrium conditions for a thin ring (a torus whose major radius much exceeded its minor radius) with a helical current (1) taking into account gravitational forces, (2) assuming the ring to be surrounded by gas, and (3) in an external magnetic field.

He applied some general considerations about the necessary conditions for a stable equilibrium from the *virial theorem* (see later discussion) given by Chandrasekhar and Fermi (1953) for the case of a magnetohydrodynamic system. He showed that scenario (1) was unstable against a "wriggling" form of perturbation. Stability was possible in scenario (2), which, as Finkelstein and Rubenstein (1964) suggested, could be relevant to ball lightning if its internal pressure is less than atmospheric pressure. This scenario is discussed later. Shafranov showed that for a minor radius of 1 cm and an excess pressure of 1 atm, the axial (ring) current would be on the order of tens of kiloamperes.

Shafranov used a magnetohydrodynamic analogy to M. Hill's (1894) stationary, spherical vortex in fluid mechanics to consider the equilibrium conditions of axially symmetric MHD configurations. Using this theorem, he reduced the problem of the equilibrium conditions for magnetohydrodynamic configurations to the theory of stationary flow of an incompressible fluid. For the case of axial symmetry, he considered general equilibrium conditions for distributed currents. He concluded that if there was no gravitation, a bounded equilibrium configuration could exist only in the presence of an azimuthal current.

The virial theorem is used extensively in a wide range of areas of physics, from the structure of molecules to astrophysical studies of stellar structure and gas clouds, star clusters, galaxies, and clusters of galaxies. The virial theorem can be used to provide the conditions necessary for stable equilibrium of a closed plasma system.

According to this theorem, the sum of the gravitational, electrical, magnetic, and internal fluid energies must be zero (Singer 1971). Witalis (1991) indicates that the virial theorem in its continuum formulation states that "the plasmoid energy contents—electromagnetic as well as gas kinetic—are all positive and together drive expansion." Equilibrium is possible, however, only in the presence of gravity because attractive forces such as gravity are associated with negative potential energies. Shafranov (1957a,b) demonstrated, and it is frequently reiterated in the literature, that in the absence of gravity, the virial theorem shows that a plasmoid, defined as a self-confined bundle of electromagnetic and material energy, cannot exist. The electromagnetic energy of a random, neutral plasma and the internal fluid energy are positive, so the virial theorem cannot be satisfied unless some other containing force, perhaps due to pressure differences with the surroundings, is invoked.[1]

Endean (1992b) offers a simple illustration of the virial theorem. When a toy balloon is inflated, work is done that results in energy being stored in the balloon in excess of that which would have been contained in the air displaced by the balloon. The energy is stored in the form of strain energy in the fabric of the balloon and in the additional kinetic energy of the gas particles because the pressure inside the balloon now exceeds that outside it. The essential point is that energy containment is possible because part of the system can withstand tensile stress. He sums up the implications of the virial theorem by saying that, for media that cannot withstand tensile stress, the average energy density of a body cannot exceed that which it would have if it were replaced by the surrounding atmosphere.

Finkelstein and Rubenstein (1964) used the difficulties posed by the virial theorem as the starting point for an analysis of the possibility of stable plasmoids being formed in the atmosphere. Acknowledging that the virial theorem indicates that self-field alone cannot confine plasma for many particle transit times, the authors indicated that confinement is, however, possible with external gas pressure,

[1] J. L. Shohet (1971) sets this as a problem in his book *The Plasma State*.

and then used the theorem to establish how much energy might be stored in such a plasmoid. They showed that the total plasma energy for a plasmoid of a 10-cm radius is at most a few kilojoules. They considered how a plasmoid might persist for several seconds despite coulomb scattering and cyclotron radiation. This analysis yielded a range of particle energies from 0.1 to 10 MeV—energies which they thought could be consistent with the generation of the plasmoid during a lightning flashover, based on estimated cloud potentials of 100 MV.

The greatest theoretical difficulty was encountered at the plasma–air boundary, where a "vacuum jacket" would be required to prevent rapid cooling of the plasma. A self-field able to provide such a jacket would use up the internal energy through ohmic heating in too short a time to account for ball lightning. Finkelstein and Rubenstein thus concluded that a plasmoid model for ball lightning could not explain ball lightning events. They did, however, show, with the above limitations, that a plasmoid of a few kilojoules of energy was allowed by the virial theorem in the presence of atmospheric pressure. This value is, of course, consistent with the estimate of ball lightning energy suggested in Chapter 11. Later, however, Powell and Finkelstein (1970) stated that the upper limit on a ball of volume 1000 cm^3 was only 100 J, and that even then the confinement time would be restricted to a few milliseconds because of ohmic dissipation at lower temperatures and thermal radiation at higher temperatures.

There were many experimental studies of plasma vortex rings, notably those of Bostick and Wells in the late 1950s and 1960s, which were related to the theoretical predictions of Shafranov and others (Bostick 1956, 1957, Harris, Theus, and Bostick 1957, Bostick, Prior, and Farber 1965, Wells 1962, 1964, 1966). Logan (1946) suggested that the magnetic field generated by rotation of charged particles in a ring, following helical paths and circular paths within the ring, could assist in confinement. Alfvén's theorem (Alfvén and Fälthammar, 1963) shows that magnetic lines of force are "frozen" into a perfectly conducting fluid. Lindberg and Jacobsen (1961) described an experiment in which a magnetized plasma was produced with both toroidal and poloidal magnetic fields. Under certain conditions, the kink instability could cause part of the toroidal field to be converted to a poloidal field so that the poloidal flux became magnetized up to five times its original value. This could offer a natural mechanism for the creation of a rotational transform. Wells and Ziajka (1978) presented an interesting and useful review that proposed the self-contained plasma vortex as a means of achieving fusion. Much more recently, Kunin and Furow (1993) have claimed the production of atmospheric toroidal plasma vortices with volumes of about 3×10^3 cm^3 and optical durations of up to 1.5 s.

K. Wolf (1915a–d) suggested that ball lightning is a rapidly rotating electron vortex ring, with the electrons ionizing air by collisions and producing a vacuum within the ball, formed by a pulse in conventional lightning. Frenkel (1940) proposed a magnetohydrodynamic vortex model for ball lightning that is described later in this chapter.

Wooding (1963) proposed that ball lightning is a plasma vortex ring. If a vortex ring were produced from plasma in a magnetic field, the outward force due to vorticity would be balanced by magnetostatic pressure. This trapped magnetic field would prevent expansion of the ring. Wooding estimated that the kinetic energy associated with the vorticity would be a few hundred joules, and pointed out that collision with a metallic object might cause an explosion by sudden disruption of the field and current distributions. In addition to the kinetic energy, the plasma would possess thermal and ionization energies amounting to several megajoules. The energy would mainly be lost by radiation, initially at a rate of about 10 MJ s^{-1}, so the plasma would persist for up to a few seconds.

There have been various attempts to circumvent the problem posed by the virial theorem. The argument presented by Finkelstein and Rubenstein (1964) has already been discussed, based on the additional confinement provided by atmospheric pressure.

In a different approach, Bergström (1973) pointed out that the virial theorem cannot be applied in the usual way to certain bodies of discrete charges, for example, ionic crystals. He indicated that these can only exist, as can self-confined plasmoids, because the thermal energy is much less than the coulomb energy. He proposed that instead of magnetic confinement, there was a strong dielectric–diamagnetic attraction that overcame repulsion of electric charge by a strong, short-range interaction. He suggested that this interaction could have a macroscopic range in thunderstorm conditions when ionization was followed by electron depletion (owing to the much greater mobility of electrons compared with ions), leaving small regions of positive charge. Acceleration of ions in the dense, positively charged cloud would provide a strong interaction in the same way that acceleration of charge by an electric field produces, by electromagnetic induction, an electric field opposed to the applied field. The estimated total electrostatic energy of a ball thus composed of a radius of 5 cm and a positive charge of 0.01 C was 20 MJ.

Muldrew (1990) presented a mathematical model for ball lightning that assumed the presence of a solid, positively charged core at its center, such as hail, a stone, or a piece of metal. In this model, a pure electron layer and a plasma layer surround the core, with the magnitude of charge of the electron layer equal to that of the core. An electromagnetic field is completely trapped by the electron and plasma layers. The ponderomotive force (radiation pressure) of the trapped field balances the electrostatic force of the electrons toward the core plus the force of atmospheric pressure. The electron temperature is sufficiently high to make absorption by electron–ion collisions small, enabling the ball to have a lifetime of seconds or more. Muldrew suggested that the large amount of energy occasionally associated with ball lightning is mainly due to the electrostatic energy of the charge on the core. The upper energy limit is determined by the size and strength of the core, and this energy could be orders of magnitude greater than the energy that can be confined by atmospheric pressure alone. Endean, who (1992b) had a high regard for this model, remarked (1993) that the suggestion of a solid body at the core was ingenious but far-fetched.

A very similar model was discussed by Alanakyan (1994). He investigated the self-localization of an electromagnetic vortex under conditions that produced a partial charge separation in the plasma that formed near the vortex. He postulated that outside the vortex there was an excess of electrons and within it there was an excess of positive charges. He showed that the presence of charges that had not been neutralized increased the energy content of the vortex without increasing the energy dissipation of the high-frequency field in the plasma. This enabled the lifetime of the vortex to be substantially increased. He examined a mechanism for confining in the interior of the vortex a positively charged solid object that could serve as ballast and retard the buoyant rise of the vortex in the atmosphere. He discussed the possible formation of a hydrodynamic vortex at the periphery of an electromagnetic vortex by transfer of momentum from a high-frequency field to the surrounding medium. This affected the nature of the motion of the vortex in a free atmosphere. He claimed that the properties of the vortex were consistent with some of the properties of ball lightning.

Witalis (1990a,b) indicated that for a magnetized plasma, the dominant quasi-neutrality plasma property was actually a severe restriction, which was usually neglected, on the motion of the charged species. He explained that quasi-neutrality alone caused the plasma total electromechanical torque tensor to act preferentially on the ions-plus-neutrals whenever the electrons were free to perform Larmor gyrations. The canonical (i.e., matter-plus-field) angular momentum of the heavy plasma species would then be conserved so that action on its matter part would have its response reaction on the magnetic vector potential field part. Starting with the premise that ball lightning originates as a positive-pulse corona discharge caused by a very strong, transient electric field, producing a room-temperature plasma by photoionization, Witalis proved that the virial theorem, instead of invalidating the concept of magnetically self-contained ball lightning, indicated pronounced two-fluid plasma properties. The virial theorem, he argued, applies to the single-fluid MHD plasma description in which the plasma mass motion is not distinguished from the motion of charge. This leads to a plasma description whose electrodynamics are essentially of a diffusive or "smoothed-out" character.

In this model, the quantum Ramsauer effect in room-temperature air followed by the proper Hall effect magnetohydrodynamic (HMHD) description yields ball lightning as a self-magnetized, self-contained whorl structure. (The Ramsauer effect is a quantum mechanical interference effect between the wave properties of free electrons and orbitally bound electrons in gas atoms or molecules.) This has rather arbitrary but about equal amounts of magnetic and kinetic energy, and a degree of ionization that depends on ball lightning size and is on the order of 0.1%.

He examined several applications of this theory, including the failure in achieving fusion plasma confinement, and succeeded in explaining several peculiar ball lightning characteristics.

While it has often been contended that the virial theorem shows that a stable plasmoid cannot be formed in air, there is experimental evidence such as the work of Andrianov and Sinitsyn (1976b) that vortex plasmoids have lifetimes considerably greater than those found in plasmas without self-containment.

A further limitation of these models is that they generally fail to explain reports of ball lightning not formed near a lightning channel or the point of impact of lightning. Perhaps, contrary to Singer's (1971) suggestion that these models cannot explain ball lightning formation inside rooms, this could be explained by plasma effects from the arc from a sideflash; however, formation within aircraft cannot be thus explained. Only Wooding's (1972) model attempts to address this question by suggesting that coherent radiation from lightning can penetrate portholes.

Plasma vortex ring models predict intense magnetic fields that would be detectable outside the ball. Blair (1973) provided anecdotal evidence of strong magnetic fields from reports of the effect of ball lightning on church bells. Similar reports were provided in earlier chapters. However, Jennison (1973) objected that, having seen ball lightning inside an aircraft (Chapter 7) at a distance of 0.5 m, with several ferromagnetic objects in his pockets, he had found no evidence of strong B fields. Evidently not all ball lightning has magnetic properties.

Yet another objection might be the special conditions that would need to prevail in order for a stationary vortex to form. Most generation mechanisms would suggest production of plasma vortex rings with very high translational velocity (see later discussion), which is inconsistent with most reports of ball lightning.

Models in which ball lightning is powered by thermonuclear fusion are described later in this chapter. Vlasov (1997) has discussed a plasma vortex ring model with a toroidal current layer creating a ring magnetic field. The current layer in the model consists of a sheath of positive ions and a monolayer of relativistic electrons that act as catalysts of nuclear reactions. More conventionally, plasma models assume that the internal energy of ball lightning is derived from recombination processes. Smirnov (1975a,b) studied the energy associated with these processes and concluded that they did not adequately explain the extended lifetime reported for ball lightning. Types of decay that may be important are dissociative recombination and three-body recombination. Typical dissociative recombination processes involving ions and electrons might include

$$e + O_2^+ \rightarrow O + O$$

$$e + N_2^+ \rightarrow N + N$$

$$e + NO^+ \rightarrow N + O,$$

whereas three-body recombination, with both positive and negative ions, might involve the typical decay processes

$$O_2^- + O_4^+ + O_2 \rightarrow 4O_2$$

$$NO^+ + NO_2^- + N_2 \rightarrow NO + NO_2 + N_2$$

A final limitation would be that electric fields cannot exceed the breakdown field strength of air. This, in turn, imposes limitations on energy density (Smirnov 1987b,c).

Some investigators have attempted to correlate ball lightning color with temperature and energy on the assumption that the ball emits blackbody radiation. If ball lightning is a plasma, this assumption may be shown to be very unlikely. Plasmas emit radiation as a result of mutual collisions. This is an important mechanism for power loss. In an electron–ion collision, the electron is accelerated so as to emit *Bremsstrahlung radiation* from free–free transitions. If the equilibrium distributions for both ions and electrons are Maxwellian and both have the same temperature, then the plasma is in thermodynamic equilibrium. The radiation emitted is blackbody, and Stefan's law enables calculation of the surface power density.

The minimum radius of a spherical plasma that can emit *Bremsstrahlung* blackbody radiation is given by $10^{47}T^{7/2}/n^2$ where T is in electron volts and n is in m^{-3}. Laboratory plasmas (and ball lightning) are always much smaller than these dimensions, hence they are not in complete thermodynamic equilibrium and are not perfect blackbody radiators (Green, Chapter 1 in Gill 1981). Radiation from ball lightning may originate mainly from molecular transitions; furthermore, an air plasma will contain impurities that will produce contributions to the emission spectrum from recombination radiation (free–bound transitions) and line radiation (bound–bound transitions).

12.3 Other Plasma Models

Neugebauer (1937, 1977) proposed that quantum-mechanical exchange forces provide the binding force or containment energy of a dense plasma ball, consisting of free electrons and positive ions. It is essential that the plasma be almost completely ionized and thus have a high charge density. If the plasma consists of equal numbers of free ions and free electrons, other electrical forces could be ignored and the exchange energy would be given by the following equation, where all symbols have their usual meanings:

$$E = -\frac{e^2 n_e h^2}{8\pi mkT}.$$

The equilibrium between thermal energy and exchange energy would be achieved at a temperature such that $E = 3/2\ kT$ and so

$$T = \left(\frac{e^2 n_e h^2}{12\pi mk^2}\right).$$

If the electron density is equivalent to the molecular concentration of gas at atmospheric pressure (i.e., ~2.7×10^{27} m^{-3}), the ball temperature has to be low (~600 K) for these exchange forces to be sufficient to provide stability. Neugebauer assumed that the decrease in electron recombination rate with increasing temperature was sufficient to allow an adequate ball lifetime. Silberg (1965) pointed out that it is difficult to see how such a structure could be generated in a thundercloud, and it is also not clear whether recombination is slowed sufficiently at the relatively low temperature of a few hundred kelvins.

Powell and Finkelstein (1970) argued that Neugebauer's model violated the correspondence principle. They commented that the density required for significant electron exchange at room temperature is that of a solid and is even greater at higher temperatures. Altschuler, House, and Hildner (1970) commented that such a ball would emit insufficient light to be visible.

Neugebauer's quantum-mechanical exchange model was revived and extended by Dijkhuis (1980, 1981, 1982, 1988, 1991b,c). Dijkhuis (1980) pointed out that many ball lightning models, including those based on chemical reactions, dc glow discharges, microwave cavities, and nuclear reactions or annihilation of antimatter, fail to explain reliable observations made in metallic airplanes, which are airtight Faraday cages. He indicated that electric charge alone could penetrate an aircraft. Electric discharge currents in ionized air magnetize the ionized region through which they pass, and the magnetic field lines are frozen into the plasma for a characteristic period known as the magnetic diffusion time τ. This time τ is directly proportional both to the permeability and to the electrical conductivity. While τ is too short for the lightning channel to account for reported ball lightning lifetimes, an increase in permeability or electrical conductivity could increase the magnetic diffusion time. Superconductivity would allow an increase of conductivity and hence τ. Charge transfer would be enhanced by plasma turbulence.

Dijkhuis deduced theoretically that the containment force is a quantum-mechanical binding force, comparable to the forces that hold the atoms in molecules together. In the formation of the ball, a phenomenon known as Bose–Einstein condensation occurs. Pressure equilibrium inside a self-confined plasma sphere is maintained by binding forces arising from gradients in boson density. The breakdown electric field in air drives electron concentration into the region of thermodynamic instability caused by exchange interactions between free electrons in the phase diagram. The electrons in the unstable region form pairs subject to Bose–Einstein statistics. A localized charge separation combined with cooling takes place within the plasma, as a result of which electrons in the form of bosons in the superconductive ground state are stored in vortex filaments only one Debye length ($\sim 10^{-8}$ m) across, running through the entire ball. This was later described (Dijkhuis 1992a,b) in terms of fractal structure (see Section 12.7).

From Feynman's equation for a charged superfluid, Dijkhuis obtained a steady-state vortex solution with self-confined electron pairs in the bosonic core. The ball is thus "a sponge-like structure of tenuous vortex cores nested in a dense ambient plasma." Radial electric fields are screened out by ions circling around the negative vortex core. Relativity theory sets an upper limit for energy storage in the vortex structure that is consistent with reported ball lightning energies. The boson model for plasma turbulence circumvents the virial constraint because the potential energy from interaction between plasma vortices is deduced to be negative.

Although the temperature in the ball is only a few thousand degrees, fusion is predicted to occur. Deuterium is the fuel for the fusion process in the ball. Very powerful local electric fields are present in the ball that would accelerate deuterium nuclei to velocities at which fusion reactions would be inevitable. Helium is formed by fusion from deuterium. A visual characteristic of a ball during fusion would be that one or more luminous rings surround it. For every 3000 or so water molecules, water contains one heavy water molecule; in other words a heavy hydrogen atom, deuterium, has replaced one of the hydrogen atoms in the water molecule. The quantity of deuterium is such that nuclear fusion can occur in ball lightning with adequate energy content, which has formed in humid air.

As the ball floats through the air, the spent fuel could be supplemented by fresh deuterium from the surrounding air, which could enter the ball by diffusion and thus reach the reaction zones because of the circulation of the plasma in the ball. In this way the ball could attain a far longer life than the usual 2–3 s. Visible cyclotron radiation would be emitted from ions because electrons in quantized orbits do not radiate. This radiation would have a nonthermal spectrum, thus explaining the absence of heat radiation from ball lightning seen at close proximity, and color would depend on core radius, which could explain occasional reports of color changes.

Dijkhuis (1982) claimed that fireball generation from a high-intensity circuit breaker arc in U.S. submarines (Silberg, 1962, 1965, 1978) was a quantum-mechanical exchange phenomenon caused by rapid cooling of electrode material evaporating from contact surfaces. He found that a formula derived for the threshold discharge current before generation of a fireball was consistent with data reported by Silberg (1978).

Dijkhuis's company, Convectron, succeeded in the 1980s in reproducing the observations reported by Silberg. Their goal was to achieve controlled thermonuclear fusion as an energy source. They built an experimental hall containing a complete submarine accumulator battery (weight 200 tonnes, short-circuit capacity 30 MW) provided with many switches. The short-circuit trials were automated by means of a control system. In 1985, the threshold current of 150,000 A was exceeded for the first time and balls with the characteristics of ball lightning (such as a constant light intensity), a diameter of 10 cm, and a life of 1 s were recorded very clearly on film.

After formation in the circuit breaker of a reactor vessel, the ball could be held in place with a stream of gas from a nozzle; the gas left the vessel via an outlet. Deuterium that had been mixed into the gas stream could enter the ball and continuously supplement the spent deuterium fuel. The reaction product, the noble gas helium, could leave the ball in the same way and be carried away by the gas stream. By placing a water jacket around the reactor vessel, the energy released might be used to make steam from water and thus electricity in the standard manner (Dijkhuis and Pijpelink 1989).

Witalis (1991) raised objections to the Dijkhuis model based on the fluid virial theorem. Allowing, however, that this theorem could be misleading owing to the averaged plasma description inherent in it, he considered a particle system formulation of the theorem. This showed that a slight excess of negative charges around each positive charge could make the electrostatic energy negative, and that a hypothetical current distribution where adjacent current elements were directed oppositely could make the magnetic energy negative. However, he argued that in either case these negative values would be insignificant compared with the positive thermal energy.

Dijkhuis (1981) replied, saying that perfect generality did not apply to the purely particle system formulation of the virial theorem. In a turbulent plasma, where electromagnetic fields may have strong local fluctuations, classical equations of motion are not applicable. In his ball lightning model, turbulence is so great that the motion of particles must be treated quantum mechanically. He maintained that "(1) the quantum-mechanical version of the virial theorem admits electromagnetic self-confinement, and (2) strong plasma turbulence leads to a self-confined boson state in ball lightning."

12.4 Other Vortex Structures

Some models suggest that vortex motion of a gas could help to preserve its spherical geometry (Coleman 1993, 1997). The Hill vortex (1894) has often been proposed as a possible configuration. Nickel (1989b) derived an expression for the lifetime τ of the Hill spherical vortex in air at atmospheric pressure $\tau \approx d^2/10$ where τ is in seconds and d in centimeters. For vortices of a diameter of 20 to 50 cm, this would provide lifetimes of 40 to 250 s and suggests that vortices of larger diameter would persist for longer. Amirov and Bychkov (1997c) found that reported ball lightning lifetimes were longer for larger diameters.

Coulvier-Gravier (1859) and Moigno (1859) proposed that ball lightning might be a vortex formed at the meeting point of two perhaps misaligned, oppositely directed strokes of ordinary lightning. Figure 12.1 shows an experiment with an electric discharge in air in which upward and downward leaders appear to be misaligned. A number of other authors have considered this possibility, particularly as a means of explaining descriptions of the rapid rotation of some ball lightning. Meissner (1930) suggested that a rapidly rotating vortex might be formed where an abrupt change in direction of a lightning channel caused a parallel but oppositely directed discharge over a short segment. He conceived the resulting structure as an evacuated center surrounded by a rotating layer of air at normal density, with the centripetal force of the rotating layer exactly balanced by the force due to atmospheric pressure. Godwin-Austin (1883) and Flint (1939a) proposed that a vortex consisting of oppositely directed streams of electrons and ions might be formed at the junction between a descending leader carrying a negative charge and a positive streamer ascending from the Earth.

Faye (1890a,b, 1891a,b) suggested that ball lightning might be formed from whirlwinds, cyclones, or tornadoes. He proposed that ball lightning is a rotating sphere, highly charged owing to friction with water droplets, hail, and other solid particles, which detaches from the lower tip of the rotating spout. This suggestion followed a report of several ball lightnings the size of billiard balls, one of which exploded on contact with the ground, which were seen with a tornado in France in 1890. There were several observations of ball lightning entering houses through stovepipes or chimneys. Other larger, red balls entered a barn and set some hay on fire before disappearing. After the event, many windows in nearby buildings were found to have holes that had evidently been melted in them. These were about 8 cm in diameter and had smooth edges. It is interesting to compare Faye's approach with that of Hands (1909, 1910) and Lagrange (1910), who explained ball lightning reports as light from whirlwinds! (Singer 1971)

Dauvillier (1965) suggested that cyclone activity could generate ball lightning. He proposed that spheres thus generated would have gyroscopic rigidity and elasticity, providing stability against bouncing from the Earth, because of a very high rotational velocity that would ionize air through molecular collision.

Voitsekhovskii and Voitsekhovskii (1974) proposed an alternative mechanism for the formation of a vortex where a region of high charge density is above one of lower charge density within an electrostatic field. This is analogous to the Rayleigh–Taylor instability in fluid mechanics.

In 1971 Singer made some general remarks concerning vortex structures whether plasmas or not (Singer 1971). He indicates that these models receive support from numerous descriptions of the rotational motion of ball lightning. Several viable formation mechanisms have been proposed. He commented, how-

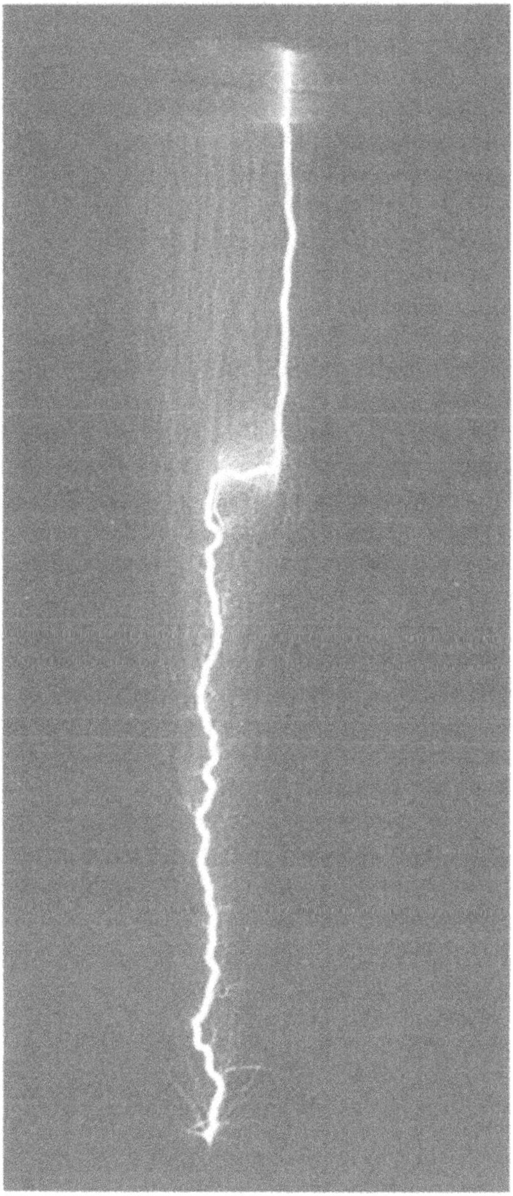

Figure 12.1. Two sharp turns in a disruptive discharge channel, probably due to the meeting of an upward-moving positive and a downward-moving negative leader. [Photograph © Dr. A. Larsson. Reproduced with kind permission from A. Larsson, "Inhibited electrical discharges in air," thesis for a Ph.D. in electricity, Institute of High Voltage Research, Uppsala University, Sweden, 1997.]

ever, that "none of the studies of ball lightning as a vortex . . . has conclusively dispelled by detailed consideration the serious difficulties on which other theories have floundered, such as the continued luminosity of the balls for long periods while they travel inside structures." He remarked that the energy of a vortex would be quickly dissipated through viscous drag and turbulence. He also (1997) pointed out the failure of these studies to explain reports of ball lightning passing through windows.

12.5 Internal Energy Models Powered by Electromagnetic Radiation

Dawson and Jones (1969) proposed that ball lightning consisted of a "radiation bubble"—a cavity bounded by highly ionized, conducting spherical walls carrying large surface currents. Resonating within the cavity are intense, high-frequency microwave fields that are the energy source for the ball and are reflected by the walls of the cavity. It was conceived that a small loop in a lightning channel or a large current in a lightning conductor might form this structure. The high frequency far exceeds the collision frequency of electrons and molecules, and the internal plasma pressure is low, thus reducing the rate of energy dissipation. The radiation pressure of the microwave energy balances the external atmospheric pressure on the sphere. Several coupled resonant modes are required to provide a uniform radiation pressure over the entire surface. These assumptions allowed energy estimates of 400 J for a sphere with a 10-cm radius and 10 kJ for a sphere with a 30-cm radius.

Jennison (1973, 1987, 1990) suggested that ball lightning is formed by a phase-locked loop of electromagnetic radiation of specific wavelength in the intense field associated with lightning activity. The standing wave thus formed would exhibit spherical configuration and excite the ambient gas to produce luminosity. The lifetime of the ball would be limited in this process, but the shape and size of the ball would be maintained by the locking of the wavelength. The typical gap between the ball and electrically conducting surfaces could result from the maintenance of equilibrium between the ball and its image in the conductor. He suggested that the diameter of ball lightning could thus be independent of air pressure, and that ball lightning might be able to exist in regions of very low pressure even though it would not then be observable from the luminosity of excited gases. Electromagnetic radiation in phase-locked loops at an exact frequency also occurs in the process of pair production, although the scale and degree of quantization of the phenomena are quite different. In support of this model, Jennison referred to events in which ball lightning had maintained constant separation from nearby surfaces; one of these was the trailing edge of the wing of an aircraft in flight.

Endean (1976) proposed a model in which ball lightning consists of electromagnetic field energy trapped in an evacuated, spheroidal cavity that is separated from the surrounding atmosphere by an ionized sheath. He revived the idea of misalignment of a descending negative leader with an ascending positive charge. He envisaged that this could establish a rotating electric dipole with a peripheral phase velocity $>c$, the speed of light, which would provide an electromagnetic pressure balance. If 10% of both the available energy and leader charge of a typical lightning discharge were made available, this could produce a ball of 10 MJ energy. The field would be internal, and a high stabilizing shear would be provided by rapid rotation of the magnetic boundary field, so plasma confinement could be achieved without instabilities. The model is, however, fundamentally flawed, as is often the case with two-dimensional ball lightning models, in that there are uncompensated end-effect pressures.

A small leakage of field through the sheath would be sufficient to maintain the ball at a constant distance from nearby structures. While this field would be insufficient to move nearby metal objects, penetration of the ball by a metal object would have significant effects. Either quiet decay by ohmic heating of the sheath and radiation through it or explosive decay by perturbation, leading to an instability, would be possible. Since the ball transports its own energy, it could drift into buildings. Endean suggests that a ball could pass through a glass window or be formed inside an airplane if a strike were close to a porthole and sufficient electric field and ionization were set up inside the aircraft. Given that the source of the internal energy of the ball is high-frequency electric field energy, an insulating window would not in any way affect the passage of the ball. Indeed, the presumed electrically conducting surrounds should have the effect of steering the ball through the window aperture. The electric plasma associated with the ball appears to pass through the window, but in fact it does not. As the electric field passes through the window, the plasma on one side extinguishes and the plasma on the other side ignites.

Endean further developed this model in subsequent papers (1978, 1992a,b, 1993, 1997a,b). He showed (1992a,b) that limits on field energy containment imposed by the tensile strength of containment structures could be avoided. He proposed two machines, one of which provided useful insights into the above model, for which there was a much-weakened theoretical virial constraint on the containment of field energy. He also (1993) presented an analytical solution for the model. He assumed that total current (conduction plus displacement) was everywhere zero, resulting in the considerable simplification of no magnetic field so that the ball would not emit electromagnetic radiation. The time-averaged electrical force is radial, depends only on radius, and is balanced by a radial plasma pressure gradient. Average energy densities could exceed the external atmospheric pressure by large factors. Thus he dealt with objections based on the virial theorem. He argued that compared with alternative explanations of ball lightning, the model had the virtues of simplicity; a straightforward, natural generation mechanism; and no special requirements. However, the central plasma pressures are so high, considerably above atmospheric pressure, that even with no conduction currents there would be enough plasma collisions to ensure an unacceptably short life (V. G. Endean, personal communication, 1998).

He claimed (1997b) that the problem of how to obtain satisfactory equilibrium configurations consistent with the virial theorem had been solved by recent work on the model, including a full analytical solution. Here plasma supports large-amplitude electron–plasma oscillations in both central core and outer shell. He showed that the central core of the electric dipole model is much more likely to consist of a vacuum electromagnetic field, with only the outer plasma shell supporting electron–plasma oscillations. Although he evolved this model independently of earlier work by Dawson and Jones, Endean later recognized that he had arrived at a mathematical solution to the general model proposed by these authors in 1969 and that the radiation bubble theory was essentially correct (Endean 1997b; V. G. Endean, personal communication, 1998).

Zheng (1990, 1992) discussed a model of ball lightning as a spherical standing wave of electromagnetic radiation trapped in a plasma shell. The shell is expelled from an air-filled cavity by the ponderomotive force. The trapped radiation energy generates plasmas continuously through ohmic heating to prevent the shell from diffusing away. This view allows a quantitative analysis for the dimension, temperature, energy, and lifetime of ball lightning.

12.6 Chemical Processes

12.6.1 Chemical Reaction

Arago (1838, 1854) suggested that ball lightning might contain compounds produced from air by the effect of lightning.

Smirnov (1975a,b, 1976a,b, 1977a–c) made a very thorough study of chemical energy as a possible source of energy for ball lightning. He rejected models using ions and excited states because the short decay times associated with them were inconsistent with the reported duration of ball lightning of several seconds. He consistently found that the internal energy of a 20-cm ball would fall to less than 0.1 mJ after 10 s. He concluded that ball lightning must be formed by a chemical process involving reactions of oxygen, ozone, and nitrogen dioxide. He argued that the luminosity of the ball thus formed could be maintained for many seconds because of the low efficiency of conversion of chemical energy to radiation energy if the ball is initially composed of ~1% ozone and at a temperature of a few hundred kelvins. The ball would end either by scattering of the chemically active medium by convection within it or by penetration into the ball of a substance that reacts explosively with the chemically active medium. Brightness and color could depend on the relative abundance of different chemical species.

Descriptions of acrid odors during or following ball lightning events may support the formation of certain gases, notably ozone or oxides of nitrogen. Dmitriev (1967a,b) detected unusually high concentrations of nitrogen dioxide, ozone, and hydrogen in samples. (His interpretation was that these were generated by the ball rather than responsible for its formation.)

Turner (1994) proposed a model based on chemical reactions, but this required the thunderstorm electric field as an energy source and is therefore discussed in Chapter 13.

Powell and Finkelstein (1970) commented that rates of reaction for decomposition of ozone or NO, or for oxidation of NO, are too great at the temperatures needed for sufficient light emission to account for observed lifetimes. As Charman (1979) points out, chemical models fail to explain why the effects seen as ball lightning are restricted to a small, spherical region of the atmosphere. If the necessary mixture of gases is produced by a cloud-to-ground discharge, then similar luminous effects should surely be seen along a substantial length of the lightning channel.

Nitrogen compounds: Some models have proposed that ball lightning is formed by reactions of oxides of nitrogen. Besnou (1852) attempted to explain the explosive behavior of some ball lightning by suggesting that it contained nitrogen triiodide formed by reactions of nitrogen with traces of iodine or related compounds initiated by electrical discharges in thunderstorms.

Oxides of nitrogen have been frequently proposed. In high-temperature oxidation, some nitrogen and oxygen from the air combine to form nitrogen (II) oxide, NO (also known as nitric oxide or nitrogen monoxide):

$$N_2 + O_2 \rightarrow 2NO$$

There is spectroscopic evidence to confirm the formation of NO in lightning discharges (Orville 1977). When this oxide cools in the presence of air, it is further oxidized to nitrogen (IV) oxide:

$$2NO + O_2 \rightarrow 2NO_2$$

Kilburn-Scott (1923) suggested that lightning generates a mass of nitric oxide, which then reacts readily with oxygen to give nitrogen dioxide (NO_2). He suggested that these concentrated oxides of nitrogen could cause immediate nitration of organic material, such as trees or haystacks, with explosive decay. This mechanism is restricted to explosion following contact with substances that may undergo nitration. [Nitrogen dioxide, nitrogen (IV) oxide, is a brown, acidic, acrid gas that is harmful if inhaled and contributes to the development of acid rain because it dissolves in water to form nitric acid.]

Chirvinskiy (1936a–d) proposed exothermic reactions of nitrous oxide with hydrogen, forming nitrogen and water.

Ozone: Ozone (O_3, trioxygen) is a colorless gas with a penetrating odor that is soluble in cold water and in alkalis. It is also a powerful oxidizing agent. Ozone is an allotrope of oxygen made up of three atoms of oxygen. It is formed when the molecule of the stable form of oxygen (O_2) is split by ultraviolet radiation or by an electrical discharge. It is produced in the stratosphere by the action of high-energy ultraviolet radiation on oxygen. Its presence there acts as a screen for ultraviolet radiation. Since lightning is an electrical discharge that emits ultraviolet light, it produces ozone, and evidence of this has been provided by lightning spectroscopy experiments (Orville 1977).

In support of ozone as a substance for ball lightning in preference to oxides of nitrogen, some authors have appealed to a number of factors, including descriptions of odors after ball lightning events and a greater than atmospheric density that would prevent air convection and cause the ball to descend. Thornton (1911a–d, 1912a,b, 1913) estimated that a sphere of ozone 50 cm in diameter could revert to oxygen with the liberation of about 10 MJ, which is consistent with estimates of energy in the "tub-of-water" event. However, more recently this has been considered an overestimate by a factor of about 25 (Pascal 1956).

Oxygen, ozone, and nitrogen dioxide: Smirnov (1975a,b, 1976a,b, 1977a,b) proposed a chemical model based on a series of reactions involving oxygen, ozone, and nitrogen dioxide. If the initial temperature is a few hundred kelvins and the initial concentration of ozone is ~1%, a transfer from chemical to radiation energy is possible with an efficiency of a few percent, with the radiation maintained for many seconds. Scattering of the chemically active medium by convection or by contamination by a reactant that produces a thermal explosion could cause the end of the ball. Differences in reported colors of ball lightning could be explained by variations in chemical composition.

The observation reported by Dmitriev (1967a,b) where gas samples were collected within 1 min of a ball lightning event offers some support for Smirnov's proposals because both ozone and nitrogen dioxide were detected.

Explosive mixtures of hydrogen and oxygen: Explosive mixtures of hydrogen and oxygen or other gases could be produced by direct impact of a previous cloud-to-ground flash on water causing electrolytic effects or thermal dissociation (de la Rive 1858, Hildebrandsson 1883, 1884a,b, 1885, 1896a,b, Schönland 1950, 1964, Benedicks 1951). The gas might possess an electric charge and be bounded by an aqueous wall. Hydrogen was also detected following a ball lightning event (Dmitriev 1967a,b), but in a concentration below the lower explosion limit (Lewis and von Elbe 1961).

Polyatomic molecules: Mathias (1924a,c,d) suggested that descriptions of a rapid descent of ball lightning and its high density at supposedly high temperatures might imply the presence of polyatomic oxygen and nitrogen molecules up to O_{12}

and $N_{12.}$ However, Lowke, Uman, and Liebermann (1969) found that the density of a high-temperature gas (3500 K) composed of 7/8 polyatomic carbon molecules (up to 17 atoms) and 1/8 air would be less than that of the atmosphere.

12.6.2 Combustion

Experiments by Nauer (1953) established that a short electrical discharge into air containing a small proportion of methane, hydrogen, propane, or benzene could generate various glowing forms that shared properties with ball lightning. Barry (1967a,b, 1968a–c, 1980a) suggested that certain regions of the troposphere contain unusually high concentrations of simple hydrocarbons generated by decaying vegetation or seepage of natural gas, but that this concentration is below that required for combustion. Under thunderstorm conditions, ionization of the hydrocarbons might occur and electrical discharges might generate more complex hydrocarbons that could tend to clump together in localized regions. Combustion could then be started by a lightning discharge into one of these regions. The required concentration of hydrocarbons would be obtained in the ordinary atmosphere only if clumping were to occur over a radius of about 100 m. A fairly large volume of the atmosphere would still be involved if the initial abundance of hydrocarbons exceeded the normal figure of about 10^{-4}%. As long as there was sufficient density of hydrocarbons, burning would continue. By using the spherical flame equation and substituting sensible values for the parameters, Barry predicted the formation of a fireball of between 6 and 130 cm in diameter. He supported these theoretical studies with experiments using mixtures of air and propane at normal temperature and pressure. With concentrations of propane between 1.4 and 1.8% (which is well below the concentrations required for combustion), a brief initial discharge lasting about 1 ms produced a small fireball. This yellow-green ball moved rapidly and randomly around the experimental chamber and had a duration of about 2 s. Barry suggested that the motion of the ball was caused by progression of the burning center from one hydrocarbon concentration to another.

In the next section we discuss the model proposed by Blok et al. (1981) in which ball lightning consists of a cloud of slowly burning, combustible aerosol particles of like charge. Ofuruton and Ohtsuki (1989, 1990) described experiments with discharge in flammable gas and/or aerosol. A luminous phenomenon with a long lifetime (2 s) was observed in 2.7% ethane and 100 cm^3 cotton fibers, and in 1.5% methane and 1.9% ethane.

Singer (1971) points out that Barry's model has the advantage over other chemical models of suggesting why the effect might be localized. Powell and Finkelstein (1970) remarked, and Barry (1980a) agreed, that such models could account for only a minority of observations where sufficient quantities of methane might be found.

12.7 Dust, Droplets, Dirty Plasmas, Aerosols, and Fractal Structures

Poey (1855) proposed that electrostatic charging of dust particles and droplets in clouds might form ball lightning. de Tastes (1884, 1885a,b) suggested that pollen and dust in the atmosphere caused light-emitting reactions similar to the light from biochemical processes such as fermentation.

Frenkel (1940) proposed that lightning could generate reactive substances in the air that would then condense as small droplets or dust. He initially envisaged a spherical film structure similar to a Leyden jar, but later rejected this in favor of a magnetohydrodynamic vortex model resembling the Hill hydrodynamic vortex. Frenkel replaced the usual circulation of fluid within the vortex with alternate layers of separated electric charge of equal magnitude, thus rendering the vortex electrically neutral. The motion of the electric charge would produce magnetic fields which, he proposed, would compress intervening shells of air (Leonov 1965a,b, Singer 1971). Stepanov, Sall, and Arutjunan (1997) proposed a model in which the ball consisted of aerosol particles with a radial field inside the ball caused by a central core and corona of opposite charge. They argued that such a structure would recover its spherical symmetry after deformation, and that the ball would be attracted to electrical conductors by inducing a mirror charge in them.

Schwegler (1951) suggested that many ball lightning events were caused by dust activated by storms, particularly dust clouds associated with volcanoes.

Cawood and Patterson (1931a,b) described an experiment involving an aerosol of charged solid particles produced by distributing dust within a chamber containing a charged electrode by means of a fan. After the fan was switched off, a spherical concentration of particles 20 cm in diameter gradually developed in the center of the chamber. The sphere could be repelled or attracted by electrodes of different polarity, but returned to the center when these electrodes were removed, probably because the charge induced on the glass envelope was the same as that of the sphere. It was shown that many of the particles had formed long ropes or chains.

Aerosol models of ball lightning have since been proposed and discussed by Blok et al. (1981), Aleksandrov, Golubev, and Podmoshenskii (1982), Mukharev (1986), and Neda, Ofuruton, and Ohtsuki (1989). Blok et al. proposed a model in which ball lightning consists of a cloud of slowly burning, combustible aerosol particles of like charge. They argued that the coexistence of noncombustible products would slow the rate of reaction, and that electrostatic forces would provide stability against a change in shape. Mukharev proposed that ball lightning is an atmospheric electrical phenomenon with an internal source of energy created in an electrified aerosol medium by an exponentially damped electromagnetic field that excites a luminous, spherical, electrode-free, coronal discharge in space. An analysis of this model was provided by Ofuruton et al. (1997c).

Bychkov (1993a,b, 1994) and Bychkov, Bychkov, and Standnik (1996) proposed that some ball lightning may consist of agglomerations of hydrocarbon polymer threads that may become highly charged by interaction with plasmas such as the lightning channel. The source of the polymeric material was postulated as organic molecules, including macromolecules, from decaying vegetation and from artificial dielectrics. Bychkov et al. (1997) reported experiments with erosive discharges in support of this model.

Gaidukov thoroughly studied the gas dynamics of ball lightning. He considered the motion of ball lightning in an air stream flowing out of a narrow opening in a flat screen. Without describing in detail the structure of ball lightning, he assumed that its material is bonded like a liquid, so that it can be characterized by a surface tension. Moreover, molecules of air do not attach to it. Using this model, he was able to account for a large number of effects associated with ball lightning, such as its passage through small holes and slits, and its capture by the trail of airplanes or helicopters. He explained the attraction of ball lightning toward narrow openings and slits in the presence of air propulsion in either direction and its subsequent flow through them as a single, purely hydrodynamic effect. Such an effect is observed when a limited volume of an ideal fluid with moderately large forces of internal tension moves in a viscous fluid of the same density. He indicated, however, that it would be necessary to postulate the absence of a viscous layer at the interface between the ball lightning and the air to explain the effects associated with the sighting of ball lightning (Gaidukov 1988a,b, 1989a,b, 1991a,b, 1992a–f, 1993a,b,e, 1994, 1997).

Smirnov (1987, 1993a, b) described a fractal cluster as a system of entangled fractal fibers formed in a gas by laser or electrical discharge action on a surface. Expansion of the weakly ionized vapor that is formed, which is accompanied by the condensation of vapor on the ions, leads to the formation of solid particles that then join into fractal aggregates. Under the action of an external electric field, the latter, in turn, join into fractal fibers out of which fractal clusters are formed. Experimental studies of ultrafine smoke particles by Forrest and Witten (1979) indicated that the particles quickly form chainlike aggregates. These fractal structures are produced within a few tens of milliseconds after the thermal explosion of materials. Smirnov (1987b,c) proposed that similar aerosols could be formed in a unipolar plasma with a high density of charged dust particles of size less than 2 μm and with a charge density of about 10^9 electron charges per cubic centimeter. This plasma is formed by breakdown of the air between the dust cloud and the ground. Clusters would be formed within about 10 to 100 ms and would be charged in the ionized air by friction.

Smirnov's analysis, based on other ball lightning models and recent scientific and extensive statistical information, suggests that the substance composing ball lightning has a sparse fractal structure. The structure envisaged is a knot of fractal fibers, similar to an aerogel with a rigid skeleton, having the density of a gas and the behavior of a solid or liquid. Its stability would be provided by its charge of about 1 μC, which would prevent its collapse. The magnitude of the associated energy is comparable with the surface energy of water. The surface electric field is about 4 kV cm^{-1} and the total electrical energy about 40 mJ. Energy is provided by chemical reactions.

Luminosity is created by thermal waves that propagate along separate fibers, using the surface energy of the structure. These form glowing hot zones with a temperature of about 2000 K. The aerosol framework could act as a catalyst for initiation of chemical reactions. The color and luminosity would be provided by admixtures of various chemicals, excited by the hot spots. The processes could continue for perhaps 15 min until the clusters discharge in the air, rendering the framework unstable. However, if there is more active matter at the hot spot, there could be earlier explosive decay. The author presented calculations of the drag force experienced by a fractal cluster in its movement through the air. The expression thus obtained supported Gaidukov's conclusion.

Singer (1997) remarked that the formation of a suitable aerogel requires very carefully controlled conditions in the laboratory, let alone in nature. He acknowledged, however, that fine glassy fibers have been observed in volcanic processes in nature, although they did not entirely fulfill the properties described by the aerogel model for ball lightning.

Dijkhuis (1992a,b) claimed log-normal relationships for a number of ball lightning parameters (see Chapter 1) and used these to argue that ball lightning theories must be scaling, like the critical phenomena near phase transitions. Fractal structure is a sure sign of this condition.

Manykin, Ojovand, and Poluectov (1993) offered a model of ball lightning based on Rydberg matter, that is, the low-density metastable state of highly excited atoms or molecules when their valence electrons undergo collective behavior before decay. This matter, they argued, could be formed in the atmosphere after a normal lightning discharge. While there is a high abundance of excited atoms and molecules, fast condensation of excited atoms occurs because the resonance dipole–dipole interaction between excited atoms is stronger than their usual van der Waals forces. The clumps of Rydberg matter, which are held in globular form by surface tension, in many respects would resemble solids, but their density would approximate that of the atmosphere. Rydberg matter is electrically conducting although it can easily be evaporated by a strong electric current that may cause explosive decay. Its energy density is about 1 J cm^{-3}.

12.8 Nuclear Processes

It has been suggested that large electric fields associated with various processes in thunderstorms might accelerate charged particles sufficiently to produce nuclear reactions (Bottlinger 1928). Dauvillier (1957) suggested that lightning might produce thermal (slow) neutrons that react with nitrogen in the atmosphere to produce ^{14}C, a radioactive isotope of which it was suggested ball lightning might be composed. However, the long half-life of this isotope (5570 years) would suggest a slow release of energy. Powell and Finkelstein (1970) remarked that this half-life is excessive. They indicate that while there are positron-emitting isotopes such as ^{12}N (10 min) and ^{15}O (2 min) that have shorter lives, the electron energy required to form them is several tens of megaelectron volts, which is considerably in excess of known electron energies in lightning strokes. They further remarked that this model does not explain the explosive mode of decay of ball lightning.

Altschuler, House, and Hildner (1970) proposed that a lightning discharge might generate protons by dissociation of atmospheric water. If these were accelerated to 1 MeV by storm potentials, despite their short mean-free path (~0.1 μm), they could cause atmospheric oxygen (^{16}O) and nitrogen (^{14}N) to form the radioactive isotopes ^{17}F (half-life 66 s) and ^{15}O (half-life 124 s) respectively, both of which are positron emitters. The annihilation of positrons with electrons in the air would produce gamma radiation. This would produce an intensely radioactive fireball, although not one of exceptional luminosity. Radial electrostatic fields caused by the emission of positrons and the subsequent annihilation process might achieve the confinement of isotopes during the formation of the ball.

These theories predict that ball lightning is radioactive. The absence of symptoms of exposure to ionizing radiation in witnesses who have reported ball lightning at close proximity offers some evidence against this (Dmitriev 1973a,b; Wooding 1976). Hill and Sowby (1970) pointed out that the radiation dose on a body at a distance of 2 m from the nuclear fireball proposed by Altschuler, House, and Hildner would be 175 rad s^{-1} from ^{15}O and 325 rad s^{-1} from ^{17}F. This would be lethal, and would also cause thermoluminescence from nearby solid objects. A study by Mills (1971) of possible thermoluminescence from a church steeple that had been struck by lightning, possibly ball lightning, yielded a null result. Other tests by Fleming and Aitken (1975) for thermoluminescence of brickwork following a nearby ball lightning observation provided data insignificantly different from control data.

Singer (1971) comments that charged particles would lose energy by collision in the atmosphere and thus not be accelerated to sufficiently high velocities to allow nuclear reactions to occur. He also indicates that peak lightning temperatures are insufficient to provide the plasma conditions necessary for thermonuclear fusion.

Ashby and Whitehead (1971) reported experiments carried out over a period of 1 year to test the hypothesis that ball lightning is caused by the annihilation of tiny fragments of meteoritic antimatter, which they postulated would come from the upper atmosphere. The hypothesis was intended to explain the high energy of ball lightning and its entry into buildings and aircraft. A very small abundance of antimatter was required to explain the frequency with which ball lightning is reported. A potential barrier was hypothesized to exist between antimatter and matter, so that annihilation would not occur when the relative velocities were low. Ashby and Whitehead suggested that emission of positrons from antimatter micrometeorites by photoemission, and by secondary emission caused by recoiling fragments, would cause the dust particles to become negatively charged. These would then be repelled and accelerated toward the ground by the foul-weather electric field, thus gaining sufficient energy to overcome the postulated potential barrier. Annihilation of the antimatter on contact with matter would provide the energy of formation of ball lightning, which would be observed as a glowing region whose luminosity was provided by ionization. A microscopic particle of a radius of 5 μm and a mass of 500 pg would provide 0.5 MJ of energy, including 511 keV gamma rays from positron–electron annihilation. The experiments were designed to detect such radiation, and recorded four events of interest, one of which was directly correlated to a severe thunderstorm.

Jennison (1971) remarked that the existence of micrometeorites of antimatter would imply the presence in the solar system of large bodies composed exclusively of antimatter. Crawford (1972) suggested that these events may have been caused by extensive air showers of cosmic rays corresponding to primaries of energy $\sim 10^{16}$ eV. However, Cecchini, Diocco, and Mandolesi (1974) indicated that the observed duration and frequency of these events were excessive compared with the known properties of air showers of the required energies. The cause of the events seemingly remains unexplained.

12.9 Charge Separation

De Tessan (1859) proposed a structure for ball lightning based on the Leyden jar. (The Leyden jar was an early form of capacitor invented in the Dutch town of Leyden about 1745. It consisted of a glass jar with layers of metal foil on the inside and outside. Contact to the inner foil was by means of a loose chain hanging inside the jar.) In the ball lightning model, a spherical shell of dry air replaced the insulating glass. Opposite charges that formed on either side of the shell compressed the air. The ball was considered to be in stable equilibrium: radial forces due to atmospheric pressure, the coulomb force between opposite charges, the consequent increase in pressure of the shell, and the reduced pressure at the center of the ball were thought to be balanced. Gradual recombination of opposite charges through the imperfect dielectric shell, producing ozone, would provide visible light. The structure would be stable against contact with the ground because this would merely remove a small amount of free charge from the outer surface, but penetration by a conductor would cause explosive decay by discharge between the oppositely charged regions. Overall electrical neutrality would prevent the ball from being attracted by conductors. This model can be rejected and considered of merely historical interest in the absence of a proposed mechanism for separation of charges where recombination and other deionization processes would be expected to occur over a very short time scale (Singer 1971).

Pühringer (1965, 1967) proposed a similar model. This followed an earlier suggestion by Walter (1909a,b), who thought that ball lightning might be a charged water bubble. This was a response to an observation described by Sauter (1890, 1892a–c, 1895). Experimental studies gave no indication of luminosity from such bubbles (Walter 1909a,b).

Pühringer suggested that ball lightning consists of a sphere of positively and negatively charged water droplets, with all the particles of one polarity of charge in the center and those of opposite polarity forming the outer shell. Vaporized water acts as an insulating layer between the two charged regions and slows the rate of recombination. Singer (1971) offered objections similar to those he raised for the de Tessan model.

Sall (1992a,b) hypothesized that conversion of the heat of condensation of water vapor into the radiation of a streamer corona discharge takes place in the volume of ball lightning. The efficiency of this process is high due to heat regeneration. He showed that the diverse colors of ball lightning could be connected with the polarity of the streamer corona and with scattering of light by a water-droplet aerosol.

Singer (1971) goes on to make a general criticism that such models have been unsuccessful in demonstrating that charged particles can be generated and confined in a spherical form in sufficient concentrations to account for ball lightning. No adequate explanation has been offered to suggest how charge separation could be maintained for several seconds against such deionizing processes as recombination or diffusion.

12.10 Ions

These models suggest that ball lightning is composed of molecular ions of gases. Reynolds (1931) first made this suggestion. He thought that the inside of the ball was an attenuated, charged gas, likening it to an evacuated filament lamp. This could explain apparently violent explosive decay indoors with no damage.

The distinction between ion and plasma models is hazy, but Singer (1971) points out that the former are characterized by low temperature, low charge density, or the absence of free electrons. They do not fulfill all the necessary criteria to be defined as a plasma (Chapter 11); hence many theoretical approaches of plasma physics are inapplicable here.

Hill (1960) concluded on the basis of rapid recombination times (a few milliseconds, mainly due to removal of free electrons captured by oxygen to form negative oxygen ions) and the difficulty of confining atmospheric plasmas that ball lightning is not a plasmoid. Hill acknowledged that there might be several different physical mechanisms responsible for phenomena reported as ball lightning. He thought that rather than a plasmoid, it was more likely that ball lightning is a region containing a strongly nonuniform distribution of space charge in the form of highly ionized gas, the ionization being mainly found in molecular ions, ionic clusters, etc., with few free electrons. The form of the electrical charge present and the absence of electrons were expected to increase recombination time. Inhomogeneity could be enhanced by the presence of dust, water vapor, and other foreign material. He likened this model to a "miniature thundercloud." A large fraction of the energy content of the ball would be in the form of ionization. Relatively weak electrostatic forces would confine the ball, and luminosity would come mainly from internal corona discharge, from molecular recombination, ordinary combustion, etc. These processes would be very slow compared with recombination times for a plasma and would be associated with quiet disappearance. Contact with metals would cause premature decay by discharge between space charge regions. A rapid increase in internal discharge rate or a mechanical shock wave could produce explosive decay.

Hill attributed the rarity of ball lightning to the difficulty of producing the necessary initial conditions for formation of the ball. He thought that the high degree of ionization required at low kinetic temperature might be achieved if there is a relatively extended transition period in which the energy input of the original stroke diminishes slowly. He later remarked (1970) that the total energy stored would depend on the degree of ionization, which would probably vary from one case to another. The observed radiation would not depend on a high temperature because it is recombination radiation in molecular plasma.

Powell and Finkelstein (1970) remarked that Hill's "miniature thundercloud" could not exceed the above limitations on energy imposed by the virial theorem (an upper limit of 100 J in 1000 cm^3), and that there was no way of accounting for continued separation of charges in moist air. Hill (1970) replied, writing that the stored ionization energy proposed by the model would not be subject to constraints imposed by the virial theorem on magnetohydrodynamic considerations. Powell and Finkelstein (1970) agreed, but said they had discounted ionization energy because of the very rapid rate of recombination.

Crew (1972) suggested a role for water in explaining charge separation in the Hill (1960) model. Electromagnetic forces compress air within the lightning channel so that it is sheathed with a region of low pressure. If the surrounding air is humid, drops of water will condense in this sheath. They are repelled by radiation pressure. Sonic shock or other effects then shatter the drops so that they become charged. Crew speculates that smaller drops, which are mainly negatively charged, are accelerated farther from the channel than larger, mostly positive drops. The water then evaporates from the heat of the flash, leaving an inner sheath of positive charge and an outer sheath of negative charge that would act as a Faraday cage except at the ends or at a discontinuity such as a bend in the channel. Here a discharge could begin and progress along the channel with fairly steady brightness.

Charman (1979) expresses the view that Crew's mechanism does not explain why luminous spheres are not formed along the entire lightning channel.

Singer (1971) commented that it had not been demonstrated that attachment of electrons to form negative ions substantially reduces the rate of recombination. Rather, he suggests that recombination between ions of opposite charge may occur more rapidly than between ions and electrons. Therefore, he reasoned, it would be necessary for solid particles or water droplets present to have a large inhibiting effect on recombination. He commented too on the problem of providing high degrees of ionization with large ionization energies at low temperatures.

Stakhanov (1973, 1974a,b, 1975, 1976a,b) suggested that recombination could be inhibited if ions were surrounded by completely filled hydrate shells consisting of water molecules. The water molecules have a large dipole moment and are attracted to positive and negative ions, forming solvate shells around them. The luminosity of ball lightning would be caused by recombination of the hydrated ions (clusters). Two different mechanisms for cluster recombination are possible, depending on heat-transfer conditions: one is explosive and the other is relatively slow. Mesenyashin (1993) also suggested that ball lightning is an electrically charged shell of water molecules.

Chapter 13

Models Based on an External Energy Source

External energy models have a number of advantages over internal ones. They do not necessarily encounter difficulties with the virial constraint discussed in the previous chapter because they do not require a high energy density in order to have a long duration (Endean 1997a,b). Effectively, a ball powered by the electric field of a thunderstorm has a large energy resource at its disposal and, simplistically, could survive for as long as the electric field was elevated. However, Endean indicates that these models need to account for high-energy ball lightning events with explosive decay and that there are formidable problems in arranging the required high input power while continuing to maintain equilibrium.

13.1 Direct Current Discharge Models

Many experiments in the nineteenth and early twentieth centuries investigated the properties of electrical discharges at atmospheric pressure and below. Comprehensive reviews of these early experiments were presented by Singer (1971) and Barry (1980a). Many early investigators, notably Planté (1860, 1868, 1872, 1873, 1875a–f, 1876a–j, 1877a–d, 1878a–c, 1879a,b, 1884a–i, 1885a,b, 1887, 1888, 1890, 1891, 1901), attempted to relate their experimental results to descriptions of ball lightning, but it was not until a firm theoretical basis was established by Townsend (1900, 1901, 1910) that the results of these experiments could be given physical interpretation. Nauer (1953, 1956) repeated and extended much of this early work. Nasser (1971) gives a very useful review of relevant theory. Beynon (1972) provides a simple, mostly qualitative account. Rees (1973) collected many relevant historical papers.

Finkelstein and Rubenstein (1964) concluded that ball lightning must be fed by an external source of energy—the dc electric fields associated with thunderstorms—through electric currents in the air. In eliminating other models, they used the virial theorem (see Chapter 12) to show that confinement of a plasma by self-field alone was consistent with conservation laws for energy and momentum only in the presence of air pressure. However, the latter condition limited the energy content to the order of kilojoules and would require the presence of a layer insulating the ball from the air, and even then, the extended lifetime of ball lightning could not be explained. They deduced from the "tub-of-water" event (Morris 1936, Goodlet 1937) that the energy of ball lightning must be much greater.

Finkelstein and Rubenstein envisaged a ball that was a luminous, localized glow discharge surrounded by a Townsend discharge. This would require small currents. Using continuity of charge, Ohm's law, and the equation of stationary flow, and assuming conductivity to be a function of the square of the electric field, they derived a nonlinear function for conductivity. They assumed a constant low conductivity for the Townsend region and a constant higher conductivity for the glow discharge region. One solution of the nonlinear equation would be a nonlocalized Townsend discharge, but another would be a spherical discharge with a uniform field surrounded by a Townsend regime. The field would be that of a dipole parallel to the applied field. The spherical region would cause convergence of electric field lines and current that would thus maintain its conductivity. Fairly gentle convection currents could carry away the energy dissipated in the outer Townsend régime. The value of the external field would be fairly critical; if it increased sufficiently there would be breakdown (presumably ordinary lightning) through the region, whereas if it decreased, the ball would disappear quietly. This model receives strong support from the many ball lightning observations that terminate with a CG flash through the ball, or where damage following apparently explosive decay of ball lightning is of a magnitude that suggests its cause as the current of ordinary lightning. Hubert (1996) described this model as a variant of St. Elmo's fire. He pointed out that traces of organic material might enhance the corona effect. He suggested further that the phenomenon described by the Finkelstein–Rubenstein model might be observed more frequently inside old houses because more modern buildings often have a metallic superstructure that might serve as a kind of Faraday cage.

Powell and Finkelstein (1970) revived the above model. They considered that the model based on heated air [Lowke, Uman, and Liebemann (1969); see Chapter 12] and models based on metastable molecular electronic states were compatible with basic laws of physics. Powell et al. (1966) proposed a model using N_2 and O_2 in metastable, excited states that require lower temperatures for energy storage or light emission than other models.

However, both the heated air and metastable models predict stability for about a second or less, together with upward convection at about 1.2 m s^{-1}. Powell and

Finkelstein pointed out that the problem of convection would also occur with the Finkelstein and Rubenstein model unless local field anomalies held the maximum field at one position in space. Powell and Finkelstein demonstrated experimentally that if this were so, large power inputs, e.g., 10 kW for a 20-cm ball, would be required to compensate for energy lost by convective flow through the ball.

Powell and Finkelstein therefore proposed a model involving the following stages. This was based on their experimental work involving radio-frequency discharges, described later. (1) Ball lightning is residual from a CG or CC lightning flash, or is formed from a glow discharge (St. Elmo's fire). The latter might be expected where ball lightning is formed remote from ordinary lightning, e.g., when the ball appears near a structure indoors. (2) In the absence of external energy input, the ball lasts from 0.5 to 1.0 s. It is powered by internal energy (electronic excitation or dissociation of gas molecules). (3) If the field following the stroke is 100 to 200 kV m^{-1}, Townsend multiplication would continually generate positive and negative ions and electrons. The positive and negative charges would leave the ball in opposite directions, with positive ions moving downward, because of the external electric field. The greater mobility of electrons would leave the ball positively charged. Thus, in contrast to the dipole of the Finkelstein and Rubenstein model, this model describes a monopole.

The ball would consist of a central region at a fairly uniform temperature where Townsend multiplication occurs and from where light would be emitted. An intermediate region with a radial temperature gradient would surround this. This, in turn, would be surrounded by ambient air. For a given set of conditions, the model predicts a constant size of ball. (4) The positive space charge cloud leaving the ball would channel further electrons into it from a surrounding region many times its dimensions.

Processes (3) and (4) can maintain the ball for as long as the poststroke field continues. If the poststroke field is downward, it exerts a downward force on the ball, which, in equilibrium, could balance convective forces, cause the ball to move downward, or even make it move against the wind. The absence of convection would reduce thermal convective mixing and heat loss from the ball. The positively charged ball would be attracted to conductors. If the ball contacts a conductor, this may cause rapid heating and breakdown of the air and thus initiate a lightning flash. Alternatively, if the concentration of metastable molecules exceeds that for thermal equilibrium, there may be a chain reaction, causing rapid release of energy. In either case, explosive decay could result.

An interesting feature of this model is its suggestion of how ball lightning could pass intact through windows. Initially, electric field lines would pass freely through the glass. As a result, a positive charge would accumulate on one side of the glass while electrons would accumulate on the other. When the ball approaches, the glass is heated or breaks down sufficiently to become electrically conducting. The glass then serves as an electrode, and a ball, which can then detach itself from the window, is formed inside the room.

Because, as discussed in Chapter 7, predominantly metal aircraft are such effective Faraday cages, dc discharge models have difficulty in explaining formation of ball lightning within aircraft. Although dc fields can penetrate the fuselage through the side windows because they are not coated or impregnated with electrically conducting material, the electric field obtained near such an aperture within the fuselage is very low compared with the field outside. A discharge outside will occur well before a discharge inside. (P. Laroche, 1998, personal communication.)

Although they agreed that it was difficult to imagine a strong dc electric field penetrating very far into an aluminum aircraft through a window, Powell and Finkelstein suggested that the above mechanism would enable a ball to be formed inside the aircraft, which would then be cut off from external power and decay within a second or so.

Fractal discharge, a form of electrical discharge that contains fractal aggregates, was discussed by Smirnov (1993c) as an analog for ball lightning (see Chapter 12). He related the properties of such discharges to the results of the Powell–Finkelstein experiments.

Turner (1994) described a hybrid model involving electric fields and chemical reactions, in which ball lightning can be considered as a thermochemical heat pump powered by the electric field of a thunderstorm. The ball consists of a central plasma core surrounded by a cooler, intermediate zone, in which recombination of most or all of the high-energy ions takes place. A zone in which temperatures are low enough for any ions present to become extensively hydrated surrounds the intermediate zone. Hydrated ions can also form spontaneously in the inner, hotter, parts of this hydration zone. Essential to the model is a region near the surface of the ball in which thermochemical refrigeration can take place. Once the ball is established, energy is supplied not only by electric fields and, possibly, electromagnetic fields, but also by the production of nitric acid from nitrogen and oxygen, and by hydration of ions. He showed that the reaction will be endothermic and can refrigerate its surface if N_2O^- and H_3O^+ ions become hydrated by more than about five water molecules before they can combine at the edge of the ball. The surface refrigeration allows the condensation of water in quantities sufficient to counteract the buoyancy of the hot plasma. The in-flow of N_2 and O_2 produces both nitrous and nitric acids, the latter being dissolved in the water droplets. The flow of gas inward past these droplets (and past those condensed around an excess of H_3O^+ ions) provides an effective surface tension for the ball that could explain its shape and mechanical stability.

Lowke (1996) proposed an electrical glow-discharge model of ball lightning, continuously varying on a microsecond scale. This coronalike discharge is sustained by electric fields resulting from dispersal of charges from a lightning strike along preferred conducting paths in the earth following a ground flash (Fig. 13.1).

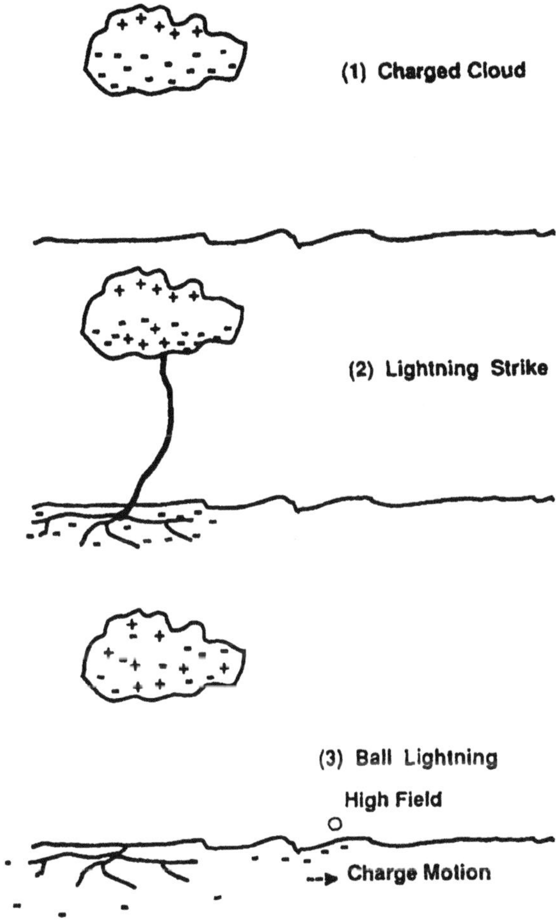

Figure 13.1. Ball lightning model due to Lowke (1996). Electric charge redistribution, during and following a lightning strike. [Reproduced from J. Lowke, "A theory of ball lightning as an electron discharge," *J. Phys. D (Appl. Phys.)* **29**(5) 1237, 1996 with the permission of IOP Publishing, Ltd.]

Arc channels and fulgurites, both of which would be highly conducting immediately after the flash, would provide the preferred paths, and the higher conductivity of water would increase the amount of charge near the surface. Charge dispersal would continue until an insulator was encountered. The dispersing charges, together with space charge distortions by ions of both polarities, would produce a highly enhanced electric field of at least 500 kV m^{-1} about 1 m above the soil, which would produce the luminosity observed as ball lightning. Although measured values of ground potential usually do not exceed 200 kV m^{-1}, it was suggested that a space charge would shield the field of the ball. The motion of the ball would be caused by changes in potential distribution over the earth's surface, for example, by a nearby subsequent lightning flash. Once generated, the ball could be sustained at lower fields. Pulsation occurs because of space-charge effects. Convective cooling owing to motion of the ball would help to maintain the glow discharge.

Explosive decay of ball lightning could be explained by formation of an arc if further ionization led to heating of the glow discharge, or by an encounter with a solid obstacle that disrupts convective flow. The ball would produce radio noise with a frequency on the order of megahertz. Dissociation of air would produce ozone and oxides of nitrogen. The motion of the charge filament in the soil beneath a window would allow ball lightning to re-form on the other side of the window. The appearance of ball lightning inside aircraft was attributed to the presence of high fields, but this aspect of the theory was not developed further.

13.2 Microwave Resonance Models

P. L. Kapitsa's (1955) deduction that a sphere of typical ball lightning dimensions (~10 cm) would glow for a maximum of 10 ms (Section 11.6) led him to the conclusion that ball lightning must be fed by an external source of energy. He proposed that this was by resonant absorption of intense radio standing waves in the microwave region. Similar suggestions that ball lightning was caused by stationary waves had been made previously by Lodge (de Jans 1912a–c), Marchant (1930), Cerillo (1943), and others, although it was Kapitsa who developed the suggestion much more formally. It has been known for some time that plasmas can absorb energy by a variety of resonance mechanisms with radiofrequency waves in the range of 100 kHz to 200 GHz (Riviere, Chapter 22 in Gill, 1981).

In developing a model based on this proposal, Kapitsa endeavored to account for the constancy of size and the absence of convective behavior of ball lightning. Conditions for resonance would depend solely on the external dimensions of the ball. The relationship between the wavelength λ and the ball diameter d is given by the characteristic oscillations of a sphere; thus, $\lambda = 3.65d$, unless ionization is weak, when $\lambda > 3.65d$. The initial scenario was the excitation of a small volume V of weakly ionized plasma, such that V is much less than $\pi d^3/6$, where d is the final diameter, by absorption of radio waves according to the resonance condition. This excitation increases the degree of ionization, and the volume of the region grows until the diameter of the ball stabilizes as d. If an increase in temperature causes expansion, there will be a deviation from the resonance condition, which will reduce the efficiency of absorption; hence the ball will cool and contract to resonance diameter. This negative feedback process maintains a stable ball size. We may thus deduce from modal reported diameters of ball lightning (10–50 cm) that wavelengths would lie in the range 37 to 183 cm; hence frequencies, from $f = c/\lambda$, would lie in the range of approximately 160 to 820 MHz. Kapitsa argued that ball lightning would be formed at antinodes in a standing-wave pattern where intensity would be greatest. The motion of the ball would follow the motion of the antinode and would not exhibit convection or depend on wind direction. If the standing wave were set up by reflection along a normal to the soil's surface, antinodes would occur on surfaces parallel to the soil at heights of $(2n + 1)\lambda/4$, where $n = 0, 1, 2. \ldots$ Near these surfaces, the raised electromagnetic field intensity would provide the necessary initial conditions and conditions for maintenance of the ball. Kapitsa argued that the ball would most often be formed close to the ground's surface at a height of $\lambda/4$, i.e., a distance from the reflecting surface and the edge of the ball equal to its radius. If similar effects were produced at higher antinodes, bead lightning might be formed. Silent decay of the ball would occur if the energy supply to the ball was terminated and the ball radiated its energy slowly. Explosive decay would occur if rapid cooling produced a weak shock wave as the sphere was filled with air.

Watson (1960) used the Mathieu equations to analyze the confinement of charged particles by a polarized electromagnetic standing wave and predicted that the particles would be bound near the nodes, not near the antinodes as Kapitsa had stated, because there was no solution of the equations at the antinodes. Tonks (1960) demonstrated that although initial ionization would be at an antinode, the ball would find an equilibrium position at a node because of radiation and atmospheric pressure. Electrons in the node region would cause further ionization by collision, provided the field strength was on the order of 1 MV m^{-1}. Experimental studies by Babat (1947) had already demonstrated the possibility of producing spherical, electrodeless plasmoid discharges at both nodes and antinodes.

Tonks (1960) and Singer (1971) discuss the difficulty of radiation pressure from an external electromagnetic wave preventing convection of a ball 10 cm in diameter with a temperature of several thousand kelvins, since the buoyancy force would be about 6 mN. It was suggested that the additional coulomb force on a charged ball attracting it to a grounded conductor might provide an additional force to stabilize against buoyancy. Tonks pointed out that the model required a substantial input power of about 0.1 MW together with a very high stability in the frequency of the generating radio frequency radiation if the ball was to be maintained stationary for some time.

Kapitsa explained the appearance of ball lightning in buildings by suggesting that an aperture such as a chimney could serve as a waveguide. It is not clear in his paper whether Kapitsa thought that a similar explanation could apply to the appearance of ball lightning inside aircraft. He refers to such an observation, then in the next sentence, writes: "In terms of our hypothesis, all these effects are explained thusly: lightning balls penetrate into closed buildings by virtue of the fact that they follow the path of short-wave electromagnetic oscillations which can propagate through apertures or along a chimney or conductor as along a waveguide." He does not develop the question of aircraft further (Kapitsa 1955).

Altschuler, House, and Hildner (1970) remarked that models based on focusing of microwaves or of electric currents into a small volume could not explain ball lightning that enters metallic enclosures such as aircraft. However, Laroche (personal communication, 1998) indicates that VHF and UHF radiofrequency fields, possibly in the frequency range of interest, can penetrate aircraft through apertures. While the main resonances within the aircraft will probably occur at lower frequencies, lower intensity resonances at higher frequencies are possible.

It was remarked in Chapter 7 that corona discharges from the extremities of aircraft, often visible as St. Elmo's fire, may generate radio interference in the gigahertz range (Boulay 1982). The majority of strikes to aircraft follow observations of corona and/or high-voltage streamers, which may then develop into leaders that allow electrical connection between charge centers in clouds. Continuing currents of 100 A or more may then flow through the aircraft, feeding the leader for periods of up to hundreds of milliseconds. These currents can produce substantial magnetic fields, with rapid rates of change that can in turn induce large voltages and currents in nearby conductors.

A lightning strike injects sufficient energy into the aircraft structure in this frequency range to excite electromagnetic resonances. For an F-106-B aircraft, these are in the high-frequency range of about 7 to 23 MHz, the lowest frequency corresponding to the fuselage half-wavelength resonance and the highest corresponding to that of the wings, with intermediate frequencies probably corresponding to a combination of the wings and tail. In laboratory scale-model tests, there was also a component at 39.8 MHz that was thought to be the second harmonic of the wing resonance (Trost and Pitts 1982). These measured frequencies fall short of those in Kapitsa's model by a factor of at least 4.

Powell and Finkelstein (1970) describe experiments by Powell et al. 1965, 1966, 1967 in which long-lived (0.5 s), bright luminosities were produced in a 250-cm resonant cavity by radiofrequency electromagnetic waves at a frequency of 75 MHz. They remarked that an aircraft fuselage provides a natural resonant cavity for radiofrequency radiation. In repeating Manwaring's experimental work, Powell and Finkelstein produced luminosities in a 15-cm Pyrex tube whose luminosity persisted for about 1 s after the power was switched off. In the open air, the luminosities persisted for about half as long, probably owing to convective mixing. These luminosities depended on gas pressure, electrode composition, and gas composition, the latter having the most significant effect. The experiment would only work with N_2, O_2, with mixtures of these, and with N_2O. Color ranged from dull blue with nitrogen, through yellow-white with intermediate brightness in air, through pure orange with N_2O, to white and very luminous with oxygen. Luminosity with N_2O persisted for up to 2 s and may have resulted from decomposition. In the other cases, the persistence of luminosity was attributed to metastable excitation of nitrogen and oxygen molecules. Metastable molecular species of nitrogen and oxygen are shown in Table 13.1. Species 1, 5, and 6 in the tables were considered to be those most likely responsible for the processes.

Kapitsa related formation of ball lightning near the end of a storm immediately after a flash of conventional lightning to the presence of ionized air that promoted the generation of radio waves and to the stimulation of oscillation by the lightning discharge. He suggested that the source of these oscillations was an oscillatory process accompanying the ionization of the atmosphere near the ground or the earth. If it were the latter, the zone of intense radio emission would be of limited extent.

Until recently, radio emissions from natural lightning in the 160- to 820-MHz range were considered to be very weak (Silberg 1961a,b, 1962; Powell and Finkelstein 1970). Pierce (1977a) stated that above 5 kHz, signal amplitudes radiated from lightning flashes have an approximately inverse dependence on frequency. Kapitsa suggested that since ball lightning is a rare phenomenon, the occurrence of corresponding radio emissions might also be rare. Subsequent field investigations of electromagnetic radiation from lightning revealed relatively narrow-band radiation at frequencies of about 0.1 to 0.2 GHz, occurring 0.1–0.4 s after the appearance of a leader stroke. The average duration of these oscillations was about 50 ns. The power spectral density (1 to 100 pW m^{-2} Hz^{-1} at a distance of 1 km from lightning) was too low to generate a plasmoid of the kind described by Kapitsa, but the investigators speculated that the power might be much greater near the ionized lightning channel. The source of the radiation was speculated as transverse plasma resonances in the lightning channel, where the magnetic field of the lightning channel amplifies magnetic Bremsstrahlung and Cerenkov radiation from electrons in the channel (Kapitsa 1969a; Kosarev et al., 1968, 1969; Ranjeloric 1969; Kosarev and Sereszhkin 1974a,b).

Since Kapitsa's work in this field, investigators have discovered lightning discharges, called superbolts (see Chapter 2), with exceptionally high optical powers (Turman 1977). These may be positive discharges at the extreme end of the energy distribution. Their radio emissions are not known and there is at present no

Table 13.1. Characteristics of Metastable Molecule Species

Species	Energy (eV)	Radiative lifetime
1. $N_2(A^3\Sigma_u^+)$	6.17	13 s
2. $N_2(^3\Delta_u)$	7.35	1 s
3. $N_2(\omega^1\Delta_u)$	8.89	1 s
4. $O_2(A^3\Sigma_u^+)$	4.43	10 s
5. $O_2(b^1\Sigma_g^+)$	1.63	8 s
6. $O_2(a^1\Delta_g^+)$	0.98	45 min

Source: Powell and Finkelstein (1970).

way to extrapolate radio emissions from optical emissions. They may, however, be possible generators for "Kapitsa waves" and it would therefore be interesting if radio spectrum data could be obtained for superbolts. Uman (1987) states that (1) positive lightning occurs more frequently in winter thunderstorms; (2) positive lightning often occurs toward the end of a storm; and (3) the median energy of positive lightning is about 7 times greater than that of negative lightning. Although Rayle's (1966) survey did not agree, Brand (1923) noted that ball lightning occurs more frequently in winter than in summer and frequently occurs toward the end of a storm.

Le Vine (1980) identified the sources of the strongest radiofrequency emissions from lightning at 3, 139, and 295 MHz as narrow bipolar pulses with an initial polarity opposite to that of return strokes that lower negative charge, thus known as narrow positive bipolar pulses (NPBPs). Willett, Bailey, and Krider (1989) reported the first wideband recordings of these pulses and determined that they radiated much more strongly than first return strokes, with more high-frequency content at frequencies from 10 to at least 50 MHz. They generally occurred as relatively isolated and infrequent pulses in intracloud flashes, but they were not associated with *K* changes or other known phenomena. They sometimes occurred either before or after the first return stroke in a CG flash. The authors identified their potential risk to aerospace vehicles whose fundamental structural resonances usually lie at high frequencies. They also determined that these pulses could occur with the opposite initial polarity.

The "Blackbeard" detector on the ALEXIS satellite detected exceptionally intense pairs of brief (a few microseconds), noiselike VHF pulses separated by a few tens of microseconds in a passband from about 25 to 100 MHz. Both pulses were equally dispersed, which is consistent with their propagation through the Earth's ionosphere. These are called transionospheric pulse pairs (Massey and Holden 1995). Although these pulses are thought to originate from thunderstorm regions, they are significantly more intense than the atmospherics produced by normal lightning activity, and may represent the tail of "normal" VHF emissions from lightning. Four possible reasons were suggested why ground-based stations have not reported corresponding observations. (1) TIPPs are a rare phenomenon that has thus not been identified as a special class of radiofrequency signal. (2) They are a rare but energetic phenomenon that is more easily detected from space than from the ground. (3) They have a different waveform when observed from space than when observed from the ground. (4) They are emitted by directional sources radiating more energy toward space than toward the ground.

TIPPs were among the signals later detected by the Forté satellite, launched in August 1997. Their instantaneous power was at least ten times greater than that of VHF signals normally associated with lightning. Although some investigators had hoped that TIPPs might provide supportive evidence for high-altitude, upward discharges related to sprites and similar phenomena, Jacobson et al. (1998) confirmed that these signals were consistent with the second pulse being generated by a ground reflection. Nonetheless, these experiments offer evidence for the existence of more intense, albeit transient, radiofrequency signals in the Kapitsa range than had previously been detected, and for the existence of both a transmitted and reflected signal as required by the Kapitsa model.

While lightning-generated radio noise at low frequencies in the atmosphere is dominated by contributions from ground flashes, above 50 kHz the influence of cloud flashes becomes significant. Between 1.5 and 12 MHz, the amplitude ratio of return stroke to cloud stroke is approximately 1:1 (Brook and Ogawa, 1977). Thus, cloud flashes should also be considered a possible source of "Kapitsa radiation." Many strikes to aircraft are apparently intracloud strikes rather than strikes from the more severe cloud-to-ground strikes (Chapter 7).

Silberg (1961a,b, 1963) was critical of some aspects of Kapitsa's model, chiefly the lack of specific detail about why this phenomenon would be selective in responding to specific radiofrequencies from the broad-band, damped emissions of a storm, and about mechanisms for plasmoid generation and containment. However, Silberg considered Kapitsa's model worthy of further development. In further evolving the model, he treated the earth as a perfectly conducting, extended plane surface and assumed that intracloud or intercloud discharges generated a radiofrequency field with a flat, discrete-band spectrum. The assumption of a discrete-band spectrum helped to circumvent the difficulty of frequency stability discussed by Tonks. Because of the dipolelike nature of the radiator, radiation from such a discharge would be linearly polarized and could be treated as plane waves in the region in question. His analysis demonstrated that under rare conditions, the electric field thus produced could support a horizontal lightning streamer and possibly a localized corona discharge. However, he pointed out that the existence of a discrete-band spectrum has not been experimentally verified. Shapiro and Watson (1963) demonstrated the possibility of using three orthogonal standing waves to confine particles.

Berger (1973) remarked that Kapitsa's theory should logically predict the generation of ball lightning-type phenomena in the vicinity of shortwave radio transmitters. However, Singer (1977) pointed out that "Kapitsa waves" have a short wavelength compared with normal transmission wavelengths, and that in the latter case, waves would disperse and become too diffuse to produce such phenomena.

Handel (1975, 1988, 1989, 1997) and Handel and Leitner (1994) proposed a maser–caviton ball lightning model in which ball lightning is a nonlinear, localized high-field soliton, known as a high-pressure caviton, forming a cavity surrounded by plasma. The source of VHF energy in the model is an atmospheric maser. The rotational energy levels of water molecules include many closely spaced pairs linked by forbidden transitions in the VHF region. It was suggested that following a sudden and extreme electric field excursion to about 10 kV m^{-1} caused by ordinary lightning, some of the forbidden transitions would be weakly allowed and a population inversion could be induced. The low rate of forbidden transitions would allow the maser action to continue for several seconds. Collisional deexcitation of molecules would be in competition with the stimulated emission. In a high-gain field mode, the latter could dominate.

The ball is formed at an antinode of the maser-generated standing wave, as in Kapitsa's model. If at a certain point the maser frequency equaled the local plasma ion frequency, resonance would increase the electric field and pump out ions from the high-field region, thus creating a resonant cavity that would lead to the formation of the caviton. Such a caviton has been shown to be a stable, quasi-spherical configuration of trapped electromagnetic field, surrounded by plasma, and moving horizontally at the level of the antinode.

Above flat ground, equipotential surfaces are planes and the atmospheric maser would therefore occupy a volume of several cubic kilometers for ball lightning formed in the open air. In cases where ball lightning was electrically shielded, the maser would occupy the space in which it was formed. In the former scenario, the sudden termination of the caviton would lead to spiking of the maser, especially if the ball is disturbed before the population inversion has been exhausted. This would be observed as a large release of energy like a large spark, but different from ordinary lightning because it is associated with large electric fields rather than large currents, and interpreted as a powerful explosion. In the open air, this electric field spike produces large ponderomotive forces on metallic objects and throws dielectric objects around. In the latter situation, the small volume of the maser would limit the stored energy to a few hundred joules, so decay of the caviton in closed spaces such as houses and aircraft would not be associated with significant energy release or explosions.

Because the model requires an extended, uniform electric field, ball lightning would not be expected to form on mountaintops. (Here, the population inversion would be limited to a cone-shaped volume.) This could explain Berger's (1973) failure to observe ball lightning in many years of lightning observation on Mt. San Salvatore, and the failure of Kapitsa's expedition to detect "Kapitsa radiation" in the Pamir mountains of central Asia.

Although the population inversion will become quickly depleted at the antinodes, it is unaffected at the nodes, and this can lead to continuous regeneration of the population inversion by the migration of Döppler shifted atoms into the spectral resonance window and by gradual air movement from nodes to antinodes.

The authors tentatively claimed that this model is unique in consistently explaining the main, well-known features of ball lightning. These include: (1) its appearance immediately after a strong electric field pulse usually caused by lightning; (2) its passage through closed windows and other dielectrics; (3) its always harmless existence in electrically shielded (e.g., metallic) enclosures, without the possibility of electrostatic-explosive demise; (4) its total absence in the vicinity of high peaks or lightning observation stations, as well as the resonant character of its positioning and motion with respect to conducting bodies; (5) its apparent lack of buoyancy in the air.

Further analysis (Handel and Leitner 1994) revealed two stable conditions, one leading to the formation of hotter, white ball lightning, and the other to the formation of colder, orange ball lightning.

Experimental support came from investigations of moist air in a wind tunnel passing between the plates of a capacitor with 80 kV across it. It was found that fast-moving, humid air passing through a large electric field generated VHF electromagnetic waves at frequencies of 207 and 247 MHz. This was explained as maser emission from the rotational levels of water molecules. Its authors also used the experiments (see later discussion) of Ohtsuki and Ofuruton to support this model.

Rice-Evans (1982) reported unexpected observations made while Rice-Evans and Franco (1980) were developing a streamer chamber for studies of cosmic rays. When the electrode configuration was at a positive point 15 cm above a grounded plate and the chamber was filled with air, a well-defined, stationary, stable, oscillating luminous cone was observed with a period of about 170 ns and hence a frequency of about 5.9 MHz. This emitted light in synchronization with the oscillations of the applied voltage. The shape of the discharge suggested self-focusing. The investigators were unable to explain the localized oscillatory condition using classical theory, but postulated that the combination of fluctuations in current and voltage, including space-charge contributions, resulted in a resonance state.

Ohtsuki and Ofuruton (1991) reported the experimental production of plasma fireballs in air at atmospheric pressure by microwave interference. The microwaves were generated by a continuous, 2.45-GHz, magnetron microwave oscillator at powers of 1 to 5 kW. These fireballs exhibited certain properties that match eyewitness observations of ball lightning, such as splitting into two parts, motion against the wind, and the ability to pass through a wall intact. Several kinds of fireball were produced, with durations from a few seconds to a few minutes. Although there should theoretically have been six antinodes in the cavity, only one fireball was formed (Fig. 13.2).

Yasui (1993, 1997) investigated these experimental results theoretically and established a model to calculate their physical parameters. The radius of a ball is one fourth of a wavelength. Its pressure was assumed to be atmospheric, and because its temperature exceeds that of the ambient atmosphere, its density would be much smaller than atmospheric. The ball is heated by joule heating and loses energy by thermal radiation, assumed to be blackbody. Free electrons are required for the formation of the ball, which would take about 0.1 s. Once the external source of energy is terminated, the ball disappears within about 0.5 s.

Subsequently, Ofuruton et al. (1997a) succeeded in producing a spherical, ellipsoidal, or crescent-shaped plasma fireball in air at atmospheric pressure that could exist independent of a metal cavity. The balls were produced using a

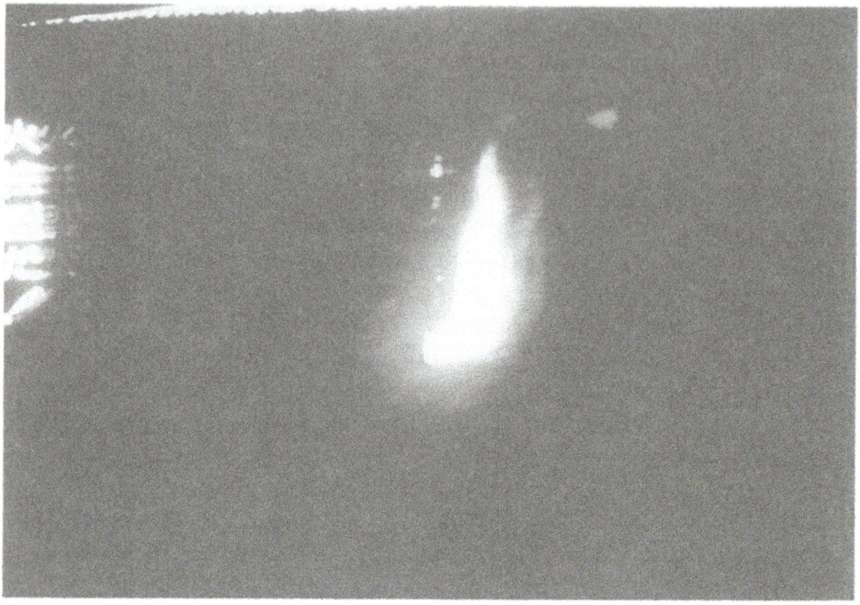

Figure 13.2. Plasma fireball produced by microwave interference. [Reproduced from Ohtsuki and Ofuruton (1991) with the kind permission of Dr. Ofuruton and *Nature*.]

microwave arrangement similar to that described above and an electrical discharge. Sometimes the balls could be generated at low microwave power. They survived for about 0.5 s and had a diameter of about 5 cm.

Golka (1994) claimed that putting a 2-in. burning candle in a home kitchen microwave oven produced cavity-formed plasmodes, which floated around for as long as the microwave energy was present.

Kamogawa and Ohtsuki (1997) proposed a related mechanism to explain observations of earthquake lights. Because of extreme stress in rocks, they argued, exo-electrons are excited and emitted, and a bulk plasmon and surface plasmon can be produced. When the plasmons travel to the surface of the Earth, they propagate as electromagnetic waves in the region 10 kHz to 100 MHz.

Hubert (1996) has commented that witnesses to such a ball would suffer risks of severe burns, and remarked that in a given environment, the focusing of waves could only occur at certain points defined by the local environmental geometry. It is difficult, then, to explain the unpredictable motion of ball lightning. He further remarks that modern electronic devices such as digital watches and radios would be damaged by an intense high-frequency field.

Chapter 14

Conclusions and Recommendations

While a large proportion of ball lightning events can be readily explained in prosaic terms, I think that there is good, consistent observational evidence that roughly spherical, luminous regions are formed in some thunderstorms and occasionally under other atmospheric conditions, and that these regions can exhibit independent motion through space. This evidence is admittedly largely anecdotal because of the very limited amount of tangible evidence in the form of traces, damage, or photographs that cannot be attributed to other causes. However, there are many credible eyewitnesses; ball lightning reports show a much higher degree of consistency than other reported phenomena such as UFOs; and one could argue that there is a much more widespread mythology associated with the latter than with ball lightning.

The limited amount of compelling photographic evidence can be explained by the transient nature of the phenomenon. The small number of cases involving severe damage or traces that cannot be explained by the effects of ordinary lightning suggests that ball lightning is a relatively low-energy phenomenon that does not interact strongly with the environment. However, while, like St. Elmo's fire, ball lightning may not be hazardous in itself, both these phenomena are often precursors of ordinary cloud-to-ground lightning flashes to the vicinity and may thus serve as a useful warning to people nearby.

Ball lightning research has made painfully slow progress in the past century and a half. This is partly because of the nature of the phenomenon, but it is also a consequence of poorly coordinated effort. The inauguration of the International Committee on Ball Lightning (ICBL) and its organization of regular international symposia on the subject is a very positive move.

Progress in evolving theoretical models for ball lightning and testing them with experiments will continue to be difficult while the empirical properties of the phenomenon being modeled have not been clearly described. In order to determine the properties of the natural phenomenon, it is essential that rapid-response investigation teams be established throughout the world. Teams should include experts with training in psychology, especially the psychology of perception, specialists in the properties of ordinary lightning, meteorologists, and physicists. Interviewing techniques, questions to be asked of witnesses, and field investigation procedures should be decided by discussion among such experts. Walker's approach of "benevolent skepticism" is to be advocated here. Information should be gathered that enables evaluation of reports and the accuracy of perception, and that also allows hypotheses to be tested. Authors of ball lightning theories should therefore also be consulted.

Telephone hotlines should be available for reporting ball lightning. Details of these should be given to meteorological offices, news media, etc. Reports with limited information content could probably simply be handled by a telephone interview and then logged for possible corroborative value, while those of greater quality and value would justify a field investigation by a team. Factors that contribute to the value of the report include its information content, its potential scientific value, reported damage or traces and their availability for analysis, time delay in reporting, qualifications and experience of the observer, and the proximity of the phenomenon to the observer. Members of the team could rapidly exchange information with one another and with teams in other countries via e-mail using a wide variety of media, including graphics and video. Reports should be correlated with meteorological and other data, for example, lightning location data (now available on the Internet).

Such investigations would enable far more conclusive evaluation of ball lightning reports and the reports could then be assigned a weighting value according to their reliability. Weighted statistics would offer far more useful relevant data about the natural phenomenon than raw statistics.

A great deal could thus be achieved on a relatively low budget. If staffing were voluntary, expenses would essentially be limited to travel and telephone calls. Within the same limited budget, ordinary video camcorders could be used to film thunderstorms. Again, Internet information about lightning activity allows forecasts of likely thunderstorm activity. Although the probability of recording a ball lightning event is most likely very low, the inexpensive and re-usable nature of the medium would perhaps make this a worthwhile activity.

The question of the nature of ball lightning has not yet been resolved, hence the title of this book. Until procedures such as those recommended here are adopted, this will probably continue to be so. Many theories are now very sophisticated, with impressive quantitative development, but one could be forgiven for thinking that they are constructed on a very weak foundation of empirical knowledge of the natural phenomenon. Matters could be improved by implementing procedures such as those given here.

Ball lightning theories discussed at the 1997 International Symposium on Ball Lightning (ISBL) included electrostatic, electrochemical, charged aerosols and fractal, nuclear, plasma and MHD, gas vortex, condensed matter, dark matter, radiation bubble, microwave interference, atmospheric maser, and relativistic models. Although the diversity of theory is still considerable, it is encouraging to see more theories being subjected to experimental tests than a decade ago. It is important, too, that ball lightning theory not develop in isolation. Research in atmospheric electricity continues to be very active and fruitful. There are many unanswered questions about thunderstorms and lightning, but progress is rapid. An example of the relevance of present research in atmospheric electricity to ball lightning is the recent discovery of high-intensity sources of VHF radiation in thunderstorms. This discovery may support models based on microwave interference and may justify a review of earlier theoretical work that was rejected at the time because radiation in this waveband was considered to be very weak.

If these recommendations are implemented, I am optimistic that questions of the existence and nature of ball lightning will be resolved within the next decade.

Appendix A: Ball Lightning Questionnaire

The questions in the questionnaire presented on the pages that follow are designed to enable an investigator to gather the information necessary to evaluate ball lightning reports, including some elements of cross-checking for accuracy, and to provide relevant data for statistical and theoretical analysis.

The design of the questionnaire was developed from the questionnaires devised by Rayle (1966) for reports of ball lightning (primarily intended for the acquisition of statistical data rather than for evaluation) and by the U.S. Air Force (Condon 1969) for reports of unidentified flying objects (primarily intended for evaluation of reports).

The use of questionnaires to acquire data about ball lightning reports has been criticized on the grounds that to ask the witness questions about ball lightning assumes that ball lightning has been observed. However, this hypothesis will generally have occurred to or been suggested to the eyewitness in order for the report to have come to the attention of the investigator.

The questionnaire is lengthy. To reduce the time required to complete it, the first four questions may be completed by the eyewitness and subsequent questions used as prompts for interview of the witness. Responses to certain questions enable the investigator to skip irrelevant sections of the questionnaire.

I offer the questionnaire as a draft. Other investigators are welcome to use it, and I would be pleased to have comments on its effectiveness or recommendations on how it might be improved.

QUESTIONNAIRE ON BALL LIGHTNING

We understand that you may have observed ball lightning, and would be very grateful if you would provide further information by completing and returning this form with details of the event. Your co-operation is very much appreciated and may be of considerable help to our research. If necessary, please attach continuation sheets to this form.

1. Please write a brief description of the event.

2. Please draw a sketch of the appearance of the ball lightning.

3. Please draw a sketch plan or map of the immediate area near the event and show the path of the ball lightning.

4. Please list any other witnesses by name, and give addresses and telephone numbers where known.

5. Date of event?	Year: Month: Day: Unknown □
6. Local time of day?	a.m. □ / p.m. □ Unknown □
7. Exact location of event ***(nearest street address or map reference)***	
8. How could the terrain in the immediate vicinity of the event best be described?	Flat □ rolling □ hilly □ mountainous □ other...................... □ unknown □
9. How could the earth's surface in the area nearest the ball lightning best be described?	Water-covered □ barren □ meadow or brush □ wooded □ built-on □ other.................. □ unknown □
10. Where were you?	within a building □ in a vehicle □ out-of-doors □ other.......... □ unknown □
11. Where were you?	below ground □ near ground level □ first floor □ other..................... □ unknown □
12. How many others that you know of saw this ball lightning?	
13. Did you observe it	through spectacles or eyeglasses? □ directly? □ through window glass? □ other?.................. □ unknown □
14. During which part of a storm did the ball appear?	Early □ middle □ late □ no storm connected □ unknown □

15. Was the storm, if any, more violent than average?	Yes □ no □ unknown □
16. Please describe the rainfall just before the event.	None □ slight □ medium □ heavy □ unknown □
17. Did you see the ball originate?	Yes □ no □ unknown □
18. Where was the ball observed?	Out-of-doors □ in a closed room □ in a room with an open window or door □ in an aircraft □ in a vehicle □ other.............. □ unknown □
19. Just before the ball appeared was there a flash of lightning?	Yes □ no □ unknown □ ***If no or unknown, please go to question 26.***
20. What was the approximate time interval between seeing this flash of lightning and hearing the thunder?	
21. How long after the lightning flash did the ball appear?	
22. Was the flash of lightning	to the ground? □ between clouds? □ within cloud? □ other.................... □ unknown □
23. Did you see the point of impact of the lightning?	Yes □ no □ unknown □
24. If the lightning flash was to ground, where did the lightning hit the ground?	Ground surface □ water surface □ tree □ power or telephone wires □ other................... □ unknown □
25. What was the distance from the point of impact of the lightning stroke to the place where the ball first appeared?	
26. When you first saw the ball, where was it?	Among clouds □ in mid-air □ contacting metal □ contacting non-metal □ contacting ground □ other.............. □ unknown □
27. Did the ball remain in contact with a solid object throughout its lifetime?	Yes □ no □ unknown □ If 'yes', what was the object?...........................
28. When you first saw it, how far away was the ball from you?	
29. What was the diameter of the ball when you first saw it?	
30. Write down the name of an object of a size comparable to the ball.	
31. Write down the name of an object which, ***if held at arm's length***, would have ***just covered or obscured*** the ball.	
32. While you were watching, did the size of the ball change?	No, size remained the same □ Yes, size decreased □ Yes, size increased □ unknown □

33. What was the shape of the ball?	Round □ elliptical □ ring-shaped □ rod-shaped □ other................. □ unknown □
34. What color(s) was the ball?	Red □ orange □ yellow □ green □ blue □ violet □ silver □ gold □ white □ black □ other.............. □ unknown □
35. If there was more than one ball, how many were there?	
36. Could you see through the ball?	Yes □ no □ unknown □
37. Which word best describes the ball?	Opaque □ translucent □ transparent □ unknown □
38. Write down the name of an everyday object which is of similar brightness to the ball, e.g., the sun, a 100-watt bulb, etc.	
39. Was the ball	brightest near the outer surface? □ brightest near the center? □ uniformly bright all over? □ unknown □
40. Did the brightness	decrease? □ increase? □ remain about the same? □ unknown □
41. Did its appearance change noticeably in any other way? ***If so, please describe.***	Yes □ no □ unknown □
42. Please describe the motion of the ball.	Mostly vertical □ mostly horizontal □ mixed □ no motion □ other................................ □ unknown □
43. What was its distance of closest approach to you?	
44. What was its maximum speed?	
45. While you were watching it, did the ball pass behind anything? ***If so, what was it, and how far away was it from you?***	Yes □ no □ unknown □
46. While you were watching it, did the ball pass in front of anything? ***If so, what was it, and how far away was it from you?***	Yes □ no □ unknown □
47. What was its minimum speed?	
48. Did the movement of the ball seem to be guided by anything? ***If so, what was it?***	Yes □ no □ unknown □
49. Did you have any impression of spinning or rotational movement within the ball?	Yes □ no □ unknown □
50. During the event, did the ball appear to pass through small holes, screens, or solid objects? ***If so, please give details.***	Yes □ no □ unknown □
51. Did you feel any heat from the ball?	Yes □ no □ unknown □

52. Did you notice any odor (smell) from the ball? ***If so, please describe.***	Yes □ no □ unknown □
53. Did you notice any sound from the ball—other than when it ended. ***If so, please describe.***	Yes □ no □ unknown □
54. Did the ball emit anything, such as sparks, lightning, etc.? ***If so, please describe.***	Yes □ no □ unknown □
55. Did the ball have halos, coronas, rays or protrusions? ***If so, please describe.***	Yes □ no □ unknown □
56. Was any unusual behavior noted concerning equipment such as radio, television, hi-fi, car engines, etc., at about this time? ***If so, please describe.***	Yes □ no □ unknown □
57. Did the ball make contact with any solid object? ***If so, what was it? If there was any resulting damage, please describe it.***	Yes □ no □ unknown □
58. Did this contact seem to bring about the end of the ball?	Yes □ no □ unknown □
59. Did you notice any particular change in shape, brightness, size, color or speed immediately before the ball ended? ***If so, please describe.***	Yes □ no □ unknown □
60. How long was the ball in sight?	
61. Was the disappearance of the ball	explosive? □ silent? □ unknown? □
62. If the disappearance of the ball was explosive, can you think of any similar sound?	
63. Where was the ball when it disappeared?	Among clouds □ in mid-air □ contacting metal □ contacting non-metal □ contacting ground □ other............... □ unknown □
64. How far was it from you when it disappeared?	
65. What was its speed when it disappeared?	
66. Was your last sight of the ball	as it disappeared or ended □ as it passed from your view □ unknown □
67. Were any photographs or video films taken?	Yes □ no □
68. Were there any traces or damage left by the ball? ***If so, please describe.***	Yes □ no □ unknown □
69. Have you seen ball lightning on any other occasion?	Yes □ no □

70. May we publish your name in connection with this event?	Yes □ no □
71. Your full name	
72. Title or degrees (if any)	
73. Your full postal address:	
74. Your telephone number	
75. Your profession	
76. Your signature	
77. Today's date	
Please return the completed form to	

Appendix B: Glossary

Definitions marked with a dagger are reproduced, with kind permission of the American Meteorological Society, from T. Glickman, *Glossary of Meteorology,* American Meteorological Society, in press.

aircraft electrification†: The accumulation of a net electrical charge on the surface of aircraft; or the separation of charge into two concentrations of opposite sign on distinct portions of the aircraft surface. Net charges appear as a result of autogenous electrification when aircraft fly through clouds of ice crystals or dust. Charge separation occurs by induction when aircraft fly through regions of strong atmospheric electrical field, as in thunderstorms. Charging may also occur by the engine exhaust carrying away a net charge leaving the aircraft charged. Development of large local charges on aircraft parts may lead to appearance of corona discharge or St. Elmo's fire and is almost always accompanied by poor radio communication due to so-called precipitation static.

air–earth conduction current† (also called fair-weather current): That part of the air–earth current contributed by the electrical conduction of the atmosphere itself. It is represented as a downward current in storm-free regions all over the world. The conduction current is the largest portion of the air–earth current, far outweighing the contributions made by the precipitation current and convection current, which are zero in storm-free regions. Its magnitude is approximately 3×10^{-12} A/m^2, or about 1800 A for the entire earth. Such observations of the vertical variation of the conduction current as have been made indicate that it is approximately uniform throughout the troposphere, a condition that is consistent with the generally accepted view that the conduction current flows from a positively charged conducting region in the lower ionosphere downward to the negatively charged earth. Only in areas of temporarily disturbed weather does the conduction current become replaced by reverse flow. Accumulating evidence points to the conclusion that the conduction current continues to exist only because of the action of thunderstorms scattered at all times over the earth, which supply the positive charge to the upper atmosphere and negative charge to the earth (see supply current).

air–earth current†: The transfer of electric charge from the positively charged atmosphere to the negatively charged earth. This current is made up of the air–earth conduction current, a point-discharge current, a precipitation current, a convection current, and miscellaneous smaller contributions. Of these, the first is by far the largest. The existence of this quasi-steady current in fair weather and the observed maintenance of the earth's net negative charge are both better established than the nature of the supply current, which must replenish the positive charge in the upper atmosphere and the negative charge on the earth.

arc discharge†: A luminous electrical discharge in a gas where the current flows more or less continuously along a narrow channel of high ionization. An arc discharge requires a continuous source of electric potential difference across the terminals of the arc. This steady-state high current discharge is to be distinguished from the low current and visually diffuse corona discharge and point discharge, as well as from the transient, high luminosity, high-current spark discharge.

atmospheric electric field: A quantitative term denoting the electric field strength of the atmosphere at any specified point in space and time. In areas of fair weather, the atmospheric electric field near the Earth's surface typically is about 100 V m^{-1} and is directed vertically in such a sense as to drive positive charges downward to the earth. In areas of fair weather, this field decreases in magnitude with increasing altitude, falling, for example, to only about 5 V m^{-1} at an altitude of about 10 km. Near thunderstorms, and under clouds of vertical development, the surface electric field varies widely in magnitude and direction, usually reversing its direction immediately beneath active thunderstorms. In areas of minimum local disturbance, a characteristic diurnal variation of electric field strength is observed. This variation is characterized by a maximum that occurs at about 19 h universal time coordinated (UTC) for all points on the Earth, and is now believed to be produced by thunderstorms which, for geographic regions, are more numerous for the world as a whole at that universal time than at any other. It is now believed that thunderstorms, by replenishing the negative charge to the Earth's surface, provide the supply current to maintain the fair-weather electric field in spite of the continued flow of the air–earth current that tends to neutralize that field.

atmospherics†: The radio frequency electromagnetic radiation originating, principally, in the irregular surges of charge in thunderstorm lightning discharges. Atmospherics are heard as a quasi-steady background of crackling noise (static) on certain radio receivers, such as AM radio. Since any acceleration of electric charge leads to emission of electromagnetic radiation, and since the several processes involved in propagation of lightning lead to very large charge accelerations, the lightning channel acts like a huge transmitter, sending out radiations with frequencies of the order of 10 kHz. Atmospherics may occasionally be detected at distances in excess of 2000 miles from their source. Advantage has been taken of this in using radio direction-finding equipment to plot cloud-to-ground lightning locations, to locate active thunderstorm areas in remote regions, and in-between weather reporting stations.

autogenous electrification†: The process by which net charge is built up on an object, such as an airplane, moving relative to air containing dust or ice crystals. The electrification is produced by frictional effects (triboelectrification) accompanying contact between the object and the particulate matter.

beaded lightning†: A particular visual variation of the end of a normal lightning flash where periodic sections of the channel appear to die out slowly because they have greater radius and hence lose heat more slowly, are seen end-on, or for other unknown reasons.

blue jet†: A narrowly collimated beam of mostly blue, with some green, light that appears to propagate upward from the tops of thunderstorms. The typical blue jet is observed to appear from the apparent top of the anvil and propagate upward, at speeds of about 100 km/s, in a narrow cone and to flare out as it reaches maximum altitude, approximately 40–50 km, such that it resembles a trumpet. Currently, there is no satisfactory theory of this phenomenon. One possible mechanism may be that jets are discharge following some sort of collimated quasi-straight ion trail emanating from the thunderstorm. Any mechanism producing blue jets will have to continue for at least 200 ms.

breakdown field†: The electric field necessary to produce breakdown.

breakdown†: The process by which electrically stressed air is transformed from an insulator to a conductor. Breakdown involves the acceleration of electrons to ionization potential in the electric field imposed by the thundercloud, and the subsequent creation of new electrons, which avalanche and expand the scale of enhanced conductivity. Breakdown precedes the development of lightning.

charge separation†: The physical process causing cloud electrification. The process can include particle collisions with selective charge transfer and particle capture of small ions at the particle scale. The process can include gravity-driven differential particle motions and convective transport of charged air parcels at the cloud scale.

cloud flash† (also called cloud flash, intracloud flash, cloud-to-cloud flash): A lightning discharge occurring between a positively charged region and a negatively charged region, both of which may lie in the same cloud. The most frequent type of cloud discharge is one between a main positive charged region and a main negative charged region. Cloud flashes tend to outnumber cloud-to-ground flashes. In general, the channel of a cloud flash will be wholly surrounded by cloud. Hence the channel's luminosity typically produces a diffuse glow when seen from outside the cloud and this widespread glow is called sheet lightning.

cloud-to-ground flash†: A lightning flash occurring between a charge center in the cloud and the ground. On an annual basis, negative charge is lowered to ground in about 95% of the flashes, the remaining flashes lowering positive charge to ground. This type of lightning flash, which can be contrasted with an intracloud flash or cloud-to-cloud flash, consists of one or more return strokes. The first stroke begins with a stepped leader followed by an intense return stroke that is the principal source of luminosity and charge transfer. Subsequent strokes begin with a dart leader followed by another return stroke. Most of the strokes use the same channel to ground. The time interval between strokes is typically 40 ms. (See return stroke, stepped leader, and dart leader.)

continuing current†: A sustained current in the lightning stroke that flows to the ground after the return stroke. Continuing currents can have a duration in excess of 100 ms with magnitudes of typically 100 A. Continuing currents occur in negative and positive cloud-to-ground flashes.

corona discharge†: A luminous, and often audible, electric discharge that is intermediate in nature between a spark discharge (with, usually, its single discharge channel) and a point discharge (with its diffuse, quiescent, and nonluminous character). It occurs from objects, especially pointed ones, when the electric field strength near their surfaces attains a value near 100,000 V/m. Aircraft flying through active electrical storms often develop corona discharge streamers from antennas and propellers, and even from the entire fuselage and wing structure. So-called precipitation static results. It is seen also, during stormy weather, emanating from the yards and masts of ships at sea (see St. Elmo's fire).

dart leader†: The leader, which, after the first stroke, typically initiates each succeeding stroke of a multiple-stroke flash lightning. (The first stroke is initiated by a stepped leader.) The dart leader derives its name from its appearance on photographs taken with streak cameras. The dart leader's brightest luminosity is at its tip, which is tens of meters in length, propagating downward at about 10^7 m/s. In contrast to stepped leaders, dart leaders do not typically exhibit branching because the preestablished channel's low gas density and residual ionization provide a more favorable path for this leader than do any alternative ones.

dielectric strength†: A measure of the resistance of a dielectric to electrical breakdown under the influence of strong electric fields; usually expressed in volts per meter. The dielectric strength of dry air at sea-level pressures is about 3,000,000 volts per meter. The exact value for air depends upon geometry of the electrodes between which the electric field is established, upon the humidity, and upon whether or not water drops are present in the air (Macky effect).

dust-devil effect†: In atmospheric electricity, a rather sudden and short-lived change of the vertical component of the atmospheric electric field that accompanies passage of a dust devil near an instrument sensitive to the vertical gradient. Such changes may be either positive or negative and the charge is probably produced by triboelectrification.

electric discharge†: The flow of electricity through a gas, resulting in the emission of radiation which is characteristic of the gas and of the intensity of the current.

electrical breakdown†: The sudden decrease of resistivity of a substance when the applied electric field strength rises above a certain threshold value (the substance's dielectric strength). For air at normal pressures and temperatures, experiment has shown that the breakdown process occurs at a field strength of about 3,000,000 V/m. This value decreases approximately linearly with pressure, and is dependent upon humidity and traces of foreign gases. In the region of high field strength just ahead of an actively growing leader in a lightning stroke, breakdown occurs in the form of a rapidly moving wave of sudden ionization (electron avalanche). The dielectric strength in a cloud of water drops is less than that in cloud-free humid air, for the water drops are distorted as a result of the Macky effect.

EOSO†: End-of-storm-oscillation; a characteristic in the electric field signature at the ground near the end of a thunderstorm. The electric field changes from upward directed (foul-weather polarity) to downward directed (fair-weather polarity) and often retains the latter polarity for several tens of minutes before returning momentarily to foul-weather polarity. Positive ground flashes and spider lightning are occasionally observed during the period of reversed-field polarity. The origin of the EOSO is only partly understood.

exogenous electrification†: The separation of electric charge in a conductor placed in a preexisting electric field. This is especially applied to the charge separation observed on metal-covered aircraft. It is the result of induction effects, and does not by itself create any net total charge on the conductor. It is to be distinguished, therefore, from autogenous electrification.

fair-weather electricity†: The distribution of ions and currents in the atmosphere and at the surface of the earth that occur during "fair weather," or in areas where there is no thunderstorm activity. This distribution sets up a downward-directed electric field referred to as a fair-weather field.

FEAWP† (field excursion associated with precipitation): A characteristic feature in electric field records made directly beneath a thundercloud in which the electric field momentarily reverses polarity during a burst of intense precipitation. The time scale for these events is of the order of minutes and of shorter duration than the EOSO. The physical origins of the FEAWP are still unresolved.

field changes†: The rapid variations of the electrical field at the Earth's surface, beneath, within, and above thunderclouds. Used to determine quantitative estimates of the charge transferred during a lightning discharge, heights of the charge centers, and many other features of thunderclouds.

forked lightning†: The common form of cloud-to-ground discharge always visually present to a greater or lesser degree, which exhibits downward-directed branches from the main lightning channel. In general, of the many branches of the stepped leader, only one is connected to ground, defining the primary, bright return stroke path, and the other incomplete channels decay after the ascent of the first return stroke.

global circuit†: The structure and combination of processes set up by the conductive earth, the conductive ionosphere, and all agents of electrification within the resistive troposphere. The so-called "dc" global circuit is characterized by the ionosphere potential and the "ac" global circuit refers to the Schumann resonances within the earth–ionosphere waveguide.

ground streamer†: An upward-advancing column of high ionization (a streamer or arc) which typically ascends from a point on the Earth's surface toward a descending stepped leader. The ground streamer usually joins the stepped leader about fifty meters above the ground, after which the upward-propagating light and current of the return stroke begin. Ground streamers occur because of the very high electric field intensities that build up directly below the descending, charged stepped leader. Often, more than one ground streamer starts up from the general area under a descending leader, but usually only one makes contact with the leader.

ground-to-cloud discharge†: A lightning discharge in which the original leader process starts upward from some object on the ground; the opposite of the more common cloud-to-ground discharge. Ground-to-cloud discharges most frequently emanate from very tall structures which, being equipotential with the earth, can exhibit the strong field intensities near their upper extremities necessary to initiate leaders.

heat lightning: Lightning or cloud illumination caused by lightning where thunder is not heard because of the distance from the thundercloud. Thunder can only be heard over a distance of about 25 km. This is not truly a distinctive form of lightning. (Uman 1987.)

intracloud flash†: A lightning discharge occurring between a positive charge center and a negative charge center, both of which lie in the same cloud. Starts most frequently in the region of the high electric field between the upper positive and lower negative space charge regions. In summer thunderstorms, intracloud flashes precede the occurrence of cloud-to-ground flashes; they also outnumber cloud-to-ground flashes. Intracloud lightning develops bidirectionally like a two-ended tree: one end of the tree is a branching negative leader, the other is a branching positive leader. Later in the flash, fast negative leaders similar to dart leaders (also called K-changes) appear in the positive end region and propagate toward the flash origin. In weather observing, this type of discharge is often mistaken for a cloud-to-cloud flash, but the latter term should be restricted to true intercloud discharges, which are far less common than intracloud discharges. Cloud discharges tend to outnumber cloud-to-ground discharges in semiarid regions where the bases of thunderclouds may be several kilometers above the earth's surface. In general, the channel of a cloud flash will be wholly surrounded by cloud. Hence the channel's luminosity typically produces a diffuse glow when seen from outside the cloud, and this widespread glow is called sheet lightning.

K changes†: The K-process is generally viewed as a "recoil streamer" or small return stroke that occurs when a propagating discharge within the cloud encounters a pocket of charge opposite to its own. In this view, the J-process represents a slowly propagating discharge that initiates the K-process. This is the case for K-changes in cloud discharges. It is reasonable to expect that cloud discharge K-changes are similar to the in-cloud portion of ground discharges.

leader†: The electric discharge that initiates each return stroke in a cloud-to-ground lightning discharge. It is a channel of high ionization that propagates through the air by virtue of the electric breakdown at its front produced by the charge it lowers. The stepped leader initiates the first stroke in a cloud-to-ground flash and establishes the channel for most subsequent strokes of a lightning discharge. The dart leader initiates most subsequent strokes. Dart-stepped leaders begin as dart leaders and end as stepped leaders. The initiating processes in cloud discharges are sometimes also called leaders but their properties are not well measured.

lightning channel†: The irregular path through the air along which a lightning discharge occurs. The lightning channel is established at the start of a discharge by the growth of a leader, which seeks out a path of least resistance between a charge source and the ground or between two charge centers of opposite sign in the thundercloud or between a cloud charge center and the surrounding air or between charge centers in adjacent clouds.

lightning current†: The current flowing in a component of the lightning flash. It is usually considered to be the current in the return stroke.

lightning discharge†: The series of electrical processes taking place within one second by which charge is transferred along a discharge channel between electric charge centers of opposite sign within a thundercloud (intracloud discharge) between a cloud charge center and the earth's surface (cloud-to-ground discharge or ground-to-cloud discharge), within two different clouds (intercloud or cloud-to-cloud discharge), or between a cloud charge and the air (air discharge). It is a very large-scale form of the common spark discharge. A single lightning discharge is called a lightning flash.

lightning flash†: The total observed lightning discharge generally has a duration less than one second. A single flash is usually composed of many distinct luminous events that often occur in such rapid succession that the human eye cannot resolve them.

lightning stroke†: In a cloud-to-ground discharge, a leader plus its subsequent return stroke. In a typical case, a cloud-to-ground discharge is made up of three or four successive lightning strokes, most following the same lightning channel.

lightning†: Lightning is a transient, high-current electric discharge whose path length is measured in kilometers. The most common source of lightning is the electric charge separated in ordinary thunderstorm clouds (cumulonimbus). Well over half of all lightning discharges occur within the thunderstorm cloud and are called intracloud discharges. The usual cloud-to-ground lightning (sometimes called streaked or forked lightning) has been studied more extensively than other lightning forms because of its practical interest (i.e., as the cause of injuries and death, disturbances in power and communicating systems, and the ignition of forest fires) and because lightning channels below cloud level are more easily photographed and studied with optical instruments. Cloud-to-cloud and cloud-to-air discharges are less common than intracloud or cloud-to-ground lightning. All discharges other than cloud-to-ground are often lumped together and called cloud discharges. Lightning is a self-propagating and electrodeless atmospheric discharge that transfers through the induction process the electrical energy of an electrified cloud into electrical charges and current in its ionized and thus conducting channel. Positive and negative leaders are essential components of the lightning. Only when a leader reaches the ground, does the ground potential wave (return stroke) affect the lightning process. Natural lightning starts as a bi-directional leader although at different stages of the process unidirectional leader development can occur. Artificially triggered lightning starts on a tall structure or from a rocket with a trailing wire. Most of the lightning energy goes into heat, with smaller amounts transformed into sonic energy (thunder), radiation, and light. (See also cloud-to-ground, intracloud, and air discharges.) Lightning, in its various forms, is known by many names such as the common streak lightning, forked lightning, sheet lightning, heat lightning, and the less common air discharge; also, the rare and mysterious ball lightning and rocket lightning. (For some detailed explanation of lightning processes, see lightning discharge and related terms.) An important effect of worldwide lightning activity is the net transfer of negative charge from the atmosphere to the earth. This fact is of great importance in one problem of atmospheric electricity, the question of the source of the supply current. Existing evidence suggests that lightning discharges occurring sporadically at all times in various parts of the earth, perhaps 100 per second, may be the principal source of negative charge that maintains the earth–ionosphere potential difference of several hundred thousand volts in spite of the steady transfer of charge produced by the air–earth current. However, there also is evidence that point discharge currents may contribute to this more significantly than lightning.

negative cloud-to-ground lightning†: A lightning flash or stroke between a cloud and ground that lowers negative charge to ground.

pilot streamer†: A relatively slow-moving, non-luminous lightning streamer, the existence of which has been postulated but not verified, to provide a physical explanation for the observed intermittent mode of advance of a stepped leader as it initiates a cloud-to-ground lightning discharge. Whereas the stepped leader descends at an average speed on the order of 10^5 m/s during its downward motion, it advances only about 50 m at a time with higher speed and then pauses for 50–100 μs before resuming its downward movement. The average downward speed has been associated with an invisible streamer, the pilot streamer, which is postulated to descend at a uniform speed only slightly in excess of ionizing speed of electrons in air and lay down a trail of weak residual ionization along which the stepped leader moves very rapidly in a pulsating manner. The idea of a pilot leader has been supplanted by more modern theory based on laboratory measurements of long spark generation.

point discharge current†: The electrical current accompanying any specified source of point discharge. In the electrical budget of the earth–atmosphere system, point discharge currents are of considerable significance as a major component of the supply current. Estimates made by B. F. J. Schonland of the point discharge current from trees in arid Southwest Africa suggest that this process accounts for about twenty times as much delivery of negative charge to the earth as do lightning discharges during typical thunderstorms. Although the great height of thundercloud bases in arid regions, such as that referred to in Schonland's study, tends to favor point discharge over lightning charge transfer, point discharge still seems more significant than lightning even in England, where Wormell found for Cambridge a ratio of about five to one in favor of point discharge over lightning charge transfer.

point discharge†: A silent, non-luminous, gaseous electrical discharge from a pointed conductor maintained at a potential that differs from that of the surrounding gas. In the atmosphere, trees and other grounded objects with points and protuberances may, in disturbed weather, be sources of point discharge current. Close to a pointed and grounded conductor that extends above surrounding objects, the local electric field strength may be many times greater than that existing at the same level far from the elevated conductor. When this local field reaches such a value that a free electron, finding itself acted upon by this field, can be accelerated (in one mean free path) to a sufficiently high velocity to ionize neutral air molecules, point discharge will begin. Different structures will yield point discharge under quite different gross field conditions, for geometry is critically important. Point discharge is recognized as a major process of charge transfer between electrified clouds and Earth, and is a leading item in the charge balance of the global electrical circuit.

positive cloud-to-ground lightning†: A positive cloud-to-ground lightning flash lowers positive charge from the cloud to the ground.

positive discharge†: A positive discharge lowers positive charge to ground via a lightning flash. The flash may be initiated in the cloud or from the ground.

positive ground flash†: A positive ground flash lowers positive charge to ground from the cloud above.

return stroke†: The intense luminosity which propagates upward from earth to cloud base in the last phase of each lightning stroke of a cloud-to-ground discharge. In a typical flash, the first return stroke ascends as soon as the descending stepped leader makes electrical contact with the earth, often aided by short ascending ground streamers. The second and all subsequent return strokes differ only in that they are initiated by a dart leader and not a stepped leader. It is the return stroke that produces almost all of the luminosity and charge transfer in most cloud-to-ground strokes. Its great speed of ascent (about 1×10^8 m/s) is made possible by residual ionization of the lightning channel remaining from passage of the immediately preceding leader, and this speed is enhanced by the convergent nature of the electrical field in which channel electrons are drawn down toward the ascending tip in the region of the streamer's electron avalanche. Current peaks as high as 300,000 ampere have been reported, and values of 30,000 ampere are fairly typical. The entire process of the return stroke is completed in a few tens of microseconds, and even most of this is spent in a long decay period following an early rapid rise to full current value in only a few microseconds. Both the current and propagation speed decrease with height. In negative cloud-to-ground flashes the return stroke deposits the positive charge of several coulombs on the preceding negative leader channel, thus charging earth negatively. In positive cloud-to-ground flashes, the return stroke deposits the negative charge of several tens of coulombs on the preceding positive leader channel, thus lowering positive charge to ground. The entire return stroke process is completed in a few tens of microseconds. In negative cloud-to-ground flashes, multiple return strokes are common. Positive cloud-to-ground flashes, in contrast, typically have only one return stroke. The return streamer of cloud-to-ground discharges is so intense because of the high electrical conductivity of the ground, and hence this type of streamer is not to be found in air discharges, cloud discharges, or cloud-to-cloud discharges.

ribbon lightning†: Ordinary cloud-to-ground lightning that appears to be spread horizontally into a ribbon of parallel luminous streaks when a very strong wind is blowing at right angles to the observer's line of sight. Successive strokes of the lightning flash are then displaced by small angular amounts and may appear to the eye or camera as distinct paths. The same effect is readily created artificially by rapid transverse movement of a camera during film exposure.

rocket lightning†: A form of cloud discharge, generally horizontal and at cloud base, whose luminous channel appears to advance through the air with visually resolvable speed, often intermittently.

rocket-triggered lightning†: A form of artificial lightning discharge initiated with a rocket trailing wire that may or may not be connected to the ground. The first phase of the discharge is a unidirectional leader starting from the tip of the wire. When the low end of the wire is not connected to ground, bi-directional leader development occurs from both ends of the wire, similar to lightning initiation from aircraft. In the case of the negative space charge overhead (usual summer thunderstorm condition), a triggered lightning may only be a positive leader or become a sequence of dart leader/return stroke processes following the initial positive leader. The latter is analogous to the subsequent return stroke process in a negative cloud-to-ground flash, with the initial positive leader being analogous to the first return stroke. In case of the positive space charge overhead (usual winter storm condition), the triggered lightning is a single negative leader.

sferics observation†: The detection of electromagnetic radiation from lightning generally in the frequency range 10–30 kHz. The physical measurement can include the electric field, the magnetic field, or both. Sferics are generally attributed to the high current phases of lightning, i.e., to return strokes and K-changes.

sferics source†: That portion of a lightning discharge that radiates strongly in the frequency interval 10–30 kHz. The physical source is generally identified with the return stroke in flashes to ground and the K-change in the case of intracloud flashes.

sheet lightning†: A diffuse, but sometimes fairly bright, illumination of those parts of a thundercloud that surround the path of a lightning flash, particularly a cloud discharge or cloud-to-cloud discharge. Thus, sheet lightning is no unique form of lightning but only one manifestation of ordinary lightning types in the presence of obscuring clouds.

space charge†: Any net electrical charge that exists in a given region of space. In electronics, this usually refers to the electrons in the space between the filament and plate of an electron tube. In atmospheric electricity, space charge refers to a preponderance of either negative or positive ions within any given portion of the atmosphere. A net positive space charge is found in fair weather at all altitudes in the atmosphere, and is largest near the earth's surface. The general downward flux of this positive space charge is known as the air–earth conduction current.

spark discharge†: That type of gaseous electrical discharge in which the charge transfer occurs transiently along a relatively constricted path of high ion density, resulting in high luminosity. It is of short duration and to be contrasted with the non-luminous point discharge and corona discharge, and also with the continuous arc discharge. The exact meaning to be attached to the term "spark discharge" varies somewhat in the literature. It is frequently applied to just the transient phase of the establishment of any arc discharge. A lightning discharge can be considered a large-scale spark discharge.

spider lightning†: Lightning with extraordinary lateral extent near a cloud base where its dendritic structure is clearly visible. This lightning type is prevalent beneath the stratiform anvil of mesoscale convective systems and is often associated with positive ground flashes. This discharge form is also referred to as "sheet" lightning.

sprite†: Weak luminous emissions that appear directly above an active thunderstorm and are coincident with cloud-to-ground or intracloud lightning flashes. Their spatial structures range from small single or multiple vertically elongated spots, to spots with faint extrusions above and below, to bright groupings which extend from the cloud tops to altitudes up to about 95 km. Sprites are predominantly red. The brightest region lies in the altitude range 65–75 km, above which there is often a faint red glow or wispy structure that extends to about 90 km. Below the bright red region, blue tendril-like filamentary structures often extend downward to as low as 40 km. High-speed photometer measurements show that the duration of sprites is only a few milliseconds. Current evidence strongly suggests that sprites preferentially occur in decaying portions of thunderstorms and are correlated with large positive cloud-to-ground flashes. The optical intensity of sprite clusters, estimated by comparison with tabulated stellar intensities, is comparable to a moderately bright auroral arc. The optical energy is roughly 10–50 kJ per event, with a corresponding optical power of 5–25 MW. Assuming that optical energy constitutes 1/1000 of the total for the event, the energy and power are on the order of 10–100 MJ and 5–50 GW, respectively. Early research reports for these events referred to them by a variety of names, including "upward lightning," "upward discharges," "cloud-to-stratosphere discharges," and "cloud-to-ionosphere discharges." Now they are simply referred to as sprites, a whimsical term that evokes a sense of their fleeting nature, while at the same time remaining nonjudgmental about physical processes that have yet to be determined. (See also blue jets.)

stepped leader†: The initial leader of a lightning discharge; an intermittently advancing column of high ionization and charge that establishes the channel for a first return stroke. The peculiar characteristic of this type of leader is its step-wise growth at intervals of about fifty to one hundred microseconds. The velocity of growth during the brief intervals of advance, each only about one microsecond in duration, is quite high (about 5×10^7 m/s), but the long stationary phases reduce its effective speed to only about 5×10^5 m/s. To help explain its mode of advance, the concept of a pilot streamer was originally suggested but has been supplanted by analogy to recent work on long laboratory sparks.

streak lightning†: Ordinary lightning, of a cloud-to-ground discharge, that appears to be entirely concentrated in a single, relatively straight lightning channel.

streamer†: A sinuous channel of very high ion density which propagates itself though a gas by continual establishment of an electron avalanche just ahead of its advancing tip. In lightning discharge, the stepped leader, dart leader, and return stroke all constitute special types of streamers.

stroke density†: The areal density of lightning discharges over a given region during some specified period of time, as number per square mile or per square kilometer per year.

superbolt†: A lightning discharge of extraordinary peak luminosity when observed from space.

supply current†: The electrical current in the atmosphere which is required to balance the observed air–earth current of fair-weather regions by transporting positive charge upward or negative charge downward. The problem of accounting for the supply current has been for many years a key problem of the field of atmospheric electricity and has received much attention. A quasi-steady current of about 1800 amperes for the earth as a whole is estimated to be required to balance the air–earth current. C. T. R. Wilson suggested in 1920 that the thunderstorms present in widely scattered regions of the earth at any one time might be responsible for the supply current. Although this suggestion has not been fully confirmed, there is growing conviction that this is correct. When one considers an average over many storms, thunderstorm lightning transports negative charge downward to earth, as does point discharge in the regions below thunderstorms. Also, positive ions flow upward above active thunderstorms.

thunderbolt†: In mythology, a lightning flash accompanied by a material "bolt" or dart; this is the legendary cause of the damage done by lightning. It is still used as a popular term for a lightning discharge accompanied by thunder.

thunderstorm charge†: The existence of regions of net charge in a thunderstorm. During transient collisions of ice crystals with riming graupel pellets charge is transferred. The separating particles then carry equal and opposite charges; the larger (often negative) particles fall while the smaller ones (often positively charged ice crystals) are carried up in the updraft to produce a vertical electric field that eventually produces lightning. There is no complete understanding of the charge transfer process, but possible processes include charges on the surface layers of the particles, charges on dislocations in the ice lattice, temperature differences along surface features that may be broken off during collisions, and contact potential differences between the surfaces of the interacting particles.

thunderstorm dipole†: The simplest representation of the electrostatic structure of an electrified cloud with overall charge neutrality. Ordinary thunderstorms are characterized by upper positive charge and lower negative charge.

thunderstorm tripole†: A refinement of the simpler dipole representation for the electrostatic structure of isolated electrified clouds. The tripole structure includes the lower positive charge center that appears in many observations. The tripole picture is generally attributed to C. C. Simpson.

triboelectrification†: A process of charge separation that involves the rubbing together of dissimilar material surfaces. The triboelectric series is a classification scheme for the ordering of the tendency for positive charge acquisition in rubbing. The detailed physical mechanism in triboelectrification is a long-unsolved problem.

VHF source†: Typically, a source of very high frequency radiation. Lightning discharges are one example of VHF sources in the atmosphere.

Appendix C: Bibliography

INTRODUCTION

This bibliography builds on the monumental catalog of about 2000 references prior to 1980 compiled by Dr. James Dale Barry and included in his book, *Ball Lightning and Bead Lightning*, brought out by Plenum Publishing in 1980. The book is now out of print, so Dr. Barry very generously allowed me to update the bibliography for inclusion here and Plenum has kindly given me permission to use the information in the original. The number of ball lightning references is now somewhat in excess of 2400 and nearly 20% of the publications on ball lightning have appeared in the 19 years since the publication of Barry's book. All ball lightning references known to Barry until 1980 and to the present author at the time of writing are included. The bibliography also includes references to other relevant sources, marked "NBL," many of which are used in this book.

Most modern references are readily available in university libraries. Some journals are now available in on-line format, and abstracts and sometimes entire articles can be accessed via the Internet. Earlier references are often accessible in the libraries attached to science museums. As Barry wrote,

> others are more difficult to locate because of indexing variations and changes in titles and affiliations. The title abbreviations used by many authors and journals are not always recognizable by the investigator since a standard abbreviation has not been consistently used over the years. Many abbreviations used are not in agreement with the Chemical Abstracts Service Source Index Standard.
>
> A number of journal abbreviations encountered during this investigation were unfamiliar and led to some difficulty in locating the reference. A few are discussed here to aid other researchers in their investigations. *Ann. Physik* is used for *Annalen der Physik*, which is also frequently referenced as *Ann. Poggendorff*, *Poggendorffs Annalen*, *Wied. Ann.*, *Wiedemanns Annalen*, and *Drudes Annalen*. *Bull. Assoc. Sci. Fr.* is used for the *Bulletin de la Société Astronomique de France*, which was also referenced as

L'Astronomie as the name was changed in 1911. Some confusion exists as a result of many errors and inconsistencies in volume and year values printed. *Le Cosmos* was used to refer to the *Revue des Sciences et leurs Applications, Paris*, and also referenced as *Kosmos* and *Le Monde*. The journal should not be confused with *Kosmos—Handweiser für Naturfreunde*. *Elektro. Zeit.* was used for the *Elektrotechnische Zeitschrift*, which was occasionally referenced as *ETZ* and *Elekt. Zeit.*, but references to *Elekt. Zeit.* prior to 1880 do not refer to *Elektrotechnische Zeitschrift*.

Gao or *Gaea* refers to *Der Ursprung der Meteoriten Geo (Gaea)*. *Isis* refers to *Gesellschaft Isis in Dresden*. The *Journal de Physique* has had several titles, *Journal de Physique, Téorique et Appliquée* and *Journal de Physique et le Radium*, and abbreviations of each name have been used. *Kleins Wochen.* was used for *Kleins Wochenschrift für Astronomie, Meteorologie und Geographie*. *Orion* is used for the *Schweizerische Astronomische Gesellschaft, Societé Astronomique de Suisse*. *Sitz. Akad. Wiss.* refers to the *Sitzungsberichte, Akademie der Wissenschaften in Wien, Mathematisch—Naturwissenschaftliche Klasse*. *Symon's Met. Mag.* was used to refer to the [*Meteorological Magazine*]. *Zeit. Deut. Met. Ges.* was occasionally used to refer to the *Zeitschrift der Deutschen Meteorologischen Gesellschaft*, which is one of the proper titles of the *Meteorologische Zeitschrift*. *Zh. Russ. Fiz. Khim. Obsh.* was used to refer to the *Zhurnal Russkago Fiziko-Khimicheskago Obshestva*.

Barry personally inspected and verified most of the pre-1980 references. Where he was unable to do so, either the reference is marked with an asterisk or no title is given.

I have the database stored on disk and welcome any additions or corrections.

BIBLIOGRAPHY

Abbadie, A.d' (1852), "Rapport sur une communication de M. Antoine d'Abbadie relative aux orages d'Ethiope," *C. R. Hebd. Séances Acad. Sci.* **34** 894.

Abbott, C. G. (1934), "Remarkable lightning photographs," *Smithsonian Miscellaneous Collections* **92, No. 12**.

Adamson, D. F. (1890), "Notes on a curious lightning flash," *Electrician* **25** 445.

Alanakyan, Yu. R. (1994), "Energy capacity of an electromagnetic vortex in the atmosphere," *Zh. Eksp. Teor. Fiz.* **105(3)** 601.

Alcantara, P. (1889), "Prie communiquer Académie chute foudre globulaire, Minas, 16 Septembre," *C. R. Hebd. Séances Acad. Sci.* **109** 496.

Alcantara, P. (1890a), *Naturwiss. Rund.* **22** 284.

Alcantara, P. (1890b), "La foudre en boule," *C. R. Hebd. Séances Acad. Sci.* **111** 496.

Aleksandrov, V. Ya., and Aleksandrov, V. Ya. (1982), "Aerosol model of ball lightning" (English translation), *Sov. Phys. Tech. Phys.* **27(10)** 1221.

Aleksandrov, V. Y., Golubev, E. M., and Podmoshenskii, I. V. (1982), "Aerosol nature of ball lightning" (in Russian), *Zh. Tekh. Fiz.* **52(10)** 1987.

Aleksandrov, V. Ya., Podmoshenskii, I.V., and Sall, S.A. (1988a), "Simulation of the motions of ball lightning" (in Russian), *Pis'ma v Zhurnal Tekhnicheskoi Fizika* **14(7, 8)** 639.

Aleksandrov, V. Ya., Podmoshenskii, I.V., and Sall, S.A. (1988b), "Simulation of the motions of ball lightning" (English translation), *Soviet Technical Physics Letters* **14(4)** 285.

Aleksandrov, V. Ya., Podmoshenskii, I.V., and Sall, S.A. (1990a), "Experiments on the action of an electric field on a gas-discharge model of ball lightning" (in Russian), *Zh. Tekh. Fiz.* **60(1)** 73.

Aleksandrov, V. Ya., Podmoshenskii, I. V., and Sall, S. A. (1990b), "Experiments on the action of an electric field on a gas-discharge model of ball lightning" (English translation), *Sov. Phys. Tech. Phys.* **35(1)** 42.

Alexander, W. H. (1907), "A possible case of ball lightning," *Mon. Weather Rev.* **35**.

Alexeff, I., and Rader, M. (1995), "Possible precursors of ball lightning-observation of closed loops in high-voltage discharges," *Fusion Technology* **27(3)** 271.

Alfvén, H. (1981), *Cosmic Plasma*, p. 88, D. Reidel, Dordrecht, The Netherlands. (NBL)

Alfvén, H., and Fälthammar, C.-G. (1963), *Cosmical Electrodynamics*, 2nd ed., p. 124, Clarendon Press, Oxford, England. (NBL)

Algeo, J., and Pyles, T. (1966), "Comments in the Anglo-Saxon Chronicle Jan 793 A.D.", in *Problems in the Origin and Development of the English Language*, p. 126, Harcourt Brace, New York.

Aliverti, G., and Lovera, G. (1950), "Uber einige in oberitalien Berichtete Kugelblitze," *Arkiv foer Meteorol. Geophysik und Bioklimat* **A3** 77.

Allen, J. R. (1993), "Goodness, gracious, great balls of fire!" unpublished paper, Physics Department, University of Nottingham, England.

Allibone, T. E. (1944), "Multiple lightning strokes," *Amer. Meteorol. Soc. Q. J.* **70** 161.

Allibone, T. E. (1977), "The long spark," in R. H. Golde (ed.) *Lightning*, Vol. 1, p. 7, Academic Press, New York. (NBL)

Alt, E. (1901), "Kugelblitz," *Meteorol. Zeit.* **18** 573.

Altschuler, M. D. (1969), "Atmospheric electricity and plasma interpretations of UFO's," in E. U. Condon (ed.) *Scientific Study of Unidentified Flying Objects*, Section VI, Chapter 7, p. 723, Bantam Books, New York.

Altschuler, M. D., House, L. L., and Hildner, E. (1970), "Is ball lightning a nuclear phenomenon?," *Nature* **228** 545.

Amirov, A. K., and Bychkov, V. L. (1994a), "ANOVA of several parameters of the SKB data bank on ball lightning," *Physica Scripta* **50(1)** 93.

Amirov, A. K., and Bychkov, V. L. (1994b), "ANOVA of the parameters of ball lightning from the SKB data bank in correspondence to its location (indoors/outdoors)," *Physica Scripta* **50(5)** 588.

Amirov, A. K., and Bychkov, V. L. (1995a), "Correlation analysis of ball lightning distribution over colors of the SKB data bank," *Physica Scripta* **52(2)** 222.

Amirov, A. K., and Bychkov, V. L. (1995b), "Ball lightning diameter-lifetime statistical analysis of SKB databank," *Physica Scripta* **51(3)** 413.

Amirov, A. K., and Bychkov, V. L. (1995c), "A study of the distribution function of the parameters of ball lightning," *J. Met.* **20(199)** 159.

Amirov, A. K., and Bychkov, V. L. (1996), "Influence of atmospheric thunderstorm conditions on the life time of ball lightning," *Physica Scripta* **53(2)** 252.

Amirov, A. K., and Bychkov, V. L. (1997a), "Influence of thunderstorm atmospheric conditions on the properties of ball lightning" (in Russian), *Zh. Tekh. Fiz.* **67(4)** 117.

Amirov, A. K., and Bychkov, V. L. (1997b), "Influence of thunderstorm atmospheric conditions on the properties of ball lightning" (English translation), *Sov. Phys. Tech. Phys.* **42(4)** 430.

Amirov, A. K., and Bychkov, V. L. (1997c), "Observational and computer assisted analysis of data from SKB data base," in Y. H. Ohtsuki and H. Ofuruton (eds.) *Proceedings: 5th International Symposium on Ball Lightning (ISBL97)*, p. 42.

Amirov, A. K., Bychkov, V. L., and Strizhev, A. Yu. (1995), "Principles of creating and processing data banks: meteorological applications—illustrated with regard to ball-lightning data processing," *J. Met.* **20(197)** 85.

Amirov, A. K., Bobkov, S. E., Bychkov, V. L., Emelin, S. E., Klimov, A. I., and Semenov, V. S. (1997), "Modern theoretical and experimental approaches to the problem of ball lightning," in Y. H. Ohtsuki and H. Ofuruton (eds.) *Proceedings: 5th International Symposium on Ball Lightning (ISBL97)*, p. 52.

Andersen, W. H. (1965), "Reply," *J. Geophys. Res.* **70** 1291.

Andersen, W. H. (1966), "Energy source for ball lightning," *J. Geophys. Res.* **71** 680.

Anderson, F. J., and Freier, G. D. (1972), "A report on ball lightning," *J. Geophys. Res.* **77** 3928.

Anderson, R. B. (1977), "Measuring techniques" in R. H. Golde (ed.) *Lightning*, Vol. 1, p. 13, Academic Press, New York. (NBL)

Anderson, R. B. (1982), "Lightning strikes to aircraft: operational data analysis," in *Proceedings: International Aerospace Conference on Lightning and Static Electricity*, Public Relations Section, Culham Laboratory, Abingdon, Oxon., UK.

Anderson, R., Bveinbjorn, S., Blanchard, D. C., Gathman, S., Jonasson, S., Moore, C. B., Survilas, H. J., and Vonnegut, B. (1965), "Electricity in volcanic clouds," *Science* **14** 1179. (NBL)

Andrew, W. L. (1927), "Ball lightning at Cattewater," *Meteorol. Mag.* **62** 186.

Andrianov, A. M., Sinitsyn, V. I. (1976a), CIS Preprint No. IAE-2719, I. V. Kurchatov Institute of Atomic Energy, Moscow.

Andrianov, A. M., Sinitsyn, V. I. (1976b), "Production of stable plasma vortices in the atmosphere," *Pis'ma Zh. Eksp. Trov. Fiz* **24(2)** 67.

Andrianov, A. M., Sinitsyn, V. I. (1977a), "Erosion-discharge model for ball lightning," *Sov. Phys. Tech. Phys.* **22(11)** 1342.

Andrianov, A. M., Sinitsyn, V. I. (1977b), "Ispol'zovanie erozionnogo razriada dlia modelirovaniia odnogo iz vozmuzhnykh vidov sharovoi molnii," *Zh. Tekh. Fiz.* **47** 2318.

Angstrom, C. A. (1883), *Ovfersigt kongl. Vetenskaps-Akadem. Forhandel. (Stockholm)* **40** 87.

Aniol, R. (1954), "Der Kugelblitz, eine Bibliographie," *Meteorol. Rund.* **7** 220.

Anon. (1670), "A narrative of divers odd effects of a dreadful thunder-clap at Strallund in Pomerania 19-29 June 1670," *Phil. Trans. Roy. Soc. London* **5** 2084.

Anon. (1692), "Some remarkable effects of a great storm of thunder and lightning at Portsmouth, Oct. 23 1685," *Phil. Trans. Roy. Soc. London* **16** 1212.

Anon. (1755), "An account of the death of Mr. George William Richmann, Professor of Experimental Philosophy, A Member of the Imperial Academy of Sciences at Petersburg, transl. from High Dutch," *Phil. Trans. Roy. Soc. London* **49** 61.

Anon. (1769), *Cour. Phys. Expr. Math.* **3** 405.

Anon. (1773), *Phil. Trans. Roy. Soc. London* **2** 231.

Anon. (1783a), *The Gentleman's Magazine* **53** 708.

Anon. (1783b), *The London Magazine* **53** 170.

Anon. (1800), "A short view of the new electrical expts. performed by Dr. van Marum," *Phil. Mag.* **8** 313.

Anon. (1836), *Naturwiss.* **3** 264. *

Anon. (1846), "Feuerkugeln" (summary), *Fortschr. Phys.* **2** 202.

Anon. (1847), "Beobachtete Phenomene-Feuerkugeln" (summary), *Fortschr. Phys.* **3** 169.

Anon. (1853), *Annals of Nottinghamshire* **2** 47.

Anon. (1853), *Annals of Nottinghamshire* **2** 157.

Anon. (1857), *Klima. Beit. Ges. Stur. Berlin* **7** 221.

Anon. (1860a), "Cas de foudre en boule à Brighton," *Cosmos* **17** 29.*
Anon. (1860b), "Coups de foudre mystérieux; foudre en boule," *Cosmos* **17** 718.*
Anon. (1860c), "Cas de foudre en boule à Brighton" (summary), *Fortschr. Phys.* **16** 624.
Anon. (1860d), *Le Cosmos* **16** 281.
Anon. (1861a), *Le Cosmos* **18** 1.
Anon. (1861b), *Le Cosmos* **18** 758.
Anon. (1862), *Le Cosmos* **21** 141.
Anon. (1864), "Coups de foudre mystérieux; foudre en boule" (summary), *Fortschr. Phys.* **16** 624.
Anon. (1866), *Le Cosmos* **21** 199.
Anon. (1876), "Killed by a meteor," *Scientific American* **34** 183.
Anon. (1877), "Ball lightning," *Nature* **15** 539.
Anon. (1878a), "Beobachtungen über Kugelblitze," *Ausland* **51** 771.*
Anon. (1878b), "Kugelblitz beobachtet in Dänemark," *Gaea* **14** 61.*
Anon. (1878c), "Kugelblitz beobachtet in Dänemark" (summary), *Fortschr. Phys.* **34(6)** 1156.
Anon. (1878d), "Beobachtungen über Kugelblitze" (summary), *Fortschr. Phys.* **34(6)** 1163.
Anon. (1878e), "Globular lightning," *J. Franklin Inst.* **105** 203.
Anon. (1878f), "Singular phenomenon at Lower Cumber," *R. Met. Soc. Q. J.* **4** 281.
Anon. (1880a), "Globular lightning," *Electrician* **5** 1.
Anon. (1880b), *Nature* **22** 466.
Anon. (1883), *Knowledge* **3** 64.
Anon. (1884a), "La foudre en boule," *Ciel et Terre* **5** 115.
Anon. (1884b), "La foudre en boule," *Ciel et Terre* **5** 365.
Anon. (1884c), *English Mech.* **1** Apr.
Anon. (1884d), "Physics notes," *Nature* **30** 569.
Anon. (1884e), *Ver. Kayser. Monate Berlin Akad.* Nov p. 613.
Anon. (1885), "Feu St. Elme," *Ciel et Terre* **6** 70. (NBL)
Anon. (1885), "Uber den gegenwartigen Stand der Kugelblitz-frage" (summary), *Meteorol. Zeit.* Mar p. 118.
Anon. (1885a), "Uber den gegenwartigen Stand der Kugelblitz-frage" (summary), *Amer. J. Meteorol.* **2** 142.
Anon. (1885b), "Special correspondence," *Electrician* **14** 433.
Anon. (1885c), "Un cas curieux de foudre en boule" (summary), *Fortschr. Phys.* **41(3)** 652.
Anon. (1885d), "Globular lightning," *J. Franklin Inst.* **119** 143.
Anon. (1885e), "Un cas curieux de foudre en boule," *La Nature* **8** 190.*
Anon. (1885f), "Artificial ball lightning," *Meteorol. Mag.* **20** 26.
Anon. (1885g), "Mass of fire falls into sea," *Science* **5** 242.
Anon. (1886), "Feu St. Elme," *Ciel et Terre* **7** 407. (NBL)
Anon. (1886a), "Globular lightning," *Electrician* **17** 293.
Anon. (1886b), "Fatal discharge of globular lightning," *Electrician* **17** 333.
Anon. (1886c), "La foudre globulaire," *L'Astronomie* **5** 310.*
Anon. (1886d), "Foudre en boule," *L'Astronomie* **5** 432.*
Anon. (1886e), "La foudre globulaire," *La Météorologie* **34** 310.
Anon. (1886f), "Literaturbericht," *Meteorol. Zeit.* **3** 515.
Anon. (1887a), "Note sur un coup de foudre," *Bull. Soc. Electr.* 350.*
Anon. (1887b), "Note sur un coup de foudre," *C. R. Hebd. Séances Acad. Sci.* **104** 143.
Anon. (1887c), "Globe lightning in France," *Electrician* **19** 134.
Anon. (1887d), "Globular lightning," *Electrician* **19** 325.
Anon. (1887e), "Curious effects of lightning," *Electrician* **19** 431.
Anon. (1887f), "Note sur un coup de foudre," *Elettr. Milano* 474.*
Anon. (1887g), "Note sur un coup de foudre," *Fortschr. Elektr.* **1** 392.*

Anon. (1887h), "Note sur un coup de foudre" (summary), *Fortschr. Phys.* **43(3)** 573.
Anon. (1887i), *L'Astronomie* **6** 107.
Anon. (1887j), "Note sur un coup de foudre," *La Lumière Electrique* **24** 476.*
Anon. (1887k), *La Nature* **15** 90.
Anon. (1887l), "Perlenblitz," *La Nature* **3** Sep.
Anon. (1887m), "Foudre en boule," *La Météorologie* **35** 18.
Anon. (1887n), "Foudre en boule," *La Météorologie* **35** 19.
Anon. (1887o), "Ancien exemple de foudre en boule," *La Météorologie* **35** 221.
Anon. (1887p), "Ball lightning," *Nature* **37** 187.
Anon. (1887q), "Globular lightning," *Nature* **37** 324.
Anon. (1887r), "Note sur un coup de foudre," *Rev. Intern. Electr.* **4** 107.*
Anon. (1887s), "Globular lightning," *Science* **10** 324.
Anon. (1887t), *Zeit. Angew. Meteorol.* **4** 137.
Anon. (1887u), *R. Met. Soc. Q. J.* **13** 306.
Anon. (1888a), "Strange lightning phenomenon at Geneva," *Electrician* **22** 5.
Anon. (1888b), "Globe lightning," *Electrician* **22** 117.
Anon. (1888c), "Eclairs en boule à St. Petersburg," *La Nature* Oct, p. 350.*
Anon. (1888d), "Eclair en boule observé à Montvilliers en 1720," *La Nature* 16 Dec, p. 6.*
Anon. (1888e), *Le Cosmos* **37** 95.
Anon. (1888f), "Foudre globulaire," *La Météorologie* **36** 173.
Anon. (1888g), "Bemerkenswerthe Blitzschlage," *Meteorol. Zeit.* **5** 158.
Anon. (1888h), *Mitt. Naturwiss. Ver., Steiermark* **25** 228.
Anon. (1888i), "Notes," *Nature* **38** 238.
Anon. (1888j), *Sitz. Genfer Phys. Gesell.* **20** Dec.
Anon. (1888k), *Zeit. Angew. Meteorol.* **5** 94.
Anon. (1888l), *Zeit. Angew. Meteorol.* **5** 165.
Anon. (1888m), *Zeit. Angew. Meteorol.* **5** 166.
Anon. (1888n), *Zeit. Angew. Meteorol.* **5** 204.
Anon. (1888o), *Zeit. Angew. Meteorol.* **5** 208.
Anon. (1888p), *Zeit. Angew. Meteorol.* **5** 214.
Anon. (1889a), *Abhandl. Natur. Ver. Bremen* **13** 312.
Anon. (1889b), *Arch. Sci. Phys. Natur.* **21**.
Anon. (1889c), "Globular lightning," *Electrician* **22** 271.
Anon. (1889d), *Elektro. Zeit.* **1** 7.
Anon. (1889e), *Elektro. Zeit.* **1** 23.
Anon. (1889f), *Elektro. Zeit.* **1** 28.
Anon. (1889g), "Sur un cas de foudre globulaire" (summary), *Fortschr. Phys.* **45(3)** 505.
Anon. (1889h), *Kosmos* **38** 142.
Anon. (1889i), *Kosmos* **38** 182.
Anon. (1889j), "Foudre en boule," *L'Astronomie* **8** 312.*
Anon. (1889k), "Sur un cas de foudre globulaire," *La Lumière Electrique* **33** 143.*
Anon. (1889l), *La Nature* **17** 1.
Anon. (1889m), *La Nature* **17** 334.
Anon. (1889n), "Foudre globulaire à Montfort-l'Amaure," *La Nature* **17** 383.*
Anon. (1889o), "Eclairs en boule à St. Petersbourg," *La Météorologie* **37** 23.
Anon. (1889p), "Foudre globulaire à Montfort-l'Amaury," *La Météorologie* **37** 192.
Anon. (1889q), "Coups de foudre en Belgique en 1887," *La Météorologie* **37** 228.
Anon. (1889r), "Notes," *Nature* **41** 303.
Anon. (1889s), *Staat. Wurt. Friedrichshafen* **163**, 16 Juli.
Anon. (1889t), *Uber Land und Meer* **32** 127.

Anon. (1889u), "Kugelblitz," *Zeit. Angew. Meteorol.* **6** 68.
Anon. (1889v), "Kugelblitz," *Zeit. Angew. Meteorol.* **6** 69.
Anon. (1889w), "Meteorologische Notizen und Korrespondenzen," *Zeit. Angew. Meteorol.* **6** 119.
Anon. (1890a), "Reproduction du phénomène de la foudre en boule," *Ciel et Terre* **11** 424.
Anon. (1890b), *Ebenda* **2** 298.
Anon. (1890c), "Notes," *Electrician* **26** 2.
Anon. (1890d), "A shower of lightning," *Electrician* **5** 74.
Anon. (1890e), "Globular lightning experiments," *Electrician* **26** 194.
Anon. (1890f), "Globular lightning," *Electrician* **22** 413.
Anon. (1890g), "Foudre globulaire," *La Nature* **18** 167.
Anon. (1890h), "Foudre en boule," *La Météorologie* **38** 246.
Anon. (1890i), *Mittlg. Natur. Ver. Steinheim* **28** 348.
Anon. (1890j), *Mittlg. Natur. Ver. Steinheim* **27** 423.
Anon. (1890k), "Notes (The Caucasus Papers)," *Nature* **42** 458.
Anon. (1890l), "Meteorologische Notizen und Korrespondenzen," *Zeit. Angew. Meteorol.* **7** 192.
Anon. (1890m), "Meteorologische Notizen und Korrespondenzen," *Zeit. Angew. Meteorol.* **7** 212.
Anon. (1890n), "Uber die Kugelblitz—Erscheinungen des Tornados von St.-Claude," *Zeit. Angew. Meteorol.* **7** 288.
Anon. (1890o), *Zeit. Elektro.* **8** 487.
Anon. (1891a), "Blitzschlag in die Kirche zu Preetz," *Elektro. Zeit.* **51** 5.*
Anon. (1891b), "Reproduction de l'éclair en boule" (summary), *Fortschr. Phys.* **47(3)** 360.
Anon. (1891c), "Kugelblitz in Berga bei Schlieber," *Gaea* **27** 627.
Anon. (1891d), "Foudre globulaire à Montfort-l'Amaury," *L'Astronomie* **5** 77.*
Anon. (1891e), *L'Astronomie* **5** 299.
Anon. (1891f), *L'Astronomie* **5** 357.
Anon. (1891g), "Reproduction de l'éclair en boule," *La Nature* **19** 122 *.
Anon. (1891h), "Foudre globulaire à Montfort-l'Amaury," *La Météorologie* **39** 99.
Anon. (1891i), *Mitt. Naturwiss. Ver. Steiermark* **28** 334.
Anon. (1891j), *Mitt. Naturwiss. Ver. Steiermark* **28** 349.
Anon. (1891k), "Notes," *Nature* **44** 327.
Anon. (1891l), *Zeit. Angew. Meteorol.* **8** 165.
Anon. (1891m), *Zeit. Angew. Meteorol.* **8** 166.
Anon. (1891n), *Zeit. Angew. Meteorol.* **8** 214.
Anon. (1892a), *Electr. Engr.* Dec 25.
Anon. (1892b), "Kugelblitz in Berga bei Schlieber" (summary), *Fortschr. Phys.* **48(3)** 367.
Anon. (1892c), "Foudre globulaire," *La Nature* Oct, p. 286.*
Anon. (1892d), "Eclairs en chapelet," *Le Cosmos* Juin, p. 319.*
Anon. (1892e), "La foudre globulaire," *Le Cosmos* Juin, p. 319.
Anon. (1892f), "Eclairs en chapelet," *La Météorologie* **40** 217.
Anon. (1892g), "Foudre globulaire," *La Météorologie* **40** 252.
Anon. (1892h), "A lightning story. Ball lightning," *Meteorol. Mag.* **28** 188.
Anon. (1892i), *Mitt. Naturwiss. Ver. Steiermark* **29** 427.
Anon. (1892j), "Notes," *Nature* **46** 548.
Anon. (1893a), "Kugelblitz," *Meteorol. Zeit.* **10** 157.
Anon. (1893b), *Tidskr. Phys. Chem.* **2** 242.
Anon. (1894a), *Kosmische Phys.* 80.
Anon. (1894b), "Eclair en boule," *La Nature* Juin, p. 47.*
Anon. (1894c), "Eclair en boule," *La Météorologie* **42** 200.
Anon. (1894d), *Zeit. Elektro.* **12** 73.
Anon. (1895), *Ver. Kgl. Meteorol. Inst. Berlin* 24.

Anon. (1896a), “Beobachtung über Kugelblitz” (summary), *Fortschr. Phys.* **52(3)** 299.

Anon. (1896b), *Le Cosmos* **45** 387.

Anon. (1896c), “Eclair en boule,” *La Météorologie* **42** 200.

Anon. (1896d), “Beobachtung über Kugelblitz,” *Zeit. Angew. Meteorol.* **13** 36.*

Anon. (1896e), “Beobachtung über Kugelblitz,” *Zeit. Angew. Meteorol.* **13** 97.*

Anon. (1896f), “Beobachtung über Kugelblitz,” *Zeit. Angew. Meteorol.* **13** 215.*

Anon. (1896g), “Beobachtung über Kugelblitz,” *Zeit. Angew. Meteorol.* **13** 216 *.

Anon. (1896h), “Beobachtung über Kugelblitz,” *Zeit. Angew. Meteorol.* **13** 262.*

Anon. (1896i), “Beobachtung über Kugelblitz,” *Zeit. Angew. Meteorol.* **13** 263.*

Anon. (1896j), “Beobachtung über Kugelblitz,” *Zeit. Angew. Meteorol.* **13** 264.*

Anon. (1897a), *Le Cosmos* **46** 547.

Anon. (1897b), “Kugelblitz,” *Meteorol. Zeit.* **14** 34.

Anon. (1897c), *Mittlg. Natur. Ver. Steinheim* **34** 175.

Anon. (1897d), *Mittlg. Natur. Ver. Steinheim* **34** 176.

Anon. (1897e), *Verhdlgn. Nat. Ver. Brunn.* **36** 41.

Anon. (1897f), “Zwei Seltsame Blitzschlage,” *Zeit. Angew. Meteorol.* **14** 93.*

Anon. (1897g), *Zeit. Angew. Meteorol.* **14** 144.

Anon. (1897h), *Zeit. Angew. Meteorol.* **14** 166.

Anon. (1897i), “Ein Kugelblitz,” *Zeit. Angew. Meteorol.* **14** 167.*

Anon. (1897j), *Zeit. Angew. Meteorol.* **14** 252.

Anon. (1898a), “Gewitter” (summary), *Fortschr. Phys.* **54(3)** 313.

Anon. (1898b), “Kugelblitz” (summary), *Fortschr. Phys.* **54(3)** 314.

Anon. (1898c), “Kugelblitz,” *Naturgeschichte Mecklenburg* **52** 100.*

Anon. (1898d), “Gewittter,” *Zeit. Angew. Meteorol.* **15** 70.

Anon. (1899), *Mittlg. Natur. Ver. Steinheim* **36** 228.

Anon. (1900a), “Uber die Blitz,” *Astr. Rund.* **2** 24.*

Anon. (1900b), “Foudre en boule,” *Bull. Soc. Astron. Fr.* **14** 510.

Anon. (1900c), “L’éclair en boule,” *Cosmos* Sep, p. 287.*

Anon. (1900d), “Notes-globular lightning,” *Electrician* **45** 587.

Anon. (1900e), “Uber die Blitz” (summary), *Fortschr. Phys.* **56(3)** 253.

Anon. (1900f), “Uber den Kugelblitz,” *Meteorol. Zeit.* **17** 382.

Anon. (1900g), “Notes,” *Nature* **62** 330.

Anon. (1900h), “A short view of the new electric experiments performed by Dr. van Marum,” *Phil. Mag.* **8** 193.

Anon. (1900i), *Zeit. Angew. Meteorol.* **17** 191.

Anon. (1901a), “L’éclair en boule” (summary), *Fortschr. Phys.* **57(3)** 266.

Anon. (1901b), “Verheerung durch einen Kugelblitz” (summary), *Fortschr. Phys.* **57(3)** 266.

Anon. (1901c), “Verheerung durch einen Kugelblitz,” *Gaea* **37** 634.

Anon. (1901d), *La Nature* **29** 2.

Anon. (1901e), *La Nature* **29** 561.

Anon. (1901f), “L’éclair en boule,” *La Météorologie* **49** 53.

Anon. (1901g), “Observation de foudre en boule,” *La Météorologie* **49** 353.

Anon. (1901h), *Zeit. Angew. Meteorol.* **18** 24.

Anon. (1902a), “Kugelblitz in Altenburg” (summary), *Fortschr. Phys.* **58(3)** 304.

Anon. (1902b), “Kugelblitzbeobachtung,” *Meteorol. Zeit.* **19** 335.

Anon. (1902c), “Kugelblitz in Altenburg,” *Zeit. Angew. Meteorol.* **19** 239.*

Anon. (1904a), *Ann. Hydro. Marit. Meteorol.* **32** 390.

Anon. (1904b), “Foudre en boule,” *L’Astronomie* **18** 378.*

Anon. (1904c), *Zeit. Angew. Meteorol.* **21** 118.

Anon. (1904d), *Zeit. Angew. Meteorol.* **21** 119.

Anon. (1905a), "Sonderbare Wirkungen eines Kugelblitzes," *Gaea* **4** 57.*
Anon. (1905b), "La foudre en boule," *La Nature* **32** 258.*
Anon. (1906a), "Foudre en boule," *Bull. Soc. Astron. Fr.* **20** 100.
Anon. (1906b), *Kosmos* **56** 309.
Anon. (1907), "Nochmals der Kugelblitz 1896, Remscheid," *Prometheus* **18** 576.*
Anon. (1908), "Eclair en boule," *Ciel et Terre* **29** 399.
Anon. (1908a), *La Mat. Fulm.* **7** 113.
Anon. (1908b), *Naturwiss. Rund.* **8** 62.
Anon. (1908c), *Umschau* **19** 304.
Anon. (1909), "Foudre en boule," *Ciel et Terre* **30** 417.
Anon. (1909a), "Fireballs," *English Mechanic* Aug 13.*
Anon. (1909b), "Kugelblitz bei Treptow" (summary), *Fortschr. Phys.* **65(3)** 283.
Anon. (1909c), "Foudre en boule," *Rev. Nepholog.* **47** 370.*
Anon. (1909d), "Kugelblitz bei Treptow," *Weltall* **9** 300.*
Anon. (1909e), "Thunderstorm at Portsmouth, February 14, 1809," *R. Met. Soc. Q. J.* **35** 140.
Anon. (1910a), "Foudre en boule," *Ciel et Terre* **31** 385.
Anon. (1910b), "Foudre en boule" (summary), *Fortschr. Phys.* **66(3)** 364.
Anon. (1910c), *Le Cosmos* **59** 30.
Anon. (1911a), "Foudre en boule," *Bull. Soc. Astron. Fr.* **25** 262.
Anon. (1911b), *Le Cosmos* **60** 19.
Anon. (1911c), *Le Cosmos* **60** 692.
Anon. (1911d), "Les éclairs globulaires," *Rev. Nepholog.* **6** 539.*
Anon. (1911e), *Ver. Kcl. Meteorol. Inst. Berlin* **231** 72.
Anon. (1911f), *Zeit. Angew. Meteorol.* **28** 161.
Anon. (1912a), "Un cas remarquable de foudre en boule," *Ciel et Terre* **33** 199.
Anon. (1912b), *Zeit. Angew. Meteorol.* **29** 199.
Anon. (1913), "Meteorologische Notizen und Korrespondenzen," *Zeit. Angew. Meteorol.* **30** 213.
Anon. (1914a), *Umschau* **25** 762.
Anon. (1914b), "Kugelblitz," *Zeit. Angew. Meteorol.* **31** 215.
Anon. (1915a), "Das Wesen der Kugelblitz," *Meteorol. Zeit.* **32** 416.
Anon. (1915b), *Umschau* **26** 229.
Anon. (1916a), "Ball lightning," *Electr. Exp.* Jan, p. 68.*
Anon. (1916b), *Le Cosmos* **65** 341.
Anon. (1919a), "Ball lightning at Salina, Kansas," *Mon. Weather Rev.* **47** 728.
Anon. (1919b), "Globular lightning," *Roy. Astron. Soc. Canada J.* **13** 299.
Anon. (1921), "Ball lightning seen at St. John's Wood," *Meteorol. Mag.* **56** 195.
Anon. (1922), *Umschau* **26** 587.
Anon. (1923), "Nature of ball lightning," *Meteorol. Mag.* **58** 259.
Anon. (1924a), "Onweder Optische Verschijnselen," *Nz. in Nederland, Amsterdam* **81**.*
Anon. (1924b), "Ball lightning," *Scientific American* **130** 242.
Anon. (1925), "Ball lightning," *Roy. Astron. Soc. Canada J.* **19** 213.
Anon. (1926a), *Meteorol. Mag.* **61** 197.
Anon. (1926b), "Kugelblitze und Fohn in der Krim," *Meteorol. Zeit.* **43** 33.
Anon. (1927), "Kugelblitz," *Meteorol. Zeit.* **44** 188.
Anon. (1930a), "August 1 1907. Ball lightning at Alpena, Michigan," *Nature* **126** 153.
Anon. (1930b), "Electrical phenomenon near Weymouth," *Nature* **126** 262.
Anon. (1930c), "Historical natural events," *Nature* **126** 262.
Anon. (1931), "Ball lightning," *Science* **74** supplement 1908 10.
Anon. (1932), *Soc. Fran. Elec. Bull.* **2** 865.
Anon. (1933), "Ball lightning phenomenon," *Science* **78** supplement 2026 9.

Anon. (1934), "Ball lightning-three discharges," *Roy. Astron. Soc. Canada J.* **28** 144.
Anon. (1939), "On the problem of ball lightning," *Roy. Meteorol. Soc. Q. J.* **65** 532.
Anon. (1942), *Tekh. Molod.* **8.**
Anon. (1945a), *New York Herald Tribune* Jan 2.
Anon. (1945b), "Foo fire reports keep scientists guessing," *Science News Letter* **47** 24.
Anon. (1951), "Nabetrachting over de bolbliksem van Lutjewolde," *Hamel en Dampkring* **49(9)** 158.*
Anon. (1952), "Thunderbolts: the electric phenomenon of thunderstorms," *Meteorol. Mag.* **81** 146.
Anon. (1955a), *Zeit. Angew. Meteorol.* **9** 62.
Anon. (1955b), *Zeit. Angew. Meteorol.* **9** 378.
Anon. (1958), *Civil Aviation-Russ.* **9** 32.
Anon. (1960a), "Interesting observation of lightning," *Meteorol. Mag.* **89** 235.
Anon. (1960b), *Pravda, Moscow* Oct 30, p. 3.
Anon. (1962), "Sharovile Molnii," *Priroda* **5** 71.
Anon. (1964a), "Awaizija i Kosmonawtika," *Bildmessg. und Luftbildwes.* **47(10)** 94.*
Anon. (1964b), "Kugelblitz trifft Flugzeug," *Naturwiss. Rund.* **17** 482.
Anon. (1964c), "New theoretical model for ball lightning," *Science News Lett.* **86** 199.
Anon. (1964d), "Ball lightning due to a leak in stroke at joint," *Science News Lett.* **86** 392.
Anon. (1968), "Ball lightning," *Popular Science* **192(6)** 14.
Anon. (1969), "Photographic case studies," in E. U. Condon (ed.), *Scientific Study of Unidentified Flying Objects*, Section IV, Chapter 3, p. 396, Bantam Books, New York. (NBL)
Anon. (1972), *Ind. Ocean Mar. Obs.* **42** 52.
Anon. (1975), "Unusual weather: ball lightning in a February thunderstorm down East in Maine," *Weatherwise* **29** 84.
Anon. (1976), "Fireballs: some firey, some dark," *J. Met.* **1(10)** 312.
Anon. (1978), "Ball lightning in Canada," *J. Met.* **3(31)** 212.
Anon. (1984a), "Some recent ball lightning incidents," *J. Met.* **9(87)** 77.
Anon. (1984b), "The puzzle of ball lightning," *J. Met.* **9(88)** 112.
Anon. (1985), "Cheap fusion—or a load of balls," *New Scientist* **19** Sep, p. 27.
Anon. (1991), "Flashes of insight into ball-lightning," *Electronics World & Wireless World* **97(97)** 456.
Anon. (1992), *IEEE Conference Record—Abstracts 1992 IEEE International Conference on Plasma Science*. Cat. No. 92CH3184-9 Institute of Electrical and Electronics Engineers, New York.
Anon. (1994), "Ninth International Conference on Atmospheric Electricity (ICAE)," *J. Geophys. Res.* **99(D5)**.
Anon. (1995), "Death from lightning or 'ball lightning,' near Kiev, Russia, June 1995," *J. Met.* **20** 393.
Anon. (1996), "Ball lightning comes down to earth," *Electronics World* **102(1724)** 539.
Anon. (1997a), "Ball lightning in Lancashire and North Wales," *J. Met.* **23(228)** 139.
Anon. (1997b), "A model for ball lightning" (book review), *J. Met.* **23(228)** 147.
Anon. (1998), "Ball lightning event at South Morecombe gas field," *Marine Observer* **68(340)** 77.
Arabadji, W. I. (1947), "On the electricity of thunderstorms," *Priroda* **7** 12.*
Arabadji, W. I. (1956), "K teorii yavlenii atmosfernogo elektrischestva," *Minskogo Gosudarstvennogo Pedagogicheskogo Instituta* **5** 77.*
Arabadji, W. I. (1960), "Groza i grozovyye protsessy," *Minsk IZD-VO Belgos-Universiteta Imeni V. I. Lenina.**
Arabadji, W. I. (1976), "On the problem of ball lightning," *J. Geophys. Res.* **81(36)** 6455.
Arabadji, W. I. (1977), "Reply," *J. Geophys. Res.* **82(24)** 3500.
Arago, F. (1838), *Sur le Tonnerre,* p. 221, Annuaire au Roi par le Bureau des Longitudes, Notices Scientifiques.*
Arago, F. (1854), Oeuvres completes: Tome 4: *Le Tonnere*, pp. 38, 50, 211, Gide et J. Baudry-Ed., Paris.
Arago, F. (1859), *Thunder and Lightning,* Paris.*
Argenteuil, M. (1859), *Le Cosmos* **14** 672.

Argyle, E. (1971), "Ball lightning as an optical illusion," *Nature* **230** 179.

Argyle, M. (1838), "On thunder and lightning," *Edinburgh New Phil. J.* **26** 81.

Arnhoff, A. G. (1993), "A theoretical explanation of ball lightning," in A. G. Keul (ed.) *Progress in Ball Lightning Research: Proc. VIZOTUM*, The Vizotum Project, Salzburg, Austria.

Arnhoff, G. (1992), "Is there yet an explanation of ball lighting?," *European Trans. on Elec. Power Eng.* **2(3)** 137.

Arnhoff, G. (1997), "On the spheric radiation," in Y. H. Ohtsuki and H. Ofuruton (eds.) *Proceedings: 5th International Symposium on Ball Lightning (ISBL97)*, p. 101.

Arrhenius, S. A. (1903), *Lehrbuch der Kosmischen Physik I.*, p. 772, S. Hirzel, Leipzig.

Artsimovich, L. A. (1978), *A Physicist's ABC on Plasma*, Mir Publishers, Moscow. (NBL)

Artsimovich, L. A. (1965), *Elementary Plasma Physics*, Chapter 4, Blaisdell, New York.

Ashby, D. E. T. F., Whitehead, C. (1971), "Is ball lightning caused by antimatter meteorites?," *Nature* **230** 180.

Ashmore, S. E. (1940), "On the problem of ball lightning," *Roy. Meteorol. Soc. Q. J.* **66** 194.

Askar'gan, G. A., Rabinovich, M. S., Savchenko, M. M., Stepanov, V. K. (1967a), "Optical breakdown-fireball in the focus of a laser beam," *JETP Lett.* **5** 121.

Askar'gan, G. A., Rabinovich, M. S., Savchenko, M. M., Stepanov, V. K. (1967b), *Zh. Eksp. Teor. Fiz.* **5(5)** 150.

Aspden, H. (1973), "Ball lightning enigma," *New Scientist* **57** 42.

Aspden, H (1981), "The enigma of ball lightning," *J. Met.* **6(62)** 258.

Aspden, H (1985), "Ball lightning," *New Scientist* **108(1478)** 75.

Aspden, H. (1993), "Ball lightning and the Schuster-Wilson hypothesis," in A. G. Keul (ed.) *Progress in Ball Lightning Research: Proc. VIZOTUM*, p. 6, The Vizotum Project, Salzburg, Austria.

Assmann, J. (1902), "Wirbelgewitter" (summary), *Fortschr. Phys.* **59(3)** 279

Assmann, J. (1903a), "Wirbelgewitter," *Zeit. Angew. Meteorol.* **20** 1.*

Assmann, J. (1903b), "Wirbelgewitter," *Zeit. Angew. Meteorol.* **20** 21.*

Atkinson, B. W. (1968), "Riddle of ball lightning," *Geographical Mag.* **41** 204.

Atkyns, T. (1712), *The Ancient and Present State of Gloucestershire*, London.

Audoin, M. (1913), "Kugelblitz, beobachtet in der Tschadsee-Region," *Meteorol. Zeit.* **30** 148.

Aulov, A. (1973), "On course-ball lightning," *Izvestia* July 28, p. 6.*

Avramenko, R. F., Bakhtin, B. I., Nikolaeva, V. I., Poskacheeva, L. P., and Shirokov, N. N. (1990a), "A study of the plasma formations produced in an erosion discharge" (in Russian), *Zh. Tekh. Fiz.* **60(12)** 57.

Avramenko, R. F., Bakhtin, B. I., Nikolaeva, V. I., Poskacheeva, L. P., and Shirokov, N. N. (1990b), "A study of the plasma formations produced in an erosion discharge" (English translation), *Sov. Phys. Tech. Phys.* **35(12)** 1396.

Avramenko, R. F., Gridin, A. Yu., Klimov, A. I., and Nikolaeva, V. I. (1992a), "Experimental study of energetic compact plasma formations" (in Russian), *Teplofizika Vysokikh Temperatur* **30(6)** 1057.

Avramenko, R. F., Gridin, A. Yu., Klimov, A. I., and Nikolaeva, V. I. (1992b), "Experimental study of energetic compact plasma formations" (English translation), *High Temperature* **30(6)** 870.

Baatz, H. (1977), "Protection of structures" in R. H. Golde (ed.) *Lightning*, Vol. II, p. 19, Academic Press, New York. (NBL)

Babat, G. I. (1942), "Electrodeless discharges and certain problems associated with them," *Vestnik Elektropromtshlennosti* **2/3**.*

Babat, G. I. (1947), "Electrodeless discharges and some allied problems," *IEEJ, London* **94(3)** 27.

Babat, G. I. (1957), "Vysokochastotnoye plamya," *Yunyi Tekhnik* **7** 56.*

Babick, M. (1955), "Kugelblitzbeobachtung in Cottbus," *Zeit. Meteorol.* **9** 378.

Babinet, M. (1852), "Sur un cas de foudre globulaire," *C. R. Hebd. Séances Acad. Sci.* **35** 1.

Babinet, M. (1881), *Cours de Physique*, p. 320, Jamin and Barry (eds.), Gauthier-Villars, Paris 3.*

Bacha, M. (1890), "Meteorologische Notizen und Korrespondenzen," *Zeit. Angew. Meteorol.* **7** 192.

Badger, R. M., Wright, A. C., Whitlock, R. F. (1965), "Absolute intensities of the discrete and continuous absorption bands of oxygen gas," *J. Chem. Phys.* **43** 4345. (NBL)

Bagote, A. (1919), "Eclairs en boule," *Bull. Soc. Astron. Fr.* **33** 284.

Bailey, B. H. (1977), "Ball lightning," *Weatherwise* **30(3)** 99.

Bailey, B. H. (1984), "Ball lightning strikes twice," *Weather* **39(9)** 276.

Bailey, T. (1775), "Events," in *Ann. Nottinghamshire, History-County, Nottingham, including the Borough*, Vol. 4, p. 47, Simpkin, Marshall, London .*

Bailey, T. (1794), "Events," in *Ann. Nottinghamshire, History-County, Nottingham, including the Borough*, Vol. 4, p. 157, Simpkin, Marshall, London.*

Baker, R. M. L., Jr. (1972), "Motion pictures of UFO's," in C. Sagan and T. Page (eds.), *UFO's—A Scientific Debate*, Chapter 8, p. 190, Cornell University Press, Ithaca, New York. (NBL)

Baker-Blocker, A., Donahue, T. M., Mancy, K. H. (1977), "Methane flux from wetland areas," *Tellus* **29(3)** 245. (NBL)

Baldit, A. (1914), "Sur un éclair en boule," *C. R. Hebd. Séances Acad. Sci.* **158** 1542.

Baljashni, W. (1899a), "Wiederholung Plante'scher Versuche," *Beibl. Ann. Phys.* **23** 843.

Baljashni, W. (1899b), "Wiederholung Plante'scher Versuche," *J. Electr.* **7** 106.*

Baltin, P. (1902), *Photogr. Mitt.* **39** 390.

Balyberdin, V. V. (1965), "Estimating the energy of ball lightning," *Samoletostroyeniye I Tekhnika Vozdushnogo Flota* **3** 102.*

Balyberdin, V. V. (1966), "Internal energy of ball lightning," *Foreign Science Bull.* **2(4)** 48.

Balyberdin, V. V. (1966), "Investigation of the mechanism of formation of an ionized vortex configuration in the atmosphere," *Samoletostroyeniye I Tekhnika Vozdashnogo Flota* **5** 3.

Balyberdin, V. V. (1967), "Still another ball lightning hypothesis," *Foreign Science Bull.* **3(5)** 103.

Balyberdin, V. V. (1975a), "Verification of the case of the origin of ball lightning," *J. App. Mech. Tech. Phys.* **16(6)** 946.

Balyberdin, V. V. (1975b), "Verification of the case of the origin of ball lightning," *Zh. Prikl. Mekh. Tekh. Fiz.* **16(6)** 132.*

Baratoux, M. (1952), "Sur une observation de foudre en boule fait en avion," *La Météorologie* **16** 164.

Barette, H. (1865a), "Foudre globulaire du Lundi 17 Juillet 1865" (summary), *Fortschr. Phys.* **21** 64.

Barette, H. (1865b), "Foudre globulaire du Lundi 17 Juillet 1865," *Mondes* **8** 726.*

Barry, J. D. (1966), "Ball lightning—a natural phenomenon in atmospheric physics," M. S. Thesis, Calif. State College, Los Angeles, California.

Barry, J. D. (1967a), "Ball lighting," *J. Atmos. Terres. Phys.* **29** 1095.

Barry, J. D. (1967b), "A model for ball lightning," *Wiss. Zeit. Elektro.* **9** 202.

Barry, J. D. (1968a), "Laboratory ball lightning," *J. Atmos. Terres. Phys.* **30** 313.

Barry, J. D. (1968b), "Fireball, ball lightning and St. Elmo's fire," *Weather* **23** 180.

Barry, J. D. (1968c), "Ball lightning in the laboratory," *Wiss. Zeit. Elektro.* **12** 7.

Barry, J. D. (1969), "Laboratory ball lightning," *Priroda* **12** 62.

Barry, J. D. (1974a), "A comprehensive bibliography of ball lightning reports," Air Force Avionics Lab. Tech. Report, Wright-Patterson AFB, Ohio AFAL-TR-73-348.

Barry, J. D. (1974b), "Bibliography of ball lightning reports," *J. Atmos. Terres. Phys.* **32** 1577.

Barry, J. D. (1979), "Frequency of ball lightning reports," *J. Geophys. Res.* **84(C1)** 308.

Barry, J. D. (1980a), *Ball Lightning and Bead Lightning: Extreme Forms of Atmospheric Electricity*, Plenum Press, New York.

Barry, J. D. (1980b), "On the energy density and forms of ball lightning," *J. Geophys. Res.* **85(C7)** 4111.

Barry, J. D., and Singer, S. (1989), "Ball lightning—the continuing challenge," in Y. H. Ohtsuki (ed.), *Science of Ball Lightning (Fireball)*, p. 1, World Scientific, Singapore.

Barry, J. D., Boney, W. E., Brandelik, J. E. (1971), "Laser emission from He-air-CH_4 and He-air-C_3H_8 mixtures," *Appl. Phys. Lett.* **18** 15. (NBL)

Bass, R. W., and Golka, R. K. (1976), "Nikola Tesla's theory of ball lightning (Colorado Springs 1899)," Eyring Research Institute, Provo, Utah (unpublished).

Battan, L. J. (1964), *The Thunderstorm*, p. 101, New American Library, New York.

Bauer, C. (1938), "Ein Blitzaufnahme," *Umschau* **42** 710.

Baum, C. E. (1982), "Direct-strike lightning electromagnetic environments on aircraft exterior and their stimulation," in *Proceedings: International Aerospace Conference on Lightning and Static Electricity, Public Relations Section*, Culham Laboratory, Abingdon, Oxon., UK.

Bauman, G. (1937), "Photographische Aufnahme eine Blitzkugel in Riga," *Meteorol. Zeit.* **54** 192.

Bay, I. (1908a), "Sur un cas de foudre en boule observé le 26 Mai 1907," *Bull. Soc. Astron. Fr.* **22** 232.

Bay, I. (1908b), "Observation de foudre en boule," *C. R. Hebd. Séances Acad. Sci.* **146** 554.

Bay, I. (1908c), "Kugelblitz," *Meteorol. Zeit.* **25** 468.

Bay, I. (1908d), "Kugelblitzbeobachtung," *Prometheus* **20** 48.*

Bayes, K. D., and Kistiakowsky, G. B. (1960), "On the mechanism of the Lewis-Rayleigh nitrogen afterglow," *J. Chem. Phys.* **32(4)** 992. (NBL)

Bayet, L. (1884a), "La foudre en boule," *Ciel et Terre* **5** 115.

Bayet, L. (1884b), "Foudre en boule," *Ciel et Terre* **5** 365.

Beadle, D. G. (1936), "A curious after effect of lightning," *Nature* **137** 112.

Beasley, W. H., Uman, M. A., Edgar, B. C., Turman, B. N. (1978), "Correlated satellite optical and ground-based electrical and magnetic field observations of lightning," paper presented at AGU Spring Meeting, Abstract in EOS-Trans. AGU **59(4)** 284. (NBL)

Beck, F. X. (1920), "Das Gewitter in Saarbrucken am 26 Juni 1928," *Meteorol. Beobachtungen der Wetterwarte Saarbr.* **II** 2.*

Beck, F. X. (1927), "Nachleuchten eines Blitzes als Perlschnur," *Zeit. Angew. Meteorol.* **44** 141.

Becker, R. O. (1963), "The biological effects of magnetic fields—a survey," *Med. Electron. Biol. Eng.* **1** 293. (NBL)

Becquerel, M., and Becquerel, E. (1855), "Traite d'Electricité et de Magnétisme," *Paris* **1** 415.*

Behn, U. (1903), "Uber Photographien von Perlschnurblitzen," *Meteorol. Zeit.* **20** 379.

Bellay, M. du. (1886), *L'Astronomie* **5** 311.

Belsey, F. H. (1983), "Ball lightning," *Electronics and Power* **29(1)** 97.

Bendorf, G. (1934), *Atmosfernoye Elektrichestvo*, Leningrad.*

Benedicks, C. (1951), "Theory of lightning balls and its application to the atmospheric phenomenon called 'flying saucers'," *Ark. Geofys.* **2(1)** 1.

Benedicks, C. (1954), "Theory of the lightning-balls and its application to the atmospheric phenomenon called flying saucers," *Arkiv foer Geofysik* **2** 1.

Berg, H. (1930), "Beobachtung eines Perlschnurblitzes," *Zeit. Angew. Meteorol.* **47** 315.

Berger, K. (1967), "Novel observations on lightning discharges: results of research on Mt. San Salvatore," *J. Franklin Inst.* **283** 478. (NBL)

Berger, K. (1973), "Kugelblitz und Blitsforschung," *Naturwiss.* **60** 485.

Berger, K. (1977a), "The earth flash" in R. H. Golde (ed.) *Lightning*, Vol. 1, p. 5, Academic Press, New York. (NBL)

Berger, K. (1977b), "Protection of underground blasting operations," in R. H. Golde (ed.) *Lightning*, Vol. II, p. 20, Academic Press, New York. (NBL)

Berger, K. (1978), "Blitzstrom-Parameter von Aufwärtsblitzen," *Bull. Schweiz. Elektrotech. Ver.* **69** 353.

Berger, K. (1982), "Short survey of initiation of earth flashes to Mount San Salvatore" (unpublished) paper presented at International Aerospace Conference on Lightning and Static Electricity.

Berger, K., and Vogelsanger, E. (1965), "Messunger und Resultate der Blitforschung der Jahre 1955–1963 auf dem Mount San Salvatore," *Bull. Schweiz. Elektrotech. Ver.* **56** 2. (NBL)

Berger, K., Anderson, R. B., and Kröninger, H. (1975), "Parameters of lightning flashes," *Electra* **40** 101. (NBL)

Bergstrom, A. (1973), "Electromagnetic theory of strong interaction," *Phys. Rev.* **D3** 4394.

Bergstrom, A., and Campbell, S. (1991), "The Ashford 'ball lightning' video explained," *J. Met.* **16(160)** 185.

Berlage, H. P. (1955), "Van Bliksemballen naar Lichtgevende aardappels en weer terug," *Hamel en Dampkrig* **52** 65.*

Berrett, P. H. (1994), "Severe thunderstorm and 'ball' lightning at Birmingham Airport on 3/4 August 1994," *J. Met.* **19(192)** 265.

Bertholon, M. (1789), "De l'électricité des météores," *Lyon* **2** 27.*

Besnou, M. (1852), "De l'état de l'iode dans l'atmosphère, et de la possibilité de la formation de l'iodure d'azote dans les orages," *Mem. Soc. Sci. Nat. Cherbourg* **1** 103.*

Beveridge, W. I. B. (1950), *The Art of Scientific Investigation,* Heinemann, London. (NBL)

Beynon, J. (1972), *Conduction of Electricity through Gases,* p. 80, George G. Harrap, London.

Bhattacharya, A. B. (1982), "Ball lightning over the river surface," *Mausam* **33(3)** 394.

Bhattacharya, A. B. (1992), "Ball lightning—an extreme form of atmospheric electricity," *Vayu Mandal* **22(3, 4)** 126.

Bigelow, P. (1907), "Electrical storm in Southern California," *Mon. Weather Rev.* **35** 228.

Bjarne-Solas, M. (1959), "Foudre en boule," *Bull. Soc. Astron. Fr.* **73** 16.

Blackhouse, T. W. (1899), "Remarkable lightning flashes," *Nature* **60** 520.

Blackman, G. d'Arcy (1922), "A waterspout with fireballs," *Meteorol. Mag.* **57** 250.

Blair, A. J. F. (1973), "Magnetic fields, ball lightning and campanology," *Nature* **243** 512.

Blake, R. (1977), "The wizard of Wendover," *Mountain West Magazine* Nov, p. 25.

Blanc, E. (1877a), "Observation d'éclairs en boule, se formant et éclatant sans bruit au-dessus d'une couche de mirages," *C. R. Hebd. Séances Acad. Sci.* **84** 666.

Blanc, E. (1877b), "Ball lightning," *Nature* **15** 539.

Blok, V. R, Kogan, A. Y., Kogan, B. Y., Krochik, G. M., and Lebedev, O. L. (1981), "Stability of the ball lightning," *Zh. Tekh. Fiz.* **51(10)** 2190.

Blumhof, F. (1905a), "Kugelblitz bei Hagelwetter" (summary), *Fortschr. Phys.* **61(3)** 324.

Blumhof, F. (1905b), "Eigentumliche Lichterscheinung," *Meteorol. Zeit.* **22** 132.

Boggs, E. M. (1898), "Ball lightning," *Mon. Weather Rev.* **26** 565.

Boichenko, A. M. (1996a), "On the nature of bead lightning" (in Russian), *Fizika Plazmy* **22(11)** 1012.

Boichenko, A. M. (1996b), "On the nature of bead lightning" (English translation), *Plasma Physics Reports* **22(11)** 917.

Boll, K. (1913), "Feuerkugel," *Zeit. Angew. Meteorol.* **30** 214.

Boll, K. (1918), "Beobachtung von Kugel und Perlschnurblitzen zu Jambol (Bulgarien)," *Zeit. Angew. Meteorol.* **35** 130.

Boll, K. (1920), "Schnellbewegte Kugelblitz," *Zeit. Angew. Meteorol.* **37** 191.

Bon, G. le (1905), "L'évolution de la matière," *Paris* 154.*

Bonacina, L. C. W. (1946), "The Widecombe calamity of 1638," *Weather* **1** 122.

Bonney, A. (1887a), "Appearance of air bubbles at Remenham, Berkshire, January 1871," *Amer. J. Meteorol.* **4** 149.

Bonney, A. (1887b), "Appearance of air bubbles at Remenham, Berkshire, January 1871," *Electrician* **19** 179.

Bonney, A. (1887c), "Appearance of air bubbles at Remenham, Berkshire, January 1871," *Meteorol. Zeit.* **22** 88.

Boot, H. A. H., Self, S. A., and Shersby-Harris, R. (1958), "Containment of a fully ionized plasma by RF fields," *J. Elect. Control* **4** 434.

Borlase, W. (1753), "An account of a storm of thunder and lightning, near Ludgram in Cornwall," *Phil. Trans. Roy. Soc. London* **48** 86.

Bort, M. T. de (1896), "Eclair en chapelet," *La Météorologie* **44** 66.

Bostick, W. H. (1956), "Experimental study of ionized matter projected across magnetic field," *Phys. Rev.* **104** 292. (NBL)

Bostick, W. H. (1957), "Experimental study of plasmoids," *Phys. Rev.* **106** 404. (NBL)
Bostick, W. H., Prior, W., and Farber, E. (1965), "Plasma vortices in the coaxial plasma accelerator," *Phys. Fluids* **8** 745. (NBL)
Botley, C. M. (1925), "A fireball in 1792," *Meteorol. Mag.* **59** 185.
Botley, C. M. (1938), "The Air and its Mysteries," *Bell Publ. London* Plate X.*
Botley, C. M. (1966), "The Widecombe calamity," *Weather* **21** 318.
Botley, C. M. (1977), "Tornadoes and ball lightning," *J. Met.* **2(21)** 271.
Botley, C. M. (1978), "Ball lightning," *J. Met.* **3(31)** 215.
Bottlinger, K. F. (1928), "Zur Frage nach der Natur der Kugelblitze," *Naturwiss.* **16** 220.
Bougon, D. (1897), "La Météorologie," *Bull. Soc. Astron. Fr.* **11** 430.
Bougon, D. (1901), "Coups de foudre," *Bull. Soc. Astron. Fr.* **15** 303.
Bougon, D. (1902a), "Communications diverses—un curieux cas de foudre globulaire," *Bull. Soc. Astron. Fr.* **16** 264.
Bougon, D. (1902b), "Foudre globulaire," *Bull. Soc. Astron. Fr.* **16** 420.
Bouin, M. (1861), *Le Cosmos* **18** 758.
Boulay, J.-L. (1982), "Atmospheric electric field and aircraft potential variation in flight," in *Proceedings: International Aerospace Conference on Lightning and Static Electricity*, Public Relations Section, Culham Laboratory, Abingdon, Oxon., UK.
Bowman, J. E. (1949), "St. Elmo's fire near Glencoe," *Weather* **4** 197. (NBL)
Boyce, C. F. (1977), "Protection of telecommunication systems," in R. H. Golde (ed.) *Lightning*, Vol. II, p. 25, Academic Press, New York. (NBL)
Boyd, T. J. M. and Sanderson, J. J. (1969), *Plasma Dynamics,* Thomas Nelson and Sons, London. (NBL)
Boys, H. A. (1910), "Remarkable lightning flash," *Meteorol. Mag.* **45** 108.
Brand, W. (1917), "Jahrliche Horbarkeitsperiode des Geschutzdonners und seine grossten bisherigen Reichweiten," *Meteorol. Zeit.* **17** 58.
Brand, W. (1923), *Der Kugelblitz: Probleme der Kosmischen Physik*, Vols. II and III, H. Grand, Hamburg.
Brand, W. (1971), "Ball lightning," NASA Technical Translation F-13228, National Aeronautics and Space Administration, Washington, DC.
Brasch, A. (1928), "Vorlaufige versuche zur Erzielung extremer elektrischer Potentiale Mittles der Atmospharischen Gewitterelektrizitat," *Naturwiss.* **16** 115.
Bremen, H. (1896), *Abhandl. Naturwiss. Ver. Bremen* **13** 312.
Brereton, O. S. (1781), "Account of the violent storms of lightning at East-bourn in Sussex, September 17, 1780," *Phil. Trans. Roy. Soc. London* **71** 42.
Brester, A. (1910), *Rev. Generale Sci.* **21** 385.
Brevdel, A. (1904), "Nature Intimé de l'Electricité du Magnétisme et des Radiations," *Bruxelles-Paris* 20.*
Brew, W. (1893), "A particular discharge of lightning," *Nature* **48** 370.
Brinkhuis, B. (1975), "Bolbliksem, een raadsel," *Zenit, Utrecht* **2** 117.
Briscoe, C. E. (1985), "Hailstorm in Norfolk, 5 July 1985," *J. Met.* **10(103)** 333.
Bromley, K. A. (1970), "Ball lightning," *Nature* **226** 253.
Brook, M. (1967), "Electric currents accompanying tornado activity," *Science* **157** 1434. (NBL)
Brook, M., and Ogawa, T. (1977), "The cloud discharge," in R. H. Golde (ed.), *Lightning*, Vol. 1, p. 6, Academic Press, New York. (NBL)
Brook, M., Kitagawa, N., and Workman, E. J. (1962), "Quantitative study of strokes and continuing currents in lightning discharges to ground," *J. Geophys. Res.* **67** 649. (NBL)
Brook, M., Armstrong, G., Winder, R. P. H., Vonnegut, B., and Moore, C. B. (1961), "Artificial initiation of lightning discharges," *J. Geophys. Res.* **66** 3967. (NBL)
Brooks, C. F. (1946), "A case of ball lightning," *Amer. Meteorol. Soc. Bull.* **47** 342.
Broughton, S. (1873a), "Ball discharge in thunderstorms," *Chem. News* **27** 172.*
Broughton, S. (1873b), "Ball discharge in thunderstorms" (summary), *Fortschr. Phys.* **29** 1003.

Broughton, S. (1873c), "Manchester Literary and Philosophical Society, Communications," *Nature* **7** 415.

Broughton, S. (1873d), "Ball discharge in thunderstorms," *Proc. Manchester Soc.* **4/3**.*

Brovetto, P., Maxia, V., and Bussetti, G. (1976), "On the nature of ball lightning," *J. Atmos. Terres. Phys.* **38** 921.

Brovetto, P., Maxia, V., and Bussetti, G. (1982), "Some remarks about a ball lightning model," *J. Atmos. Terres. Phys.* **44(11)** 925.

Brown, G. H. (1957), "Ball lightning," *Meteorol. Mag.* **86** 375.

Brown, J. (1858), *The Times*, London December 1.*

Brown, S. C. (1951), "High frequency gas discharge breakdown," *Proc. IRE* **39** 1493.

Browne, T. (1964), "Account of a thunderstorm on June 28, 1665," in *Misc. Writings of Sir Thomas Browne*, p. 195, Faber and Faber, London.

Bruce, C. E. R. (1963a), "Ball lightning, stellar rotation and radio galaxies," *Engineer* **216** 1047.

Bruce, C. E. R. (1963b), "Ball lightning," *IEEJ, London* **9** 357.

Bruce, C. E. R. (1964), "Ball lightning," *Nature* **202** 996.

Bruce, C. E. R. (1967), "Lightning," *Research* **2** 12.

Bruce, C. E. R., and Golde, R. H. (1941), "The lightning discharge," *IEEJ, London* **88 II** 487.

Bruce, H. (1960), *Elect. Rev.* **167** 1070.

Bruchsal, H. (1900), *Zeit. Angew. Meteorol.* **17** 191.

Bruley-Mosley, M. (1908a), "Boule de feu," *Bull. Soc. Astron. Fr.* **22** 528.

Bruley-Mosley, M. (1908b), "La Météorologie," *L'Astronomie* **22** 528.*

Brzak, W. (1892a), "Kugelblitz" (summary), *Fortschr. Phys.* **48(3)** 360.

Brzak, W. (1892b), "Kugelblitz," *Meteorol Zeit.* **9** 355.

Buchwalder, H. (1832), *Naturwiss.* **1** 327.*

Buchwalder, H. (1891), *Lehrbuch Meteorol.* 327.

Budde, E. (1920), "Schnellbewegte Kugelblitz," *Zeit. Angew. Meteorol.* **37** 87.

Burbidge, P. W., and Robertson, D. J. (1982), "A lightning-associated phenomenon and related geomagnetic measurements," *Nature* **300** 623.

Burckhardt, F. (1895a), "Ein Blitzstrahl von 14 Juli 1894" (summary), *Fortschr. Phys.* **51(3)** 400.

Burckhardt, F. (1895b), "Ein Blitzstrahl von 14 Juli 1894," *Verh. Naturf. Ges. Basel* **11** 134.*

Burger, K. (1973), "Kugelblitz und Blitzforschung," *Naturwiss.* **60** 485.

Burgess, C. R. (1950), *Meteorology for Seamen*, p. 55, Brown, Son and Ferguson, Glasgow.

Butti, M. (1852a), "Foudre globulaire à Milan, en 1841 (Juin?)," *C. R. Hebd. Séances Acad. Sci.* **35** 193.

Butti, M. (1852b), "Foudre globulaire à Milan, en 1841," *Le Cosmos* **1** 350.*

Butz, J. S. (1968), "Great balls of fire," *Air Force Mag.* April 15, p. 149.

Bychkov, V. L. (1993a), "Ball lightning as a polymer formation," in A. G. Keul, (ed.) *Progress in ball lightning research: Proc. VIZOTUM*, p. 11, The Vizotum Project, Salzburg, Austria.

Bychkov, V. L. (1993b), "Ball lightning as a polymer formation," *J. Met.* **18(179)** 149.

Bychkov, V. L. (1994), "Polymer ball lightning model," *Physica Scripta* **50(5)** 591.

Bychkov, V. L. Bychkov, A. V, and Standnik, S. A. (1996), "Polymer fire balls in discharge plasma," *Physica Scripta* **53(6)** 749.

Bychkov, V.L., Bobkov, S. E., and Stadnik, S. A. (1997), "On one type of luminous objects of Hessdalen Valley," in Y. H. Ohtsuki and H. Ofuruton (eds.) *Proceedings: 5th International Symposium on Ball Lightning (ISBL97)*, p. 47.

Bychkov, V. L., Smirnov, B. M., and Stridjev, A. Ju (1993), "Analysis of the Russian-Austrian ball lightning data banks," *J. Met.* **18(178)** 113.

Bychkov, V. L., Bychkov, A. V., Vasiliev, M. N., and Klimov, A. I. (1996), "Evaluation of possible ball lightning energy by analysing an event involving damage to a metal pot," *J. Met.* **21(207)** 77.

Bychkov, V. L., Emelin, S. E., Klimov, A. I., and Semenov, V. S. (1997), "Approaches to ball lightning modelling with erosive discharges," in Y. H. Ohtsuki and H. Ofuruton (eds.) *Proceedings: 5th International Symposium on Ball Lightning (ISBL97)*, p. 188.
Caballero, E. (1890a), "Foudre en boule," *Ciel et Terre* **11** 250.
Caballero, E. (1890b), "Globular lightning at Pontevedra," *Electrical Engineer* **9** 187.
Caballero, E. (1890c), "Globular lightning," *Electrician* **24** 414.
Caballero, E. (1890d), "Kugelblitz in Pontevedra," *Elektro. Zeit.* **11** 66.*
Caballero, E. (1890e), "Kugelblitz in Pontevedra" (summary), *Fortschr. Phys.* **46(3)** 425.
Caballero, E. (1890f), *La Nature* **18** 167.
Caballero, E. (1890g), "Foudre en boule," *La Météorologie* **38** 245.
Caballero, E. (1890h), "Ball lightning," *Meteorol. Mag.* **25** 99.
Caballero, E. (1890i), "Kugelblitz," *Meteorol. Zeit.* **7** 158.
Caballero, E. (1890j), "Ball of fire," *Nature* **41** 303.
Caballero, E. (1890k), *Naturwiss. Rund.* **22** 284.
Caballero, E. (1890l), "Kugelblitz in Pontevedra," *Rev. Intern. Elect.* **6** 202.
Caballero, E. (1890m), "Kugelblitz in Pontevedra," *Rev. Sci.* **46** 286.*
Cabane, L. (1887), *L'Astronomie* **6** 459.
Cade, C. M. (1962), "Thunderbolts as the X-weapon," *Discovery* **23** 23.
Cade, C. M. (1964), "Radiation-the new weapon," *Science Progress* **52** 232.
Cade, C. M., and Davis, D. (1969), *Taming of the Thunderbolts*, Abelard-Schuman, New York.
Cade, M., and Davis, D. (eds.) (19 XXX), *Flame of the Heaven: An Anthology of Ball Lightning*, London.*
Cadenat, M. (1890), "Uber die Kugelblitz-Erscheinurgen des Tornados von St.-Claude," *Zeit. Angew. Meteorol.* **7** 285.
Cailleux, M. (1932), "Foudre en boule," *Bull. Soc. Astron. Fr.* **46** 569.
Callahan, P. S. (1979), "Insects as unidentified flying objects: author's reply to comment; 1," *Appl. Opt.* **18(16)** 2724.
Callahan, P. S., and Mankin, R. W. (1978), "Insects as unidentified flying objects," *Appl. Opt.* **17(21)** 3355.
Callaud, A. (1874), *Trait des Paratonnerres*, Paris.*
Campbell, S. (1981a), "Not lightning damage," *Weather* **36(3)** 66.
Campbell, S. (1981b), "How not to photograph ball lightning," *Br. J. Photog.* **128(43)** 1096.
Campbell, S. (1982), "Ball lightning at Crail—1968," *Weather* **37(3)** 75.
Campbell, S. (1983a), "The snowfall of 11 February 1983 in Mallorca," *J. Met.* **8(78)** 103.
Campbell, S. (1983b), "A close encounter with ball lightning," *J. Met.* **8(81)** 204.
Campbell, S. (1985), "Ball lightning," *New Scientist* **108(1478)** 75.
Campbell, S. (1987a), "Ball lightning exposed! Another picture puzzle," *Br. J. Photog.* **134(6645)** 1537.
Campbell, S. (1987b), "A close encounter with ball lightning" (further comment by author), *J. Met.* **12(119)** 172.
Campbell, S. (1987c), "Ball lightning at Crail—1968" (further comment by author), *Weather* **42(10)** 329.
Campbell, S. (1988a), "Russian accounts of ball lightning," *J. Met.* **13(128)** 126.
Campbell, S. (1988b), "The Smethwick ball lightning report," *J. Met.* **13(134)** 391.
Campbell, S. (1988c), "The Childerhose UFO: fact or fiction?," *Br. J. Photog.* Sept. 29, p. 72.
Campbell, S. (1991), "Fireball by Day," *Br. J. Photog.* Apr. 4, p. 22.
Campbell, S. (1992a), "Does ball lightning exist?" in M. Stenhoff (ed.) *Proceedings: Fourth TORRO Conference: Ball Lightning*, p. 15, Tornado and Storm Research Organisation (TORRO), Oxford Brookes University, Oxford, United Kingdom.
Campbell, S. (1992b), "Great balls of fire," *The Skeptic* Nov/Dec, p. 12.
Campbell, S. (1993), "Comments on ball lightning and 'greenhouse effect'," *J. Met.* **18** 259.

Campbell, S. (1997), "Ball lightning through the window glass" (Comment on Turner 1997a), *J. Met.* **22(220)** 216.

Cancani, M. (1889a), "Sopra un caso di duplice fulminazione avvenuto a Canterano il 26 Aprile 1889 e sul'esistenza dei fulmini globulari," *Atti Pontif. Accad. Rom. D. Nuovi Lincei* **5** 796.*

Cancani, M. (1889b), "Sopra un caso di duplice fulminazione avvenuto a Canterano il 26 Aprile 1889 e sul'esistenza dei fulmini globulari" (summary), *Fortschr. Phys.* **45(3)** 514.

Capron, J. R. (1883a), "Ball lightning," *Meteorol. Mag.* **18** 41.

Capron, J. R. (1883b), "Globe lightning," *Meteorol. Mag.* **18** 57.

Carlheim-Gylensköld, V. (1905a), "Des foudres en boule," *Berichte Internat. Meteorol. Direktorenkonf Innsbruck* App. XXIII 1.*

Carlheim-Gylensköld, V. (1905b), *Kgl. Zeit Meteorol. Geodyn.* Suppl. 113.*

Carpenter, D. G. (1962), "Plasma theory applied to ball lightning," Thesis, Iowa State University, Ames, Iowa.

Carpenter, D. G. (1963), "Ball lightning as a plasma phenomenon," *AIAA Student J.* **1** 25.

Carre, C. (1889), *La Lumière Electrique* **33** 143.

Carruthers, J., and Foster, B. D. P. (1947), "Ball lightning," *Meteorol. Mag.* **72** 210.

Cartwright, B. W. (1938), "Lightning," *Life Mag.* **3** Aug 23, p. 77.

Cartwright, J. (1872), *J. Soc. Tel. Engr.* **1** 372.

Cawood, W., and Patterson, H. S. (1931a), "A curious phenomenon shown by highly charged aerosols," *Nature* **128** 637. (NBL)

Cawood, W., and Patterson, H. S. (1931b), "Ball lightning, old puzzle to science, made out of smoke," *Science News Lett.* **20** 366.

Cecchini, S., Diocco, G., and Mandolesi, N. (1974), "Positron annihilation in EAS and ball lightning," *Nature* **250** 637.

Cecil, H. (1884), "Fireball," *Nature* **30** 289.

Cerrillo, M. (1943), "Sobre las posibles interpretaciones electromagneticas del fenomeno de las centellas," *Comision Implusora y Coordinadora de la Investieacion Cientifica* **1** Anuario. 151.

Cerrillo, M. (1960), "Concerning the possible electromagnetic interpretation of the phenomenon of lightning balls," Unpublished report, Raytheon Co., Wayland, Massachusetts.

Chalmers, J. A. (1949), *Atmospheric Electricity* p. 152, Clarendon Press, Oxford, England.

Chalmers, J. A. (1957), *Atmospheric Electricity* p. 255, Pergamon Press, New York.

Chalmers, J. A. (1976), *Atmospheric Electricity* p. 390, Pergamon Press, London.

Chalmers, L. (1888), *Phil. Trans. Roy. Soc. London* **10(2)** 481.*

Chalmers, L. E. (1751), "An account of an extraordinary fireball bursting at sea," *Phil. Trans. Roy. Soc. London* **46** 366.

Chamberlin, J. (1945), "The foo fighter mystery," *American Legion Mag.* **39** 9.

Chameides, W. L., Steadman, D. H., Dickerson, R. R., Rush, D. W., and Cicerone, C. J. (1977), "NO production in lightning," *J. Atmos. Sci.* **34** 143. (NBL)

Chandler, M. (1896), "Violent lightning stroke in South Devon," *Meteorol. Mag.* **31** 5.

Chandrasekhar, S. and Fermi, E. (1953), *Astrophysics J.* **118** 116. (NBL)

Charlier, L. A. (1966), "Kugelblitz-Beobachtung in Vorarlberg," *Wetter und Leben* **18** 204.

Charman, W. N. (1971a), "Perceptual effects and reliability of ball lightning reports," *J. Atmos. Terres. Phys.* **33** 1973.

Charman, W. N. (1971b), "After images and ball lightning," *Nature* **230** 576.

Charman, W. N. (1972), "The enigma of ball lightning," *New Scientist* **56** 632.

Charman, W. N. (1976), "Ball lightning photographed," *New Scientist* **69** 444.

Charman, W. N. (1979), "Ball lightning," *Phys. Reports* **54(4)** 261.

Charman, W. N. (1982), "Ball-lightning: the unsolved problem," *Weather* **37(3)** 66.

Chattock, A. P. (1922), "Globular lightning discharge," *Nature* **109** 106.

Chattock, H. (1898), *BAAS* 801.

Chauveau, B. (1922), *Electricité Atmosphérique,* Librairie Octave Doin, Paris.

Cheetham, G. R. (1965), "St. Elmo's fire," *Marine Observer* **26(172)** 81.

Chen, F. F. (1974), *Introduction to Plasma Physics,* Plenum Press, New York. (NBL)

Child, W. (1755), "On the effects of lightning at Dorking in Surrey," *Phil. Trans. Roy. Soc. London* **49** 309.

Chirvinskiy, P. N. (1936a), "Materials on observations of ball lightning," *Klimat i Pogoda* **11** 5.*

Chirvinskiy, P. N. (1936b), "Materialy po navlyudeniyam nao sharovymi molniyai covrannye V. K. Cherkasom," *Klimat i Pogoda* **5** 49.*

Chirvinskiy, P. N. (1936c), *Klimat i Pogoda* **5** 68.

Chirvinskiy, P. N. (1936d), "K teorii sharovynh molniya," *Meteorologiya I Gidrologiya* **7** 78.

Chirvinskiy, P. N. (1949), "Sharov'le molnii," *Priroda* **6** 14.*

Chirvinskiy, P. N. (1954), "Pervye nablyudateli sharovykh molny," *Priroda* **43(8)** 116.

Chown, J. B., Scharfman, W. E., and Marita, T. (1959), "Voltage breakdown characteristics of microwave antennas," *Proc. IRE* **47** 1331.

Chown, M. (1993), "Fire and water: a recipe for ball lightning," *New Scientist* **137(1865)** 18.

Clare, P. (1850), "An account of some thunderstorms and extraordinary electrical phenomena that occurred in the neighbourhood of Manchester on Tue. 16th July 1850," *Phil. Mag.* **37** 329.

Clark, L. (1872), *J. Soc. Tel. Engr.* **1** 372.

Clark, R. (1965), "Ball lightning," *Weather* **20** 134.

Claubry, G. de (1874a), "Sur le tonnerre en boule," *C. R. Hebd. Séances Acad. Sci.* **79** 137.

Claubry, G. de (1874b), "Sur le tonnerre en boule" (summary), *Fortschr. Phys.* **30** 1177.

Claubry, G. de (1874c), "Sur le tonnerre en boule," *Inst. Sci. Soc. Fr.* **2** 241 *.

Claubry, G. de (1874d), "Sur le tonnerre en boule," *Mondes* **34** 50.

Clavel, J. (1901), "Coup de foudre intéressant," *Ciel et Terre* **22** 302.

Cliff, J. S. (1977), "Insulation coordination" in R. H. Golde (ed.) *Lightning,* Vol. II, p. 24, Academic Press, New York. (NBL)

Clifford, D. W. (1982), "Characteristics of lightning strikes to aircraft," in *Proceedings: International Aerospace Conference on Lightning and Static Electricity,* Public Relations Section, Culham Laboratory, Abingdon, Oxon., UK. (NBL)

Cocking, A. (1884), "Fireball," *Nature* **30** 269.

Coleman, P. F. (1993), "An explanation of ball lightning?," *Weather* **48(1)** 30.

Coleman, P. F. (1997), "Vortex breakdown burner hypothesis of ball lightning," in Y. H. Ohtsuki and H. Ofuruton (eds.) *Proceedings: 5th International Symposium on Ball Lightning (ISBL97),* p. 176.

Colgate, S. A. (1967), "Tornadoes: mechanism and control," *Science* **157** 1431. (NBL)

Colucci, S. (1976), "Ball lightning observation," *Weather* **31** 68.

Comanesti, G. (1938), "Foudre en boule," *Bull. Soc. Astron. Fr.* **52** 265.

Condon, E. U. (ed.) (1969), *Scientific Study of Unidentified Flying Objects,* Bantam Books, New York.

Cook, M. L. (1982), "Ball lightning incident in Berkhamsted, 13 September 1981," *J. Met.* **7(65)** 18.

Cooke, B. (1742), "Fire-ball seen Dec. 11, 1741," *Phil. Trans. Roy. Soc. London* **42** 25.

Corliss, W. R. (1977), *Handbook of Unusual Natural Phenomena,* Sourcebook Project, Glenarm, Maryland.

Corliss, W. R. (1982), "Lightning, auroras, nocturnal lights, and related luminous phenomena," in *A Catalog of Geophysical Anomalies,* Sourcebook Project, Glenarm, Maryland.

Corliss, W. R. (1988), "Beyond the rainbow," *Mariners Weather Log* **32(1)** 22.

Coroniti, S. C. (1959), "Evaluation of Kapitza's article on the nature of ball lightning," Unpublished report, RD-2-29-2-6, Avco Corp., Boston, Massachusetts.

Coroniti, S. C. (ed.) (1963), "Problems of atmospheric and space electricity," in *International Conference Proceedings,* Elsevier, Amsterdam.

Coroniti, S. C., and Hughes, J. (eds.) (1968), *Planetary Electrodynamics,* Gordon and Breach, New York.

Corum, K. L., and Corum, J. F. (1990a), "High-voltage rf ball lightning experiments and electrochemical fractal clusters," *Usp. Fiz. Nauk* **160(4)** 47.

Corum, K. L., and Corum, J. F. (1990b), "RF high voltage fire ball experiments and electro-chemical fractal clusters," ICBL Article Series No. 1992/1, International Committee on Ball Lightning (ICBL), Terneuzen, The Netherlands.

Coulvier-Gravier (1859), *Récherches sur les Météores et sur les Lois qui les Regissent,* p. 185, Paris.

Courtrai, R. (1860), *Le Cosmos* **16** 281.

Covington, A. E. (1970), "Ball lightning," *Nature* **226** 252.

Cowan, E. W. (1889), "Globular lightning," *Electrician* **22** 720.

Craggs, J. D., Hopwood, W., and Meek, J. M. (1947), "On the localized afterglows observed with long sparks in various gases," *J. Appl. Phys.* **18** 919.

Craige, H. (1930), "Observations of lightning," *Science* **72** 344.

Crawford, J. F. (1972), "Antimatter and ball lightning," *Nature* **239** 395.

Crawford, W. (1895), "Lightning-chain formation," *Nature* **53** 5.

Crew, E. W. (1972), "Ball lightning," *New Scientist* **56** 764.

Crew, E. W. (1976a), "Atmospheric mysteries and lightning," *Electr. Rev.* **199(24)** 21. (NBL)

Crew, E. W. (1976b), "Fall of a large ice lump after a violent stroke of lightning," *J. Met.* **1(10)** 312.

Crew, E. W. (1981a), "Subjective aspects of ball lightning seen inside aircraft," *J. Met.* **6(56)** 43.

Crew, E. W. (1981b), "Ball lightning theories," *J. Met.* **6(62)** 260.

Crew, E. W. (1983), "Ball lightning," *Electronics and Power* **29(4)** 292.

Crew, E. W. (1992a), "Possible ball lightning in Austria," *J. Met.* **17(169)** 166.

Crew, E. W. (1992b), "Possible ball lightning in Austria," *J. Met.* **17(172)** 274.

Crew, E. W. (1995), "Ball lightning in 'More about France'" (Letter), *J. Met.* **20** 391.

Cromer, A. (1993), *Uncommon Sense—the Heretical Nature of Science,* Oxford University Press, New York. (NBL)

Crookes, W. (1892), *Chem. News* **65** 301.

Cross, G. A. (1926), "La météorologie-un cas de foudre globulaire," *Bull. Soc. Astron. Fr.* **40** 504.

Crossley, A. W. (1924), "Ball lightning," *Nature* **114** 10.

Crotsenburg, C. N. (1898), "Ball lightning," *Mon. Weather Rev.* **26** 358.

Crouch, K. E. (1982), "Detection of sparks in fuel system tests," in *Proceedings: International Aerospace Conference on Lightning and Static Electricity, Public Relations Section, Culham Laboratory, Abingdon, Oxon., UK.*

Culafic, M. R. (1965), *Grom i Zastita Od Njeg,* p. 76, Izdavacko Preduzece, Gredevinska, Knjiga, Belgrade.*

Cunisset-Carnot, M. (1897), "La foudre en boule," *Bull. Soc. Astron. Fr.* **11** 299.

Daerdaeevskii, B. L. (1957), "Redkii snimok molnii," *Priroda* **4** 113.

Daguin, L. (1887), *La Nature* Sep 3.

Dale, H. D. (1923), "Ball lightning," *Meteorol. Mag.* **58** 258.

Dalloz, P. (1928), *La Montagne* **24(210)** 81.*

Datlov, J. (1965), "The inverted ion magnetron principle and its usefulness in laboratory plasma physics and cosmic electrodynamics," *Czech J. Phys.* **B15** 858.

Dauvillier, A. (1957), "Foudre globulaire et reactions thermonucléaires," *C. R. Hebd. Séances Acad. Sci.* **245** 2155.

Dauvillier, A. (1965), "Sur la nature de la foudre globulaire," *C. R. Hebd. Séances Acad. Sci.* **260** 1707.

Dauvillier, A. (1970), "Sur le mécanisme de la foudre globulaire," *Rev. Gen. Elect.* **79** 433.

Davidov, B. V. (1958), "Redkaia fotografiia sharovi molnii," *Priroda* **47(1)** 96.

Davies, B. (1925), "Ball lightning phenomena," *Nature* **115** 640.

Davies, D. W., and Standler, R. B. (1972), "Ball lightning," *Nature* **240** 144.

Davies, P. C. W. (1971), "Ball lightning or spots before the eyes?," *Nature* **230** 576

Davies, P. C. W. (1976), "Ball lightning," *Nature* **260** 573.

Davies, P. C. W. (1987), "Great balls of fire," *New Scientist* **116(1592/1593)** 64.

Dawson, G. A., and Jones, R. C. (1968a), "Model for ball lightning," in *Planetary Electrodynamics*, S. C. Coroniti and J. Hughes (eds.), Vol. II, p. 193, Gordon and Breach, New York.

Dawson, G. A., and Jones, R. C. (1968b), "Ball lightning as a radiation bubble," paper presented at Fourth International Conference on the Universal Aspects of Atmospheric Electricity.

Dawson, G. A., and Jones, R. C. (1969), "Ball lightning as a radiation bubble," *Pure Appl. Geophys.* **75** 247.

Dayton, L. (1991), "Microwaves cook up ball lightning to order," *New Scientist* **129(1761)** 24.

Decerfz, C. (1847), "Sur un cas de foudre qui a frappé la Prison de La Chatre," *C. R. Hebd. Séances Acad. Sci.* **25** 85.

Decharme, C. (1884a), "Notes relatives à un orage qui à éclair à Amiens, le 24 février dernier," *C. R. Hebd. Séances Acad. Sci.* **98** 606.

Decharme, C. (1884b), "Memoires presentes-foudre globulaire," *C. R. Hebd. Séances Acad. Sci.* **98** 658.

Decharme, C. (1884c), *La Lumière Electrique* **11** 551.

Decker, F. (1921), "Beobachtung von Kugelblitz," *Meteorol. Zeit.* **38** 183.

Dember, H., and Meyer, U. (1912a), "Kugelblitz am 12 Mai 1912 bei Dresden" (summary), *Fortschr. Phys.* **68(3)** 310.

Dember, H., and Meyer, U. (1912b), "Kugelblitz am 12 Mai 1912 bei Dresden," *Meteorol. Zeit.* **29** 384.

Dendy, R. (ed.) (1993), *Plasma Physics: An Introductory Course*, Cambridge University Press, Cambridge, England. (NBL)

Denning, W. F. (1894), "Meteors and electrical phenomenon," *Meteorol. Mag.* **29** 18.

Derham, W. (1729), "Of the meteor called ignis fatuus from observations made in England," *Phil. Trans. Roy. Soc. London* **26** 204.

Dessens, J. (1965), "Quelques tornades Francaises recentes," *J. Récherches Atmosphériques* **2** 91.

Dewan, E. M. (1954), "Eyewitness accounts of Kugelblitz," paper AFCRL-125, U.S. Air Force Cambridge Research Laboratory.

Dewan, E. M. (1964), "Attempted explanations of ball lightning," paper No. 67, AFCRL-64-927, U.S. Air Force Cambridge Research Laboratory.

Dewar, R. A. (1983), "Ball lightning: fifth state of matter," *Spec. Sci. Technol.* **6(3)** 267.

Dibner, B. (1977), "Benjamin Franklin" in R. H. Golde (ed.) *Lightning*, Vol. I, p. 2, Academic Press, New York. (NBL)

Diendorfer, G. (1993), "Lightning models as description of cloud to ground discharges," in A. G. Keul (ed.) *Progress in Ball Lightning Research: Proc. VIZOTUM*, p. 14, The Vizotum Project, Salzburg, Austria.

Dijkhuis, G. C. (1979), "Thermonuclear energy from ball lightning," in *Proc. 14th Intersociety Energy Conversion Engineering Conf.*, Vol. II, p. 1614.

Dijkhuis, G. C. (1980), "A model for ball lightning," *Nature* **284(5752)** 150.

Dijkhuis, G. C. (1981), "Reply to E. A. Witalis," *Nature* **290(5802)** 160.

Dijkhuis, G. C. (1982), "Threshold current for fireball generation," *J. Appl. Phys.* **53** 3516.

Dijkhuis, G. C. (1985a), "Method and apparatus for creating and maintaining a self-supporting plasma ball," European Patent 0019 668 B1.

Dijkhuis, G. C. (1985b), "Het fase-diagram van bolbliksem: model en toetsing," *Nederlands Tijdschrift voor Natuurkunde* **B51** 125, 128.

Dijkhuis, G. C. (1988), "Scaling law for fusion power from ball lightning," in J. Janiszewski, W. Moron, and W. Sega (eds.) *Proceedings of the International Wroclaw Symposium on Electromagnetic Compatibility*, p. 21.

Dijkhuis, G. C. (1990), "Cold fusion process in dense boson plasma," European Patent 0 422 057 B1.

Dijkhuis, G. C. (1991a), "Boundary layer model and calculation for horizontal thundercloud electrification preceding natural and rocket-triggered lightning," in H. Kikuchi (ed.) *Environmental and Space Electromagnetics*, p. 113, Springer-Verlag, Berlin.

Dijkhuis, G. C. (1991b), "2nd International Symposium on Ball-Lightning," *Usp. Fiz. Nauk.* **161(1)** 187.

Dijkhuis, G. C. (1991c), "State equation and phase diagram for fractal growth in ball lightning" in H. Kikuchi (ed.) *Environmental and Space Electromagnetics*, p. 535, Springer-Verlag, Berlin.

Dijkhuis, G. C. (1992a), "Statistics and structure of ball lightning," in M. Stenhoff (ed.) *Proceedings: Fourth TORRO Conference: Ball Lightning*, p. 61, Tornado and Storm Research Organisation (TORRO), Oxford Brookes University, Oxford, United Kingdom.

Dijkhuis, G. C. (1992b), "Statistics and structure of ball lightning," in *Proceedings of the Third International Symposium on Ball Lightning*.

Dijkhuis, G. C. (1993a), "Medical effects in a recent ball lightning case from Amsterdam," in A. G. Keul, (ed.) *Progress in Ball Lightning Research: Proc. VIZOTUM*, p. 18, *The Vizotum Project, Salzburg, Austria.*

Dijkhuis, G. C. (1993b), "The 3rd International-Symposium on the Ball-Lightning," *Usp. Fiz. Nauk.* **163(5)** 124.

Dijkhuis, G. C. (1994), "Verhulst dynamics and fractal stretching of transition layer vorticity," in H. Kikuchi (ed.) *Dusty and Dirty Plasmas, Noise and Chaos in Space and in the Laboratory*, Plenum Press, New York.

Dijkhuis, G. C. (1995), "Random walk and fractal deformation of transition layer vorticity," in F. Fauchais (ed.) *Proceedings of the International Seminar on Heat and Mass Transfer under Plasma Conditions*, Begell House, New York.

Dijkhuis, G. C. (1996a), "Report of the Fourth International Symposium on Ball Lightning," *J. Met.* **21** 105.

Dijkhuis, G. C. (1996b), "Constructions for scale-free and kink-free vortex stretching," *Physica B* **228** 144.

Dijkhuis, G. C. (1997), "Helix string field in ball lightning," in Y. H. Ohtsuki and H. Ofuruton (eds.) *Proceedings: 5th International Symposium on Ball Lightning (ISBL97)*, p. 144.

Dijkhuis, G. C., and Pijpelink, J. (1989), "Performance of high-voltage test facility designed for investigation of ball lightning," in Y. H. Ohtsuki (ed.) *Science of Ball Lightning (Fireball)*, p. 325, World Scientific, Singapore.

Dixon, F. E. (1934), "Lightning," *The Peterborough Citizen* Sep 27.

Dixon, F. E. (1955), "Photography and ball lightning," *Weather* **10** 98.

Djebari, B., and Hamelin, J. (1982), "Electromagnetic pulse emitted by lightning," in *Proceedings: International Aerospace Conference on Lightning and Static Electricity*, Public Relations Section, Culham Laboratory, Abingdon, Oxon., UK. (NBL)

Dmitriev, M. T. (1956), "Sharovaya molniia vio vremya zimnei groz'l," *Priroda* **7** 109.

Dmitriev, M. T. (1967a), "Chemical analysis of ball lightning," *Foreign Science Bull.* **3(9)** 30.

Dmitriev, M. T. (1967b), "Priroda sharovoi molnii," *Priroda* **6** 98.

Dmitriev, M. T. (1969a), "Stability mechanism for ball lightning" (in Russian), *Zh. Tekh. Fiz.* **39(2)** 387.

Dmitriev, M. T. (1969b), "Stability mechanism for ball lightning" (English translation.), *Sov. Phys. Tech. Phys.* **14(2)** 284.

Dmitriev, M. T. (1970), "New data on ball lightning," *Wissenschaft. Zeit. Tech. Hochsch. ILM* **16** 87.*

Dmitriev, M. T. (1971a), "Ball lightning: new observations and new hypotheses," *NASA-TT-F-13, 931*, National Aeronautics and Space Administration, Washington, DC.

Dmitriev, M. T. (1971b), "Sharovyye molnii: novyye nablyudeniya I novyye gipotezy," *Priroda* **6** 50.

Dmitriev, M. T. (1973a), "Novyye problemyy sharovyi molnii," *Priroda* **4** 60.

Dmitriev, M. T. (1973b), "New problems of ball lightning" (in Russian), *Priroda* Extracts 60.

Dmitriev, M. T., Bakhtin, B. I., and Martynov, V. I. (1981), "The thermal factor of ball lightning," *Zh. Tekh. Fiz.* **51(12)** 2567.

Dmitriev, M. T., Deriugin, V. M., and Kalinkevich, G. A. (1972), "Ob opticheskom izlucheniii sharovoi molnii," *Zh. Tekh. Fiz.* **42** 2187.

Dmitriev, M. T., Deriugin, V. M., and Kalinkevich, G. A. (1973), "Optical emission from ball lightning," *Sov. Phys. Tech. Phys.* **17(10)** 1724.

Dmitriev, V. G., Emel'yanov, V. N., Kashintsev, M. A., Kulikov, V. V., Solov'ev, A. A., Stel'makh, M. F., and Cherednichenko, O. B. (1979), "Non-linear perception of infra-red radiation in the 800-1355 nm range with human eye," *Sov. J. Quantum Electron.* **9(4)** 475.

Dobelmann, V. (1911), "Foudre en boule," *Bull. Soc. Astron. Fr.* **25** 262.

Dobson, H. J. E. (1925), "Ball lightning at Norwood," *Meteorol. Zeit.* **60** 162.

Dolbear, A. E. (1887), "Globular lightning," *Nature* **11** 38.

Dolbear, A. E. (1888), "Globular lightning," *Science* **11** 38.

Dolezalek, H. (1951), "Einige Kugelblitz-beobachtungen," *Geophys. Pure Appl. (Milan)* **20** 183.

Dolezalek, H. (1977), "Comment on 'On the Problem of Ball Lightning' by W. I. Arabadji," *J. Geophys. Res.* **82(24)** 3498.

Dolgopolov, N. (1957), "Budushcheye sharovoi molnii," *Znaniye-Sila* **3** 17.*

Dorofeev, S. B., Sinitsyn, V. I., and Smirnov, B. M. (1982), "Decomposition of ozone absorbed by activated carbon, and processes occurring in ball lightning," *High Energy Chemistry* **16(3)** 206.

Doublet, J. M. G. (1899a), *L'Eclairage Electrique* **20** 393.

Doublet, J. M. G. (1899b), *L'Eclairage Electrique* **21** 142.

Drake, F. D. (1972), "On the abilities and limitations of witnesses of UFO's and similar phenomena," in C. Sagan and T. Page (eds.) *UFO's—A Scientific Debate*, Cornell University Press, Ithaca, New York. (NBL)

Drobyshevski, E. M. (1997), "Ball lightning: terrestrial manifestation of dark electric matter of the universe," in Y. H. Ohtsuki and H. Ofuruton (eds.) *Proceedings: 5th International Symposium on Ball Lightning (ISBL97)*, p. 157.

Ducros, M. H. (1904), "Foudre; orages," *Bull. Soc. Astron. Fr.* **18** 476.

Dudley, L. P. (1942), "Lightning and kindred phenomena," *Electrician* **129** 507.

Dufay, J., and Mao-Lin, T. (1949), "Spectra of lightning," *Ann. Geophys.* **5(2)** 137.

Dumas, M. (1866), *Bull. Soc. Encouragement* **13** 153.

Dunn, M. (1894), *Les Nouveautes Photographiques*, pp. 253–259, Libraire Illustre, Paris.*

Dyck, C. van (1887), "Globular lightning," *Nature* **11** 110.

Dyck, C. van (1888), "Globular lightning," *Science* **11** 110.

Eaton, H. S. (1887a), "Note on a display of globular lightning at Ringstead Bay, Dorset, on Aug 17 1876," *Electrician* **19** 179.

Eaton, H. S. (1887b), "Note on a display of globular lightning at Ringstead Bay, Dorset, on Aug 17 1876," *Meteorol. Mag.* **22** 87.

Eaton, H. S. (1888), "Note on a display of globular lightning at Ringstead Bay, Dorset, on Aug 17 1876," *Amer. J. Meteorol.* **4** 148.

Eckardt, B. (1914), "Merkwurdige Blitzform," *Meteorol. Zeit.* **31** 548.

Edlund, M. (1883), *Ofver. Kongl. Vet. Akad. Forh.* **40** 86.

Egely, G. (1989a), "Hungarian ball lightning observations in 1987," in Y. H. Ohtsuki (ed.) *Science of Ball Lightning (Fireball)*, p. 19, World Scientific, Singapore.

Egely, G. (1989b), "Physical problems and physical properties of ball lightning" in Y. H. Ohtsuki (ed.) *Science of Ball Lightning (Fireball)*, p. 81, World Scientific, Singapore.

Egely, G. (1993), "Analysis of Hungarian ball lightning observations," in A. G. Keul (ed.) *Progress in Ball Lightning Research: Proc. VIZOTUM*, p. 22, The Vizotum Project, Salzburg, Austria.

Ekval, H. N. (1957), "Cougnard deionizers: new mystery in lightning protection," *Electrical World* **147(2, 11)** 11.

Eleccion, M. (1970), "Ball lightning," *IEEE Trans. Aero. Elect. Sys.* **6** 83.

Eletskii, A. V., and Smirnov, B. M. (1991), "Kinetics of prebreakdown phenomena in atmospheric air," *J. Phys. D (Appl. Phys.)* **24(12)** 2175.

Eliot, L. (1773), "Account of the effects of lightning at Steeple Ashton and Holt, in the county of Wilts. on the 20th June 1772," *Phil. Trans. Roy. Soc. London* **68** 231.

Eliseev, A. A. (1941), *M. V. Lomonosov*, Leningradscoe Otdelenie, Leningrad.

Elliot, G. (1904), "Ball lightning," *Meteorol. Mag.* **39** 133.

Emelin, S. E., Pirozerskii, A. L., Semenov, S. V., and Skvortsov, G. E. (1997a), "Propagation characteristics of the dynamic state in a capillary discharge jet" (in Russian), *Pis'ma v Zhurnal Tekh. Fiz.* **23(19–20)** 54.

Emelin, S. E., Pirozerskii, A. L., Semenov, S. V., and Skvortsov, G. E. (1997b), "Propagation characteristics of the dynamic state in a capillary discharge jet" (English translation), *Sov. Tech. Phys. Letters* **23(10)** 758.

Emelin, S. E., Semenov, V. S., Bychkov, V. L., Belisheva, N. K., and Kovshik, A. P. (1997a), "Some objects formed in the interaction of electrical discharges with metals and polymers" (in Russian), *Zh. Tekh. Fiz.* **67(3)** 19.

Emelin, S. E., Semenov, V. S., Bychkov, V. L., Belisheva, N. K., and Kovshik, A. P. (1997b), "Some objects formed in the interaction of electrical discharges with metals and polymers" (English translation), *Sov. Phys. Tech. Phys.* **42(3)** 269.

Endean, V. G. (1976), "Ball lightning as electromagnetic radiation," *Nature* **263** 753.

Endean, V. G. (1978), "Ball lightning," in *5th International Conference on Gas Discharges*, p. 116.

Endean, V. G. (1992a), "Electromagnetic-field energy containment," *IEEE Proceedings A (Science Measurement and Technology)* **139(4)** 137.

Endean, V. G. (1992b), "Electromagnetic field energy models—some recent developments," in M. Stenhoff (ed.) *Proceedings: Fourth TORRO Conference: Ball Lightning*, p. 75, Tornado and Storm Research Organisation (TORRO), Oxford Brookes University, Oxford, United Kingdom.

Endean, V. G. (1993), "Spinning electric dipole model of ball lightning," *IEEE Proceedings* **140A(6)** 474.

Endean, V. G. (1997a), "Development of the radiation bubble model of ball lightning," *J. Met.* **22(217)** 98.

Endean, V. G. (1997b), "The virial theorem with an extension for statically electrified gaseous spheres," in Y. H. Ohtsuki and H. Ofuruton (eds.) *Proceedings: 5th International Symposium on Ball Lightning (ISBL97)*, p. 96.

Epinal, M. (1821), *Ann. Chem. Phys.* **19** 86.*

Epstein, H. (1911), *Zeit. Physik* **71** 28.

Eriksson, A. J. (1977a), "Video-tape recording of a possible ball lightning event," *Nature* **268** 35.

Eriksson, A. J. (1977b), "An unusual lightning flash," *Weather* **32** 102.

Errera, L., and Uecle, A. (1904), "Un cas interéssant d'éclair en boule," *L'Astronomie* **18** 99.*

Esnault, G. (1934), "Etude de coup de foudre globulaire du 29 Avril 1930 sur la Ferme de l'Echellerie Commune de Wieil-Bauge pres Bauge (Maine-et-Loire)," *La Météorologie* **10** 552.*

Espee, M. de (1852), "Corréspondance," *C. R. Hebd. Séances Acad. Sci.* **35** 400.

Espert, M. (1852), "Tonnerre en boule," *C. R. Hebd. Séances Acad. Sci.* **35** 192.

Ette, A. I. I. (1966), "Laboratory studies of point-discharge from multiple points in irregular configuration," *J. Atmos. Terres. Phys.* **38** 982. (NBL)

Evans, C. (1973), "Cults of Unreason," *Harrap*. (NBL)

Evans, H. (1993), "Ball lightning in the wider context," in A. G. Keul (ed.) *Progress in Ball Lightning Research: Proc. VIZOTUM*, p. 27, The Vizotum Project, Salzburg, Austria.

Evenson, K. M. (1968), "Ball lightning research at Highland Lookout Mountain," Douglas Missile and Space Systems Division. Paper DAC-60941, 51-1-R8.

Evershed, S. (1889), "Globular lightning," *Electrician* **22** 443.

Evrard, F. (1885), "Note sur les observations des coups de foudre en Belgique," *Mouvement Industriel, Belge* **3** 21.*

Evrard, F. (1889), "Coups de foudre en Belgique pendant l'année 1887," *Ciel et Terre* **10** 270.

Evrard, F., and Lambotte, L. (1889), "Cinquieme note sur les observations des coups de foudre en Belgique," *Bull. Soc. Belg. Electr.* **6** 34.*

Evrard, F., and Lambotte, L. (1891a), "Sur les observations des coups de foudre en Belgique" (summary), *Fortschr. Phys.* **47(3)** 355.

Evrard, F., and Lambotte, L. (1891b), "Sur les observations des coups de foudre en Belgique," *La Lumière Electrique* **41** 192.

Evrard, F., and Lambotte, L. (1891c), "Sur les observations des coups de foudre en Belgique," *Zeit. Angew. Meteorol.* **8** 319.*

Evteev, B. F. (1982a), "Regularities of aircraft electrification in clouds and precipitation at a presound flight speed," in *Proceedings: International Aerospace Conference on Lightning and Static Electricity*, Public Relations Section, Culham Laboratory, Abingdon, Oxon., UK.

Evteev, B. F. (1982b), "Study of electrification of isolated bodies in a flow of drops," in *Proceedings: International Aerospace Conference on Lightning and Static Electricity*, Public Relations Section, Culham Laboratory, Abingdon, Oxon., UK.

Fahrig, E. (1889), "Globular lightning," *Meteorol. Mag.* **24** 10.

Fahrig, E. (1890), "Globular lightning," *Electrician* **22** 413.

Falkner, M. F. (1964), "An occurrence of ball lightning," *Meteorol. Mag.* **93** 95.

Faraday, M. (1833), "Experimental Researches in Electricity," *Phil. Trans. Roy. Soc. London* **123** 23.

Faraday, M. (1839), "Experimental Researches in Electricity," *B. Quaritch, London* **1** 523.

Fay, M. H. (1890), "Globular lightning," *Electrician* **26** 5.

Faye, H. (1877), *Annuaire au Roi par le Bureau des Longitudes* 483.

Faye, H. (1890a), "Sur les boules de feu où globes électriques du tornado de St.-Claude, d'après le rapport de M. Cadanet," *C. R. Hebd. Séances Acad. Sci.* **111** 492.

Faye, H. (1890b), "Sur les boules de feu où globes électriques du tornado de St.-Claude" (summary), *Fortschr. Phys.* **46(3)** 424.

Faye, H. (1891a), *L'Astronomie* **10** 22.

Faye, H. (1891b), *Gaea* **27** 185.

Fehr, U. (1963), "Ball of fire, a laboratory illuminated cloud phenomenon," Master's thesis, Hebrew University, Jerusalem, Israel.

Felsher, M. (1970), "Ball lightning," *Nature* **227** 982.

Ferguson, R. M. (1866), "Electricity," *London and Edinburgh J.* 97.*

Ferran, A. (1896a), "Foudre en boule," *Le Cosmos* Oct, p. 354.*

Ferran, A. (1896b), "Foudre en boule," *La Météorologie* **44** 236.

Fessenden, R. A. (1899), "Nature of the lightning discharge," *Electrical World Eng.* **35** 551.

Feynman, R. P., Leighton, R. B., and Sands, M. (1965), *The Feynman Lectures on Physics,* Vol. 3, p. 10, Addison-Wesley, Reading, Massachusetts. (NBL)

Fieux, R., and Hubert, P. (1976), "Triggered lightning hazards," *Nature* **260** 188. (NBL)

Fieux, R., Gary, G., and Hubert, P. (1975), "Artificially triggered lightning above land," *Nature* **257** 212.

Figuier, L. (1870), *Les Grandes Inventions,* Paris.*

Figuier, L. (1884), "Meraviglie e conquiste delle scienza l'elettricita e le sue applicazioni," *Milan* **1**.*

Findeisen, W. (1940), "Uber die Entstehung der Gewitter Elektrizitat," *Meteorol. Zeit.* **57** 201. (NBL)

Finkelstein, D. (1968), "Research directed toward understanding behavior and detection of plasmas in magnetic fields," paper AFCRL-68-0210, U.S. Air Force Cambridge Research Laboratory.

Finkelstein, D., and Rubenstein, J. (1964), "Ball lightning," *Phys. Rev.* **135** A390.

Fischer, E. (1981), "Ball lightning—a combustion phenomenon," *Naturwiss.* **68** 568.

Fitzgerald, M. (1878), "Notes on the occurrence of globular lightning and waterspouts in Co. Donegal, Ireland," *Roy. Meteorol. Soc. Q. J.* **4** 160.

Flammarion, C. (1874), *The Atmosphere,* pp. 426, 440, Harper and Brothers, New York.

Flammarion, C. (1899), "Les victimes de la foudre et ses caprices," *Bull. Soc. Astron. Fr.* **13** 145.

Flammarion, C. (1904a), "Les victimes de la foudre et ses caprices," *Bull. Soc. Astron. Fr.* **18** 153.

Flammarion, C. (1904b), "Foudre en boule," *Bull. Soc. Astron. Fr.* **18** 378.

Flammarion, C. (1904c), "Les caprices de la foudre," *Bull. Soc. Astron. Fr.* **18** 421.

Flammarion, C. (1905), *Thunder and Lightning* (translated by Walter Mostyn), Chatto and Windus, London.

Flammarion, C. (1908), "Les mystères de la foudre," *Bull. Soc. Astron. Fr.* **22** 440.

Flammarion, C. (1911), "Foudre en boule," *Bull. Soc. Astron. Fr.* **25** 262.

Flammarion, C. (1927), *Les Phénomènes de la Foudre,* p. 117, Paris.*

Fleming, S. J., and Aitken, M. J. (1975), "Radiation dosage associated with ball lightning," *Nature* **252** 220.

Flint, H. T. (1939a), "Ball lightning," *Roy. Meteorol. Soc. Q. J.* **65** 532.

Flint, H. T. (1939b), "A note on a theory proposed by T. H. Neugebauer," *Roy. Meteorol. Soc. Q. J.* **65** 532.

Flora, D. (1954), *Tornadoes of the United States,* University of Oklahoma Press, Norman, Oklahoma. (NBL)

Floyd, F. G. (1887), "Globular lightning," *English Mech.* **45** 159.

Fontenay, G. de (1901), "Coups de foudre," *Bull. Soc. Astron. Fr.* **15** 303.

Fontvielle, W. de (1867), *Eclairs et Tonnere,* pp. 51, 67, Hachette et Cie, Paris.*

Fontvielle, W. de (1868), *Thunder and Lightning,* p. 32, Sampson, Low, Son and Marston, London.*

Forrest, S. R., and Witten, T. A. (1979), "Long range correlations in smoke-particle aggregates," *J. Phys. A: Math. Gen.* **12(5)** L109. (NBL)

Forster, T. (1823), Researches about Atmospheric Phaenomena, p. 214, Harding, Mavor and Lepard, London.

Foster, B. D. P. (1947), "Ball lightning," *Meteorol. Mag.* **72** 210.

Fouchet, J. (1964), "Le méchanisme complexe de l'éclair et le mystère de la foudre en boule," *La Nature* **92(3347)** 81.*

Fougières, G. (1896), *Dictionnaire des Antiquités Grecques et Romaines,* Ch. Daremberg et Edm. Saglio, Paris.*

Francois, H. (1911), "La Météorologie-curieux coup de foudre en boule," *Bull. Soc. Astron. Fr.* **25** 333.

Frank-Kamenetsky, D. A. (1963), *Plazma-Chetvertoye Sostoyaniya Veshchestva,* Moscow.*

Franklin, B. (1751), *Experiments and Observations on Electricity made at Philadelphia,* E. Cave, London. (NBL)

Franklin, B. (1774), "A letter of Benjamin Franklin, Esq., to Mr. P. Collinson, F.R.S., concerning an electrical kite," *Phil. Trans. Roy. Soc. London* **47** 565. (NBL)

Freier, G. D. (1960), "The electric field of a large dust devil," *J. Geophys. Res.* **65** 3504. (NBL)

Freiherr, J. A. (1896), "Kugelblitz in Carlskrona," *Meteorol. Zeit.* **13** 476.

Fremholt, G. (1877a), "Ein Kugelblitz in Dänemark" (summary), *Fortschr. Phys.* **33** 1289.

Fremholt, G. (1877b), "Ein Kugelblitz in Dänemark," *Gaea* **13** 565.

Frenkel', Y. I. (1940), "O priroda sharovoi molnii," *Zh. Eksp. Teor. Fiz.* **10** 1424.

Frenkel', Y. I. (1949), *Teoriya Yavleniy Atmosfernogo Elektrichestva,* Moscow-Leningrad.*

Friedrich, W. (1926), "Elmsfeuerbeobachtung auf dem Kahlen-Asten," *Meteorol. Zeit.* **43** 229. (NBL)

Friesenhof, G. (1910), "Ein merkwurdiger Blitz," *Meteorol. Zeit.* **27** 560.

Fritsch, V. (1962), "Der Nachweis und die geoelektrische Untersuchung eines Blitznestes," *Gerlands Beitrage zur Geophysik* **71** 300. (NBL)

Fromholt, H. (1884), *Köln. Zeit.* 17 Apr.

Fry, E. (1906), "A remarkable lightning discharge," *Nature* **74** 198.

Fuess, P. (1931), "Kugelblitz," *Zeit. Angew. Meteorol.* **48** 255.

Fulks, J. R. (1962), "On the mechanics of a tornado," National Severe Storms Project Report No. 4, U.S. Dept. of Commerce, Washington, DC. (NBL)

Funder, L. (1939), "Ermittlung, Ursachen und Bedeutung des Ionengehaltes der Grossen Wetter," *Beitr. zur Geophys.* **54** 370.

Gabrakov, S. (1965), "Kulbovidnata mulniia," *Khid. Meteorol.* **15** 59.*

Gaidukov, N. I. (1988a), "Equations of motion of ball lightning in the field of a point source" (in Russian), *Doklady Akad. Nauk. SSSR* **301(5)** 1076.

Gaidukov, N. I. (1988b), "Equations of motion of ball lightning in the field of a point source" (English translation), *Sov. Phys. Doklady* **33(8)** 571.

Gaidukov, N. I. (1989a), "Motion of ball lightning in an air flow through a wide circular opening in a flat screen" (in Russian), *Zh. Tekh. Fiz.* **59(2)** 88.

Gaidukov, N. I. (1989b), "Motion of ball lightning in an air flow through a wide circular opening in a flat screen" (English translation), *Sov. Phys. Tech. Phys.* **34(2)** 181.

Gaidukov, N. I. (1991a), "Hydrodynamic model of the passage of ball lightning through a narrow slot in a flat screen" (in Russian), *Zh. Tekh. Fiz.* **61(11)** 49.

Gaidukov, N. I. (1991b), "Hydrodynamic model of the passage of ball lightning through a narrow slot in a flat screen" (English translation), *Sov. Phys. Tech. Phys.* **36(11)** 1223.

Gaidukov, N. I. (1992a), "Equations of motion of ball lightning in a rocket airstream" (in Russian), *Izvestiya Akad. Nauk. SSSR, Mekhanika Zhidkosti i Gaza* **27(1)** 174.

Gaidukov, N. I. (1992b), "Equations of motion of ball lightning in a rocket airstream" (English translation), *Trans. Fluid Dynamics* **27(1)** 132.

Gaidukov, N. I. (1992c), "Hydrodynamic model for the streaming of ball lightning through a narrow opening in a flat barrier" (in Russian), *Zh. Tekh. Fiz.* **62(2)** 27.

Gaidukov, N. I. (1992d), "Hydrodynamic model for the streaming of ball lightning through a narrow opening in a flat barrier" (English translation), *Sov. Phys. Tech. Phys.* **37(2)** 127.

Gaidukov, N. I. (1992e), "Motion of ball lightning in time-dependent air currents" (in Russian), *Zh. Tekh. Fiz.* **62(6)**.*

Gaidukov, N. I. (1992f), "Motion of ball lightning in time-dependent air currents" (English translation), *Sov. Phys. Tech. Phys.* **37(6)** 729.

Gaidukov, N. I. (1993a), "Hydrodynamic interaction of ball lightning and unsteady airflows" (in Russian), *Izvestiya Akad. Nauk. SSSR, Mekhanika Zhidkosti i Gaza* **28(1)** 192.

Gaidukov, N. I. (1993b), "Hydrodynamic interaction of ball lightning and unsteady airflows" (English translation), *Fluid Dynamics* **28(1)** 145.

Gaidukov, N. I. (1993c), "Magnetic field of ball lightning" (in Russian), *Teplofizika Vysokikh Temperatur* **31(6)** 1038.

Gaidukov, N. I. (1993d), "Magnetic field of ball lightning" (English translation), *High Temperature* **31(6)** 954.

Gaidukov, N. I. (1993e), "Hydrodynamic model of the interaction of ball lightning with the airstream of an aircraft in flight" (in Russian), *Zh. Tekh. Fiz.* **63(9)** 16.

Gaidukov, N. I. (1993f), "Equations of motion of ball lightning in the magnetic field of a hanging conductor" (in Russian), *Zh. Tekh. Fiz.* **63(10)** 20.

Gaidukov, N. I. (1994), "Steady-state motion of the ball lighting observed by MT Dmitriev at the Onega River" (in Russian), *Zh. Tekh. Fiz.* **64(2)** 38.

Gaidukov, N. I. (1997), "Continuous flow past ball lightning in the air flow from a flying vehicle" (English translation), *High Temperature* **35(1)** 164.

Gaillard, D. (1912), "Phénomènes électriques dans la région du Tchad," *Bull. Soc. Astron. Fr.* **26** 436.

Galczak, J. (1982), "Studies of lightning-attachment testing using aircraft models," in *Proceedings: International Aerospace Conference on Lightning and Static Electricity*, Public Relations Section, Culham Laboratory, Abingdon, Oxon., UK.

Galli, I. (1890), "Kugelblitz," *Zeit. Angew. Meteorol.* **7** 238.

Galli, I. (1891), *Mem. R. Accad. Sci. Ist. Bologna* **1** 315.

Galli, I. (1902), *Zeit. Angew. Meteorol.* **19** 239.

Galli, I. (1907), "Communicazione relativa ai parafulmini ed ai fulmini ascendenti e globulari," *Atti. Pontif. Accad. Rom. D. Nuovi Lincei*.*

Galli, I. (1908), "Della protezione offera dai parafulmini edi alcuni questi sulla natura del fulmine," *Mem. Pontif. Accad. Rom. D. Nuovi Lincei* 36.*

Galli, I. (1909a), "Di un recente fulmine a Roccagorga e di altri casi egualmente notevoli," *Atti. Pontif. Accad. Rom. D. Nuovi Lincei.**

Galli, I. (1909b), "Di alcuni fulmini recenti e antichi nel Lazio e Roma," *Atti. Pontif. Accad. Rom. D. Nuovi Lincei.**

Galli, I. (1910a), "Notizie di alcuni fulmini," *Atti. Pontif. Accad. Rom. D. Nuovi Lincei.**

Galli, I. (1910b), "Di una recente fulminazione a cori, con appendica di observazioni sur lampi," *Atti. Pontif. Accad. Rom. D. Nuovi Lincei.**

Galli, I. (1910c), "I principali caratteri dei fulmini globulari," *Mem. Pontif. Accad. Rom. D. Nuovi Lincei* **28** 54.*

Galli, I. (1910d), "Raccolta e classifiezione di fenomeni luminosi osservati nei terremoti," *Bull. Soc. Sismol. Ital.* **14** 69.*

Galli, I. (1910e), "Raccolta e classifiezione di fenomeni luminosi osservati nei terremoti," *Bull. Soc. Sismol. Ital.* **14** 221.*

Galli, I. (1911a), "I principali caratteri dei fulmini globulari," *Ciel et Terre* **32** 238.

Galli, I. (1911b), "Gli effetti meccanici dei fulmini globulari," *Mem. Pontif. Accad. Rom. D. Nuovi Lincei* **29** 9.*

Galli, I. (1911c), "Ein Zylinderblitz," *Rev. Nepholog.* **6** 538.*

Galli, I. (1911d), "Ein Zylinderblitz," *Zeit. Angew. Meteorol.* **28** 161.*

Galli, I. (1912), "Gli effetti fisici e chimici dei fulmini globulari," *Mem. Pontif. Accad. Rom. D. Nuovi Lincei* **30** 17.*

Galli, I. (1913a), "Altri fulmini globulari inediti," *Atti. Pontif. Accad. Rom. D. Nuovi Lincei* **66** 11.*

Galli, I. (1913b), "Effetti dei fulmini globulari sugli alberi e sulle erbe," *Mem. Pontif. Accad. Rom. D. Nuovi Lincei* **31** 225.*

Galli, I. (1913c), "Altri fulmini globulari inediti" (summary), *Fortschr. Phys.* **69(3)** 320.

Galli, I. (1914), "Fulmini globulari nel 1914," *Atti. Pontif. Accad. Rom. D. Nuovi Lincei* **68** 177.*

Galli, I. (1915a), "Fulmini globulari nell'anno 1915," *Atti. Pontif. Accad. Rom. D. Nuovi Lincei* **69** 59.*

Galli, I. (1915b), "Effetti dei fulmini globulari sugli alberi e sulle erbe" (summary), *Fortschr. Phys.* **71(3)** 255.

Galli, I. (1916a), "Osservazioni inedite o quasi sconoscuite di fulmini globulari e di effetti rari," *Atti. Pontif. Accad. Rom. D. Nuovi Lincei* **16** Mar.*

Galli, I. (1916b), "Fulmini globulari nel 1914" (summary), *Fortschr. Phys.* **72** 274.

Galli, I. (1917), "Fulmini globulari nel'anno 1915" (summary), *Fortschr. Phys.* **73(3)** 136.

Galli, I. (1919), "Osservazioni Italiane di fulmine globulare di lampi rari nel 1917 e 1918: Nota XXI," *Atti. Pontif. Accad. Rom. D. Nuovi Lincei* **15** Jun.*

Galli, I. (1920a), "La foudre globulaire," *Ciel et Terre* **36** 73.

Galli, I. (1920b), "Osservazioni inedite o quasi sconoscuite di fulmini globulari e di effetti rari," *Ciel et Terre* **36** 76.

Galli, I. (1920c), "La foudre globulaire et les travaux," *Ciel et Terre* **36** 140.

Gallop, J. W., Dutt, T. L., and Gibson, H. (1960), "Forces on charged particles of a plasma in a cavity," *Nature* **188** 397.

Galton, E. M. G. (1971), "Ball lightning," *The Times*, London **58** May 17, p. 13.*

Gardner, M. (1952), *Fads and Fallacies in the Name of Science*, Dover, New York. (NBL)

Garfield, E. (1976), "When citation analysis strikes ball lightning," *Current Comments* **20** 5.

Garreau, R. (1939a), "Académie des Sciences-coup de foudre violent," *Bull. Soc. Astron. Fr.* **53** 424.

Garreau, R. (1939b), "Observation d'en coup de foudre en boule," *C. R. Hebd. Séances Acad. Sci.* **209** 60.

Gathemann, H. (1889a), "Eigenartiger Blitzschlag," *Ann. Hydro. Marit. Meteorol.* **17** 39.

Gathemann, H. (1889b), "Kugelblitz," *Zeit. Angew. Meteorol.* **6** 90.

Gavache, M. R. (1951), "Foudre globulaire?," *Bull. Soc. Astron. Fr.* **65** 474.

Geddes, A. E. M. (1939), *Meteorology,* 2nd ed. Blackie and Son, London.

Geerk, J., and Kleinwachter, H. (1960), "Gasentladung in Brennpunkt eines Radar paraboloids," *Zeit. Phys.* **159** 378. (NBL)

Geist, F. (1949), "Eine Kugelblitzbeobachtung in Oberbayern," *Umschau* **19** 255.

Geller, R., and Lucarain, M. (1961), "Oscillations ioniques stationnaires dans les plasmoides de haut fréquence en absence de champ magnétique," *C. R. Hebd. Séances Acad. Sci.* **253** 1542. (NBL)

Genz, H. (1952), *Zeit. Erd. Unter.* **4** 186.

Gerasimenko, T. F. (1956), "Sharovaia molniia vo vremia zimmei grozy," *Priroda* **7** 109.

Gerharz, R. (1979), "Coffers re-opened," *Optical Spectra* **13(17)** 13.

Gerlach, W. (1927), "Uber die Beobachtung eines Kugelblitzes," *Naturwiss.* **15** 522.

Gersten, A. (1963), "Plasma ball of fire," Thesis, Hebrew University of Israel, Jerusalem, Israel.

Gervais, M. R. V. (1934), "La météorologie-la foudre en boule," *Bull. Soc. Astron. Fr.* **48** 277.

Gezekus, N. A. (1899), *Sharovidnyye I Plamenoobraznyye Molnii kak Osobyye Tipy Razryadov Atmosfernogo Elektrichestva,* St. Petersburg.*

Gezekus, N. A. (1900a), "Elektricheskoye plamya I sharovidnaya molniya," *Zh. Fiz. Khim. Obsh.* **32** 127.

Gezekus, N. A. (1900b), *O Sharovoya Molnii,* St. Petersburg.*

Gibbs-Smith, C. H. (1971), "On fireballs," *Nature* **232** 187.

Gilbert, O. (1907), *Die Meteorologischen Theorien des Griechischen Altertums,* p. 619, Leipzig.*

Gilbey, J. C. M. (1984), "Orange-coloured ball lightning in one room and blue flashes from a white centre in another room during a thunderstorm in Wakefield," *J. Met.* **9(92)** 245.

Gill, R. D. (ed.) (1981), *Plasma Physics and Nuclear Fusion Research,* Academic Press, New York. (NBL)

Gillett, H. G. D. (1983), "Ball lightning—all in the mind," *Nature* **301(5900)** 460.

Gilmore, G. (1919), "Globular lightning," *Nature* **103** 284.

Ginsburgh, I., and Bulkley, W. (1976), "Ball lightning," *Nature* **259–270**.

Gladkov, K. (1958), "Zagadka sharovoi molnii," *Tekh. Molod.* **26** 28.*

Gladyshev, G. P. (1989), "The high-temperature physico-chemical processes in the lightning storm atmosphere (a physico-chemical model of ball lightning)," in Y. H. Ohtsuki (ed.) *Science of Ball Lightning (Fireball),* p. 242, World Scientific, Singapore.

Gladyshev, G. P., and Smirnov, B. M. (1989), "International symposium on ball lightning (Tokyo, July 4 to 6, 1988)" (in Russian), *Usp. Fiz. Nauk.* **157(2)** 364.

Glickman, T. (1999), *Glossary of Meteorology* (in press), American Meteorological Society. (NBL)

Gobel, P. (1889a), "Kugelblitz," *Zeit. Angew. Meteorol.* **6** 68.

Gobel, P. (1889b), "Kugelblitz," *Zeit. Angew. Meteorol.* **6** 237.

Gockel, A. (1895), *Das Gewitter,* Gorres-Gesellschaft, Bonn, Germany.*

Gockel, A. (1905a), "Das Gewitter" (summary), *Fortschr. Phys.* **61(3)** 318.

Gockel, A. (1905b), "Das Gewitter," *Naturwiss. Rund.* **20** 447.*

Gockel, A. (1909), "Uber Kugelblitz," *Meteorol. Zeit.* **26** 458.

Goderus, A. (1889), "Foudre en boule," *Anon., 'Les Orages', Ciel et Terre* **9** 193.

Godlonton, R. (1896), "A remarkable discharge of lightning," *Nature* **53** 273.

Godwin-Austin, W. H. (1883), "Lightning phenomenon," *Nature* **28** 173.

Gold, E. (1952), "Thunderbolts: the electric phenomenon of thunderstorms," *Nature* **169** 561.

Golde, R. H. (1962), "Thunderstorms," *IEEJ, London* **8** 242.

Golde, R. H. (1967), "The lightning conductor," *J. Franklin Inst.* **283** 451. (NBL)

Golde, R. H. (1973), *Lightning Protection,* Edward Arnold, London. (NBL)

Golde, R. H. (1977a), "Lightning currents and related parameters" in R. H. Golde (ed.) *Lightning,* Vol. I, p. 9, Academic Press, New York. (NBL)

Golde, R. H. (1977b), "The lightning conductor," in R. H. Golde (ed.) *Lightning*, Vol. II, p. 17, Academic Press, New York. (NBL)

Golde, R. H. (ed.) (1977), *Lightning,* Academic Press, New York.

Golka, R. K. (1991), "How to create ball lightning," in *Proceedings of the 1991 International Aerospace and Ground Conference on Lightning and Static Electricity*, NASA Conference Publication, Vol. 2, p. 3106, National Aeronautics and Space Administration, Washington, DC.

Golka, R. K. (1994), "Laboratory-produced ball lightning," *J. Geophys. Res.* **99(D5)** 10679.

Golka, R. K., and Bass, R. W. (1977), "Tesla's ball lightning theory, a BGK-wave solution-pair Langmuir–Debye sheath surface-tension hypothesis, the Farnsworth effect, and the 'hydrotron' electrostatic-internal self-contained plasmoid concept," in *Annual Controlled Fusion Theory Conference,* paper J3.

Goodlet, B. L. (1937), "Lightning," *IEEJ, London* **81** 1.

Gordon, W. (1742), "Account of a fireball seen Dec. 11 1741," *Phil. Trans. Roy. Soc. London* **42** 58.

Gouw, M. J. E. ter (1907), "Boule de feu," *Bull. Soc. Astron. Fr.* **21** 313.

Gposu, S. (1960), *Electrotechnica* **8(2)** 57.

Graf-Pfeil, L. (1886a), "Der Ursprung der Meteoriten," *Gaea* **22** 229.*

Graf-Pfeil, L. (1886b), "Die Hischberger Kugelblitze," *Gaea* **22** 474.*

Graham, E. (1996), "Great balls of fire! Ball lightning in Dublin, Ireland, on 31 August 1992," *J. Met.* **21(214)** 412.

Graham, E. (1997), "Plasmodes in a microwave oven," *J. Met.* **22(219)** 196.

Gray, D. E. (ed.) (1957), *American Institute of Physics Handbook,* pp. 2–125, McGraw-Hill, New York. (NBL)

Greaves, C. (1879), "Discussions on waterspouts and globular lightning," *Roy. Meteorol. Soc. Q. J.* **4** 165.

Grefen, H. (1884), *Schr. Nat. Ver. Sch. Holst.* **6** 613.

Gregg, A. (1941), *The Furtherance of Medical Research,* Oxford University Press, Oxford, England. (NBL)

Gridnev, A. G. (1997), "Phase transitions in the plasma recombinated in a gaseous atmosphere," *Atm. Oceanic Opt.* **10(11)** 842.

Grigor'ev, A. I. (1993), "Some statistical data on birth and death of ball lightning," in A. G. Keul (ed.) *Progress in ball lightning research. Proc. VIZOTUM*, p. 30, The Vizotum Project, Salzburg, Austria.

Grigor'yev, A. I. (1978), "Radiation, electrical, and magnetic properties of ball lightning," *Sov. Met. Hydr.* **8(8)** 91.

Grigor'yev, A. I. (1986), "Correlation between duration and typical linear dimension of ball lightning," *Sov. Met. Hydr.* **6** 78.

Grigoriev, A. I., and Bychkov, V. L. (1994), "Report of the Conference 'Long-lived plasma formations and poorly-explored forms of natural electrical discharges in the atmosphere'," *J. Met.* **19** 243.

Grigorev, A. I., and Grigoreva, I. D. (1989a), "Correlations between some properties of ball lightning" (in Russian), *Zh. Tekh. Fiz.* **59(2)** 79.

Grigorev, A. I., and Grigoreva, I. D. (1989b), "Correlations between some properties of ball lightning" (English translation), *Sov. Phys. Tech. Phys.* **34(2)** 176.

Grigorev, A. I., and Grigoreva, I. D. (1992), "Estimating the volume-averaged energy density of the ball lightning," *Elektron. Obrabotka Materialov* **No. 5** 29.

Grigorjev, A. I., Grigorjeva, I. D., and Shirjajeva, S. O. (1989), "Statistical analysis of the ball lightning properties," in Y. H. Ohtsuki (ed.) *Science of Ball Lightning (Fireball)*, p. 88, World Scientific, Singapore.

Grigorev, A. I., Grigoreva, I. D., and Shiryaeva, S. O. (1992a), "Some characteristics of the initiation and termination of ball lightning" (in Russian), *Pis'ma v Zhurnal Tekh. Fiz.* **18(23–24)** 46.

Grigorev, A. I., Grigoreva, I. D., and Shiryaeva, S. O. (1992b), "Some characteristics of the initiation and termination of ball lightning" (English translation), *Sov. Tech. Phys. Letters* **18(12)** 781.

Grigor'ev, A. I., Grigor'eva, I. D., and Shiraeva, S. O. (1992c), "Ball lightning penetration into closed rooms: 43 eyewitness accounts," *J. Scientific Exploration* **6** 261.

Grigorev, A. I., Grigoreva, I. D., and Shiryaeva, S. O. (1995a), "Stability of ball lightning against its unneutralized self-charge" (in Russian), *Zh. Tekh. Fiz.* **65(2)** 1.

Grigorev, A. I., Grigoreva, I. D., and Shiryaeva, S. O. (1995b), "Stability of ball lightning against its unneutralized self-charge" (English translation), *Sov. Phys. Tech. Phys.* **40(2)** 117.

Grigorev, A. I., Grigoreva, I. D., Mukhina, E. I., and Shiryaeva, S. O. (1991a), "Regularities of electrohydrodynamic disintegration of ball lightning," *Elektron. Obrabotka Materialov* **No. 6** 31.

Grigorev, A. I., Shiryaeva, S. O., Grigoreva, I. D., Lazaryants, A. E., and Mukhina, E. I. (1991b), "The possibility of ball lightning splitting in two" (in Russian), *Zh. Tekh. Fiz.* **61(4)** 25.

Grigorev, A. I., Shiryaeva, S. O., Grigoreva, I. D., Lazaryants, A. E., and Mukhina, E. I. (1991c), "The possibility of ball lightning splitting in two" (English translation), *Sov. Phys. Tech. Phys.* **36(4)** 385.

Grigoreva, I. D., Grigorev, A. I., Poluektov, P. P., and Shiryaeva, S. O. (1994), "On size distribution of ball lightning," *Akad. Nauk. Moldavskoi SSR* **No. 5** 33.

Grobe, G. (1927a), "Wieder ein Kugelblitz auf dem Brocken," *Meteorol. Zeit.* **44** 312.

Grobe, G. (1927b), "Eine Kugelblitzartige Erscheinung," *Meteorol. Zeit.* **44** 391.

Gronemann, J. (1896a), "Twee Merkwaardige Bliksems" (summary), *Fortschr. Phys.* **52(3)** 301.

Gronemann, J. (1896b), "Twee Merkwaardige Bliksems," *Nat. Tijdschr. Ned. Indie* **3** 269.*

Grossmann, D. (1849), "Feuerkugeln" (summary), *Fortschr. Phys.* **5** 455.

Grunsky, C. E. (1907), "Electric storm in Southern California," *Mon. Weather Rev.* **35** 228.

Gubichev, V. A. (1966), "Possible mechanism of ball lightning formation," *Ul'yanovskiy Gosudorstvennyy Pedagogicheskiy Institut Imeni I. N. Ul'yanova, Uchenyye Zapiski Seriya Tekhnicheskaya* **20(4)** 203.

Gubichev, V. A. (1968), "A new theory of ball lightning," *Foreign Science Bull.* **4(6)** 27.

Gucinski, H. (1978), "Observation of ball lightning," *Weatherwise* **31(4)** 167.

Guillemin, A. (1883), *Le Monde-Physique* **3** 492.

Guillon, M. (1948), "Foudre en boule," *Bull. Soc. Astron. Fr.* **62** 323.

Gunther, S. (1897), *Geophysik.* Stuttgart.*

Gurevich, A. V., and Pitaevskii, L. P. (1964a), "Recombination coefficient in a dense low-temperature plasma," *Zh. Eksp. Teor. Fiz.* **46** 1281.

Gurevich, A. V., and Pitaevskii, L. P. (1964b), "Recombination coefficient in a dense low-temperature plasma," *Sov. Phys. JETP* **19(4)** 870.

Haag, D. (1914), "Kugelblitz," *Zeit. Angew. Meteorol.* **31** 240.

Hagenau, M. (1896a), "Rathselhafte Blitzerscheinungen," *Gartenlaube* **24** 395.*

Hagenau, M. (1896b), "Rathselhafte Blitzerscheinungen" (summary), *Fortschr. Phys.* **52(3)** 299.

Hagenguth, J. H. (1951), "The lightning discharge," in T. F. Malone (ed.) *Compendium of Meteorology*, p. 136, American Meteorological Society, Boston.

Hagenguth, J., and Anderson, J. G. (1952), *AIEEE* **71(3)** 641.

Hahn, P. (1892), "Eigenthumlicher Kugelblitz," *Ann. Hydro. Marit. Meteorol.* **20** 257.

Haidinger, W. von. (1868a), "Elektrische Meteore am 20 Oktober 1868 in Wien beobachtet," *Sitzber. Math. Naturwiss. Kgl. Akad. Weiss. Wien* **II, 58** 761.

Haidinger, W. von. (1868b), "Die sudwestlichen Blitzkugeln am 20 Oktober 1868," *Sitzber. Math. Naturwiss. Kgl. Akad. Weiss. Wien* **II, 58** 955.*

Haidinger, W. von. (1868c), "Ein kugelformiger Blitz, am 30 August 1865 gesehen zu Feistritz bei Peggau in Steiermark," *Sitzber. Math. Naturwiss. Kgl. Akad. Weiss. Wien* **II, 58** 1045.

Haidinger, W. von. (1869a), "Ein kugelformiger Blitz, am 30 August 1865 gesehen zu Feistritz bei Peggau in Steiermark" (summary), *Fortschr. Phys.* **25** 961.

Haidinger, W. von. (1869b), "Elektrische Meteore am 20 Oktober 1868 in Wien beobachtet" (summary), *Fortschr. Phys.* **25** 961.

Haidinger, W. von. (1869c), "Die sudwestlichen Blitzkugeln am 20 Oktober 1868" (summary), *Fortschr. Phys.* **25** 969.

Haidinger, W. von. (1869d), "Elektrische Meteore am 20 Oktober 1868 in Wien beobachtet," *Inst. Sci. Soc. Fr.* **37** 88.*

Haidinger, W. von. (1869e), "Ein kugelformiger Blitz, am 30 August 1865 gesehen zu Feistritz bei Peggau in Steiermark," *Inst. Sci. Soc. Fr.* **37** 112.*

Haidinger, W. von. (1902), *Photogr. Mitt.* **39** 761.

Hall, R. L. (1972), "Sociological perspectives on UFO reports," in C. Sagan and T. Page (eds.) *UFO's—A Scientific Debate*, Chapter 9, p. 213, Cornell University Press, Ithaca, New York. (NBL)

Halluitte, K. (1909a), "Orage sur mer," *Bull. Soc. Astron. Fr.* **23** 461.

Halluitte, K. (1909b), "Orage sur mer," *C. R. Hebd. Séances Acad. Sci.* **148** 1066.

Halluitte, K. (1909c), "Orage rémarquable sur l'Atlantique Sud dans la nuit du 17 au 18 février 1909 par 33 degré lat. S. et 48 degré long" (summary), *Fortschr. Phys.* **65(3)** 284.

Halluitte, K. (1909d), "Orage rémarquable sur l'Atlantique Sud," *La Météorologie* **57** 99.

Halluitte, K. (1909e), "Gewitter auf dem Meere," *Meteorol. Zeit.* **26** 380.

Haltermann, H. (1896a), "Uber Elmsfeuer auf See," *Jahrbuch der Astronomie und Geophysik* **7** 374.

Haltermann, H. (1896b), "Le feu St.-Elme en mer," *Ciel et Terre* **17** 482.

Hamilton, C. W. (1960), "Sustained, localized, pulsed-microwave discharge in air," *Nature* **188** 1098. (NBL)

Hamilton, G. (1896), "Merkwurdige Blitzschlage in Schweden 1," *Meteorol. Zeit.* **13** 475.

Hamilton, W. (1795), "An account of the late eruption of Mt. Vesuvius," *Phil. Trans. Roy. Soc. London* **85** 73.

Hampton, D. L., Heavner, M. J., Wescott, E. M., and Sentman, D. D. (1996), "Optical spectral characteristics of sprites," *Geophys. Res. Letts.* **23** 82. (NBL)

Han Ming, Wu and Yunming, Chen. (1989), "Magnetohydrodynamic equilibrium of plasma ball lightning," *Phys. Fluids B (Plasma Physics)* **1(8)** 1753.

Han-Ming, Wu and Oakes, M. E. (1991), "Magnetohydrodynamics equilibrium of a self-confined elliptical plasma ball," *Phys. Fluids B (Plasma Physics)* Vol. 3, Issue 8, Part 1, 2113.

Handel, P. H. (1975), "Maser theory of ball lightning," *Bull. Amer. Phys. Soc., Ser. II* **20** 26.

Handel, P. H. (1988), "Maser-caviton ball lightning mechanism," in *Proceedings: 8th International Conference on Atmospheric Electricity*, p. 177, International Association of Meteorology and Atmospheric Physics, International Commission on Atmospheric Electricity, Tucson, Arizona.

Handel, P. H. (1989), "New approach to ball lightning," in Y. H. Ohtsuki (ed.) *Science of Ball Lightning (Fireball)*, p. 254, World Scientific, Singapore.

Handel, P. H. (1997), "Theory of the stationary nonlinear ball lightning system of fireball and atmospheric maser," in Y. H. Ohtsuki and H. Ofuruton (eds.) *Proceedings: 5th International Symposium on Ball Lightning (ISBL97)*, p. 114.

Handel, P. H., and Leitner, J. F. (1994), "Development of the maser-caviton ball lightning theory," *J. Geophys. Res.* **99(D5)** 10689.

Hands, A. (1901), "Ball lightning," *Roy. Meteorol. Soc. Q. J.* **35** 301.

Hands, A. (1909), "Fireballs," *English Mech.* **90** 40.

Hands, A. (1910), *Ciel. et Terre*, **31** 383.

Hann, J. (1906), *Naturwiss.* **2** 481.*

Hann, J. von (1926), *Lehrbuch der La Météorologie*, p. 661, Verlag von Chr. Herm. Tauchnitz, Leipzig.

Hannibal, L. S. (1985), "Ball lightning," *Radio Electronics* **56(9)** 16.

Hanson, A. W. (1982), "Recent experimental work on lightning attachment point location tests," in *Proceedings: International Aerospace Conference on Lightning and Static Electricity*, Public Relations Section, Culham Laboratory, Abingdon, Oxon., UK.

Hapke, L. (1893), "Merkwurdige Blitz," *Meteorol. Zeit.* **10** 227.

Hare, A. T. (1889a), "Globular lightning," *Electrical Engineer* **8** 443.

Hare, A. T. (1889b), "Globular lightning," *Nature* **40** 415.

Hare, C. (1878), "Globular lightning," *Nature* **19** 5.

Hare, R. (1828), "On the causes of the inadequate protection afforded by lightning rods, in some cases, and the means of insuring their perfect competency: also, a refutation of the prevalent idea, that metals are peculiarly attractive of electricity," *Amer. J. Sci.* **13** 322.

Hare, R. (1840), "Notices of tornadoes," *Amer. J. Sci. Arts* **38** 73.

Harper, R. (1993), "Lightning and ball lightning incident in Wales," *J. Met.* **18(179)** 157.

Harris, E. G., Theus, R. B., and Bostick, W. H. (1957), "Experimental investigations of the motion of plasma projected from a button source across magnetic fields," *Phys. Rev.* **105** 46. (NBL)

Harris, Sir W. Snow (1843), On the Nature of Thunderstorms, J. W. Parker, London.

Harrison, E. R. (1958), "Spherical plasmoids in low pressure electrodeless discharges," *J. Electron. Control* **5** 319. (NBL)

Hartmann, W. K. (1969a), "Analysis of UFO photographic evidence," in E. U. Condon (ed.) *Scientific Study of Unidentified Flying Objects*, Section III, Chapter 2, p. 75, Bantam Books, New York. (NBL)

Hartmann, W. K. (1969b), "Processes of perception, conception and reporting," in E. U. Condon (ed.) *Scientific Study of Unidentified Flying Objects*, Section VI, Chapter 2, p. 567, Bantam Books, New York. (NBL)

Hartmann, W. K. (1972), "Historical perspectives: photos of UFO's," in C. Sagan and T. Page (eds.) *UFO's—A Scientific Debate*, Chapter 2, p. 11, Cornell University Press, Ithaca, New York. (NBL)

Hasenauer, L. (1913), "Perlschnurblitz," *Meteorol. Zeit.* **30** 350.

Hatch, A. J. (1958), "Plasma studies in a low pressure high frequency discharge," in *Proceedings: Second International Conference on the Peaceful Uses of Atomic Energy: Controlled Fusion Devices*, Vol. 32, p. 333. (NBL)

Hawkins, A. C. (1944), "An observation of ball lightning," *Amer. J. Meteorol.* **25** 364.

Hawkins, A. C. (1946), "An observation of ball lightning," *Amer. J. Meteorol.* **47** 5.

Hayden, E. (1888), "Globular lightning," *Science* **11** 110.

Head, A. P. (1995), "Ball lightning at Bristol on 12 January 1993" (Letter), *J. Met.* **20** 214.

Heen, P. de (1903), "Problème de la theorie mécanique de l'electricité," *Bruxelles* 36.*

Heidke, P. (1932), "Perlschnurblitz in Hamburg am 13 Juli 1932," *Meteorol. Zeit.* **49** 437.

Helmholtz, H. von (1896), *Vort. Reden. Braunschweig.* **1** 289.

Helmont, J. B. von (1652), *Ortus Medicinae*, p. 74, Elsevier, Amsterdam.*

Hennig, R. (1895a), "Uber ungewöhnliche Blitzentladungen" (summary), *Fortschr. Phys.* **51(3)** 398.

Hennig, R. (1895b), "Uber ungewöhnliche Blitzentladungen," *Naturwiss. Woch.* **10** 65.*

Hentschel, K. H. (1993), "A selection from 130 ball lightning reports" in A. G. Keul (ed) *Progress in Ball Lightning Research: Proc. VIZOTUM*, p. 32, The Vizotum Project, Salzburg, Austria.

Herbert, M. (1847), "Cas de foudre en boule," *C. R. Hebd. Séances Acad. Sci.* **25** 34.

Herman, J. (1964), "On the electrical properties of blowing snow," *Ann. Geophysique, Paris* **20(3)** 235. (NBL)

Hermant, A., and Hubert, P. (1988), "Ball lightning photographed?," *Atmos. Res.* **22(3)** 275.

Hernqvist, K. G. (1956), "Discharge mechanism of mercury pool arcs," *J. Appl. Phys.* **27** 1226.

Hernqvist, K. G., and Johnson, E. O. (1955), "Retrograde motion in gas discharge plasmas," *Phys. Rev.* **98** 1576.

Herrich, E. (1954), "Ein Kugelblitz-erlebnis," *Kosmos* **50** 265.

Hertz, H. (1883), "Versuche über die Glimmentladung," *Ann. Physik.* **255** 782.

Hesehus, N. (1876a), *Zh. Fiz. Khim. Obsh.* **8** 311.

Hesehus, N. (1876b), *Zh. Fiz. Khim. Obsh.* **8** 356.

Hesehus, N. (1899), "Uber den Kugelblitz," *Natschnoje Obosienija* **7** 1373.*

Hesehus, N. (1900a), "Uber den Kugelblitz," *Beibl. Ann. Phys.* **24** 200.
Hesehus, N. (1900b), "Uber den Kugelblitz" (summary), *Fortschr. Phys.* **56(3)** 253.
Hesehus, N. (1900c), "Uber den Kugelblitz," *Meteorol. Zeit.* **17** 38.
Hesehus, N. (1900d), "Uber den Kugelblitz," *Prometheus* **12** 207.*
Hesehus, N. (1901), "Kugel-und Flammenblitz als besondere Entladungsformen der atmosphärischen Elektrizitat," *Physik. Zeit.* **2** 578.
Hesehus, N. (1903), "Sur la nature de l'éclair en boule," *Bull. Soc. Astron. Fr.* **17** 101.
Hildebrandsson, H. H. (1883), "Kugelblitz" (summary), *Fortschr. Phys.* **39(3)** 518.
Hildebrandsson, H. H. (1884a), *Gaea* **21** 604.
Hildebrandsson, H. H. (1884b), "Foudre globulaire," *La Nature* **12** 302.*
Hildebrandsson, H. H. (1885), "Foudre globulaire," *La Météorologie* **31** 364.
Hildebrandsson, H. H. (1896a), "Merkwurdige Blitzschlage in Schweden 1," *Meteorol. Zeit.* **13** 475.
Hildebrandsson, H. H. (1896b), "Merkwurdige Blitzschlage in Schweden 3," *Meteorol. Zeit.* **13** 476.
Hildebrandsson, H. H. (1909), "Perlschnurblitz," *Meteorol. Zeit.* **26** 44.
Hill, E. (1897), "Globular lightning," *Nature* **56** 293.
Hill, E. L. (1960), "Ball lightning as physical phenomenon," *J. Geophys. Res.* **65** 1947.
Hill, E. L. (1961), "The physical nature of ball lightning," American Institute of Aeronautics and Astronautics (AIAA) Paper No. DP-61-574, presented at AIAA Great Lakes District Meeting, Minneapolis, Minnesota.
Hill, E. L. (1970), "Ball lightning," *Amer. Scientist* **58** 479.
Hill, E. R., and Sowby, F. D. (1970), "Radiation from ball lightning," *Nature* **228** 1007.
Hill, M. J. M. (1894), "On a spherical vortex," *Phil. Trans. Roy. Soc. London* **A188** 213. (NBL)
Hill, R. D. (1977), "Thunder," in R. H. Golde (ed.) *Lightning*, Vol. I, p.11, Academic Press, New York. (NBL)
Hill, R. H. (1963), "Determination of charges conducted in lightning strokes," *J. Geophys. Res.* **68** 1365. (NBL)
Hillinger, C. (1979), "Lightning tested as new energy source," *Los Angeles Times*, Apr 29, p. 1 (part 4).
Hofmann, U. (1919), "Kugelblitz Beobachtung," *Zeit. Angew. Meteorol.* **36** 126.
Hohn, W. (1966), "Beobachtung eines Kugelblitzes," *Wetter und Leben* **18** 56.
Hohr, A. (1903), "Zwei Merkwurdige Kugelblitz-Erscheinungen," *Meteorol. Zeit.* **20** 570.
Holl, H. (1900a), "Kugelblitz" (summary), *Fortschr. Phys.* **56(3)** 253.
Holl, H. (1900b), "Kugelblitz," *Zeit. Angew. Meteorol.* **17** 191.*
Holman, J. P. (1981), *Heat Transfer*, McGraw-Hill International, Singapore. (NBL)
Holmes, M. (1934a), "Three discharges of ball lightning," *Nature* **133** 179.
Holmes, M. (1934b), "Ball lightning-three discharges," *Roy. Astron. Soc. Canada J.* **28** 144.
Holtzer, R. E., and Workman, E. J. (1939), "Photographs of unusual discharges occurring during thunderstorms," *J. Appl. Phys.* **10** 659.
Holtzer, R. E., Workman, E. J., and Snoddy, L. V. (1938), "Photographic study of lightning," *J. Appl. Phys.* **9** 134.
Homma, H. (1901), "Feuerkugel," *Meteorol. Zeit.* **18** 576.
Hoppe, E. (1885), "Uber atmosphärische und Gewitter-Elektrizität," *Meteorol. Zeit.* **2** 100.
Horn, F., and Tillman, C. (1889a), "Beobachtungen über Gewitter in Bayern, Wurttemberg, und Baden wahrend des Jahres 1889," *Beob. Meteorol. Stat. Konig. Bayern* **11** 25.*
Horn, F., and Tillman, C. (1889b), "Beobachtungen über Gewitter in Bayern, Wurttemberg, und Baden wahrend des Jahres 1889," *Deutsch. Meteorol. Jahrb. Bayern.**
Horn, F., and Tillman, C. (1890), "Beobachtungen über Gewitter in Bayern, Wurttemberg, und Baden wahrend des Jahres 1889" (summary), *Fortschr. Phys.* **46(3)** 412.
Horner, O. W. (1904), "Ball lightning," *Meteorol. Mag.* **39** 111.
Horstmann, W. (1933), "Kugelblitz," *Meteorol. Zeit.* **50** 508.
Horton, R. E. (1921), "Unusual lightning," *Mon. Weather Rev.* **49** 242.

Howland, M. (1979), "Freak lightning ball injures 29 in Colchester," *The Times Argus*, Barre-Montpelier, Vermont, Aug 6.

Hubert, P. (1975a), "Tentative pour observer la foudre en boule dans le voisinage d'éclair déclenchés artificiellement," Rapport DPH/EP/76/349, Commisariat a l'Energie Atomique, Service d'Electronique Physique, Centre d'Etudes Nucleaires de Saclay, France.

Hubert, P. (1975b), "Récherche sur la foudre au moyen d'éclairs déclenchés artificiellement," *La Météorologie* **6** 9. (NBL)

Hubert, P. (1996), "Nouvelle enquête sur la foudre en boule—analyse et discussion des résultats" Rapport PH/SC/96001, Commisariat a l'Energie Atomique, Service d'Electronique Physique, Centre d'Etudes Nucleaires de Saclay, France.

Huc, C. (1894), "Tonnerre en boule," *C. R. Hebd. Séances Acad. Sci.* **119** 474.

Hugueny, M. (1845), "Sur une trombe observé à Dijon le 25 juillet 1845," *C. R. Hebd. Séances Acad. Sci.* **21** 443.

Hulst, J. M.d', and d'Hulst, J. M. (1980), "Globular lightning observed from Puy de Dome?," *J. Récherches Atmos.* **14** 177.

Humboldt, A. von (1845), *Kosmos, entwurf einer Physichen Weltbeschreibung*, J. G. Gottascher, Stuttgart.*

Humboldt, A. von (1864), *Cosmos, a Sketch of a Physical Description of the Universe,"* p. 345, (translated by E. C. Otte), H. G. Bohn, London.

Humphreys, W. J. (1920), *Physics of the Air*, Franklin Inst., Philadelphia, Pennsylvania.

Humphreys, W. J. (1924), "Ball lightning," *Science* **60** 135.

Humphreys, W. J. (1929), *Physics of the Air*, (2nd ed.), Chapter 17 and 18, McGraw-Hill, New York.

Humphreys, W. J. (1931), "Wanted-reports of ball lightning observations," *Science News Lett.* **20** 73.

Humphreys, W. J. (1936), "Ball lightning," *Amer. Phil. Soc. Proc.* **76** 613.

Humphreys, W. J. (1940), *Physics of the Air*, (3rd ed.), Chapter 17 and 18, p. 369, McGraw-Hill, New York.

Humphreys, W. J. (1942), *Ways of Weather*, p. 242, Jacques Cattell Press, Lancaster, Pennsylvania.

Hunneman, M. E. (1937), "A fire-ball," *Science* **86** 244.

Hunt, M. (1972), "Ball lightning," *Weather* **27** 302.

Hunt, M. (1979), "Ball lightning," *Weather* **34** 413.

Ignatovich, V. K. (1992), "Electromagnetic model of ball lightning," *Laser Physics* **2(6)** 991.

Igolkin, S (1997), "Aerodynamics of hot aerosol and cluster particles," in Y. H. Ohtsuki and H. Ofuruton (eds.) *Proceedings: 5th International Symposium on Ball Lightning (ISBL97)*, p. 70.

Igolkin, S., and Savelyev, S. K. (1997), "Condensation model of the ball lightning," in Y. H. Ohtsuki and H. Ofuruton (eds.) *Proceedings: 5th International Symposium on Ball Lightning (ISBL97)*, p. 80.

Indrikson, T. (1900a), "Kugelformige elektrische Funken," *Beibl. Ann. Phys.* **23** 706.

Indrikson, T. (1900b), "Sharoobrazenya Elektricheskaya Iskry," *Zh. Fiz. Khim. Obsh.* **32** 53.

Ingle, E. H. (1971), "Lightning's effects," *New Scientist* **52** 185.

International Aerospace Conference on Lightning and Static Electricity (proceedings) (1982), Public Relations Section, Culham Laboratory, Abingdon, Oxon., UK.

Israel, H. (1950), "Das Gewitter, Ergebnisse und Probleme der modernen Gewitterforschung," in Probleme der Kosmischen Physik, p. 232, Akademische Verladsgesellschaft, p. 232, Geest und Portig, K-G, Leipzig.

Israel, H. (1957), *Atmosphaerische Elektrizität*, Akademische Verlagsgesellschaft, Geest und Portig, K-G, Leipzig. (NBL)

Israel, H. (1958), "Blitzphotographie," *Umschau* **58** 586.

Israel, H. (1960), "Kugelblitz-aufnahmen," *Physik. Blatter* **16** 348.

Israel, H. (1971), *Atmospheric Electricity*, Keter Press, Jerusalem. (NBL)

Israel, H., Toepler, M., and Trier, O. (1960), "Uber Fragen des Kugelblitzes," *Phys. Blatter.* **16** 348.

Iur'ev, A. (1960), "Secret of the burning ball," *Komsom. Pravda*, Oct 30, p. 3.

Iur'ev, A. (1961), "Secret of the burning ball," AFCRL-21, U.S. Air Force Cambridge Research Laboratory, Cambridge, Massachusetts AD-251773.

Iwanovsky, A. (1965a), "The energy of ball lightning," *Foreign Science Bull.* **1(12)** 19.

Iwanovsky, A. (1965b), "Soviet research on ball lightning," *Foreign Science Bull.* **1(10)** 30.

Iwanovsky, A. (1968), "A new theory of ball lightning," *Foreign Science Bull.* **4(6)** 21.

Jackel, W. (1904a), "Kugelblitz in Laesgen, kreis Grunberg in Schlesien am 23 April 1904" (summary), *Fortschr. Phys.* **60(3)** 310.

Jackel, W. (1904b), "Kugelblitz in Laesgen, kreis Grunberg in Schlesien am 23 April 1904" (summary), *Zeit. Angew. Meteorol.* **21** 119.*

Jacobson, A. R., Knox, S. O., Franz, R., amd Enemark, D. C. (1998), "Forté observations of lightning radio-frequency signatures: capabilities and basic results," Report No. LA-UR-98-2046, Los Alamos National Laboratory, Los Alamos, New Mexico. (NBL)

Jaeger, A. (1892), "Uber einige elektrische Erscheinungen," *Meteorol. Zeit.* **9** 359.

Jaeger, A. (1961), "Kugelblitz über Oberschefflenz kreis Mosbach/Baden am 16 Mai 1960, um 22 uhr 30," *Meteorol. Rund.* **14** 31.

Jakobsson, S. (1973), "Hnatteldingar," *Reykjavik, Icelandic Met. Soc., Vedrid* **18** 31.

Jakovlev, V. G., and Balyberdin, V. V. (1968), "On some features of the behaviour of ball lightning in the atmosphere" (in Russian), *Harkov, Respub. Mezv. Nauc.-Tehn. Sbornik, Samolet. Tehn. Vozd. Flota. Vyp.* **16** 59.

James, A. (1992), "Fatalities attributed to ball lightning," in M. Stenhoff (ed.) *Proceedings: Fourth TORRO Conference: Ball Lightning*, p. 7, Tornado and Storm Research Organisation (TORRO), Oxford Brookes University, Oxford, United Kingdom.

James, A. (1995), "TORRO Ball Lightning Division Report," *J. Met.* **20** 309.

James, A. (1996), "Ball lightning at Ruislip, Middlesex, 26 July 1995," *J. Met.* **21(210)** 256.

James, A. (1997), "Ball lightning at West Harnham, Salisbury, 7 June 1996," *J. Met.* **22(217)** 108.

James, A., and Meaden, T. (1997), "Ball lightning at Tewkesbury, Glos On 7 June 1996," *J. Met.* **22(217)** 106.

Jamin, M. (1774), *Cours de Physique Expérimentale et Mathématique,* Vol. I, p. 470, Sigaud de la Ford, Leyden.*

Janezic, E. (1908), "Interessanter Kugelblitz," *Meteorol. Zeit.* **25** 52.

Janhunen, P. (1991), "Magnetically dominated plasma models of ball-lightning," *Ann. Geophys.-Atmos. Hydrosph. Space Sci.* **9(6)** 377.

Jans, C. de (1909a), "Les tentatives d'explication de la foudre globulaire," *Bruxelles* 16.*

Jans, C. de (1909b), "Les tentatives d'explication de la foudre globulaire," *Bull. Soc. Electr. Belge* **26** 123.*

Jans, C. de (1910), "Coup d'oeil retrospectif sur les essais d'explication de la foudre globulaire," *Ciel et Terre* **31** 499.

Jans, C. de (1911a), "Coup d'oeil retrospectif sur les essais d'explication de la foudre globulaire," *Ciel et Terre* **32** 155.

Jans, C. de (1911b), "Coup d'oeil retrospectif sur les essais d'explication de la foudre globulaire," *Ciel et Terre* **32** 255.

Jans, C. de (1911c), "Coup d'oeil retrospectif sur les essais d'explication de la foudre globulaire," *Ciel et Terre* **32** 301.

Jans, C. de (1911d), "Coup d'oeil retrospectif sur les essais d'explication de la foudre globulaire" (summary), *Fortschr. Phys.* **67(3)** 342.

Jans, C. de (1912a), "Coup d'oeil retrospectif sur les essais d'explication de la foudre globulaire," *Bruxelles* 8.*

Jans, C. de (1912b), "Coup d'oeil retrospectif sur les essais d'explication de la foudre globulaire," *Ciel et Terre* **33** 18.

Jans, C. de (1912c), "Coup d'oeil retrospectif sur les essais d'explication de la foudre globulaire," *Ciel et Terre* **33** 143.

Jans, C. de (1912d), "Un cas remarquable de foudre en boule," *Ciel et Terre* **33** 299.

Jans, C. de (1912e), "Coup d'oeil retrospectif sur les essais d'explication de la foudre globulaire" (summary), *Fortschr. Phys.* **68(3)** 318.

Jeanrenard, H., and Burnand, D. (1906), "Feuerkugel" (summary), *Fortschr. Phys.* **62(3)** 287.

Jeanrenard, H., Burnand, D., and Bolzheim, A. von (1906), "Feuerkugeln," *Meteorol. Zeit.* **23** 330.

Jefferson, G. T. (1963), "A photograph of a sub-sun," *Meteorol. Mag.* **92** 254. (NBL)

Jeffreys, H. (1921), "Results of the ball lightning enquiry," *Meteorol. Mag.* **56** 208.

Jennings, P. R. (1990), "Mysterious scorching of a curtain during a violent squall on 21 February 1990 Was ball lightning involved?," *J. Met.* **15(148)** 162.

Jennings, R. C. (1962), "Path of a thunderbolt," *New Scientist* **13** 156.

Jennison, R. C. (1969), "Ball lightning," *Nature* **224** 895.

Jennison, R. C. (1971), "Ball lightning and after-images," *Nature* **230** 576.

Jennison, R. C. (1972), "Ball lightning," *Nature* **236** 278.

Jennison, R. C. (1973), "Can ball lightning exist in a vacuum?," *Nature* **245** 95.

Jennison, R. C. (1987), "The non-particulate nature of matter and the universe," in L. Kostro, A. Posiewnik, J. Pykacz, and M. Zukowski (eds.) *Problems in Quantum Physics: Gdansk '87. Recent and Future Experiments and Interpretations.* p. 163, World Scientific, Singapore.

Jennison, R. C. (1990), "Relativistic phase-locked cavity model of ball lightning," in *Physical Interpretations of Relativity Theory: Proceedings*, Vol. II, p. 359, British Society for the Philosophy of Science, London.

Jennison, R. C. (1992), "Concerning the assessment of ball lightning reports," in M. Stenhoff (ed.) *Proceedings: Fourth TORRO Conference: Ball Lightning*, p. 25, Tornado and Storm Research Organisation (TORRO), Oxford Brookes University, Oxford, United Kingdom.

Jennison, R. C. (1997), "Ball lightning: a general critique," in Y. H. Ohtsuki and H. Ofuruton (eds.) *Proceedings: 5th International Symposium on Ball Lightning (ISBL97)*, p. 6.

Jennison, R. C., Lobeck, R., and Cahill, R. J. (1990), "A video recording of ball lightning," *Weather* **45(4)** 151.

Jensen, J. C. (1933a), "Ball lightning," *Physics* (now *J. Appl. Phys.*) **4** 372.

Jensen, J. C. (1933b), "Ball lightning," *Scientific Monthly* **37** 190.

Jensen, J. C. (1934), "Ball lightning," *Nature* **133** 95.

Jesse, O. (1892a), "Blitz von Ausserordentlich grossem durchmesser," *La Météorologie* **42** 295.

Jesse, O. (1892b), "Blitz von Ausserordentlich grossem durchmesser," *Meteorol. Zeit.* **9** 434.

Johnson, E. O. (1955), "Studies of externally heated hot cathode arcs, Part I, The low-voltage form of the ball-of-fire mode (the low-voltage arc)," *RCA Review* **16** 499. (NBL)

Johnson, J. C. (1954), "Lightning forms," in *Physical Meteorology*, pp. 300–315, Wiley, New York.

Johnson, L. H. (1958), "Ball lightning," *Marine Observer* **28** 11.

Johnson, P. O. (1965), "Ball lightning and self-containing electromagnetic fields," *Amer. J. Phys.* **33** 119.

Johnstone, J. (1890), "Note," *Electrician* **25** 472.

Jolivet, C. (1895a), "Eclairs en boule," *Le Cosmos* **31** 167.*

Jolivet, C. (1895b), "Eclairs en boule," *La Météorologie* **43** 171.

Jones, A. T. (1910), "A laboratory illustration of ball lightning," *Science* **31** 14, 144.

Jones, H. S. (1936), "Ball lightning," *London Daily Mail*, November 5.

Jones, I. (1977), "Giant ball lightning," *J. Met.* **2(21)** 271.

Jones, R. C. (1968), "Return stroke core diameter," *J. Geophys. Res.* **73** 809.

Jones, W. N. (1932), "Ball lightning," *Nature* **130** 545.

Joubert, J. (1881), *Traite. Elem. Elect.* 586.

Joule, B. J. B. (1878), "On a remarkable flash of lightning," *Nature* **18** 260.

Joule, J. P. (1850), "On a remarkable appearance of lightning," *Phil. Mag.* **37** 127.

Kadomtsev, B. B. (1990), "On the ball lightning phenomenology," *Comments on Plasma Physics and Controlled Fusion* **13(5)** 277.

Kadomtsev, B. B. (1991), "Ball lightning as a phenomenon of self-organization," *J. Moscow Phys. Soc.* **1(4)** 335.

Kahler, K. (1924), *Zeit. Angew. Meteorol.* **41** 95.

Kaiser, J. (1924), "Ball lightning," *Science* **60** 293.

Kaiser, R., and Lortz, D. (1995), "Ball lightning as an example of a magnetohydrodynamic equilibrium," *Phys. Rev. E* **52(3)B** 3034.

Kales, J. W. (1900), *Western Electrician* **26** 129.

Kamogawa, M., and Ohtsuki, Y. H. (1997), "Plasmon decay model for the luminous phenomena in the earthquake," in Y. H. Ohtsuki and H. Ofuruton (eds.) *Proceedings: 5th International Symposium on Ball Lightning (ISBL97)*, p. 87.

Kamogawa, M., Ohtsuki, Y. H., Ofuruton, H., Kondo, N., Kashima, H., Honda, S., Tokumoto, N., Tabuchi, K., Muramatsu, S., Okada, T., Ikeda, M., and Aizima, S. (1997), "Measurement of electromagnetic noises for ball lightning existence," in Y. H. Ohtsuki and H. Ofuruton (eds.) *Proceedings: 5th International Symposium on Ball Lightning (ISBL97)*, p. 167.

Kapitza, P. L. (1955), "O priroda sharovoi molnii," *Doklady Akad. Nauk. SSSR* **101** 245.

Kapitza, P. L. (1958), "Uber die Natur des Kugelblitzes," *Phys. Blatter.* **14** 11.

Kapitza, P. L. (1961), "The nature of ball lightning," in D. J. Ritchie (ed.) *Ball Lightning*, Consultants' Bureau, New York.

Kapitza, P. L. (1962), "Elektronika bol'shnik moshchnostei," *Usp. Fiz. Nauk.* **78** 181.

Kapitza, P. L. (1968a), "Sharovaya moliniya I radiozluchenie lineinykh molnii," *Zh. Tekh. Fiz.* **38** 1829.

Kapitza, P. L. (1968b), "Ball lightning and radio emission from linear lightning" (in Russian), *Zh. Tekh. Fiz.* **38** 1829.

Kapitza, P. L. (1969a), "Ball lightning and radio emission from linear lightning," *Sov. Phys. Tech. Phys.* **13(11)** 1475.

Kapitza, P. L. (1969b), "Free plasma filament in a high frequency field at high pressure" (in Russian), *Zh. Eksp. Teor. Fiz.* **57(6)** 1801.

Kapitza, P. L. (1970a), "Free plasma filament in a high frequency field at high pressure," *Sov. Phys. JETP* **30** 973.

Kapitza, P. L. (1970b), "A thermonuclear reactor with a plasma filament freely floating in a high-frequency field," *Sov. Phys. JETP* **31** 199.

Kapitza, P. L. (1970c), "A thermonuclear reactor with a plasma filament freely floating in a high-frequency field" (in Russian), *Zh. Eksp. Teor. Fiz.* **58** 377.

Kapuscha, C. (1896), "Merkwurdige Blitz," *Meteorol. Zeit.* **13** 279.

Karuve, A. de (1858), *Trait. Electr. Th. Appl.* **3** 197.

Kassner, C. (1943), "Kugelblitz," *Zeit. Angew. Meteorol.* **60** 59.

Kaubler, R. (1963), "Beobachtung eines Kugelblitzes," *Physik. Blatt.* **19(8)** 374.

Kayser, H. (1884a), *Monat. Ber. Berlin Akad.* Nov 613.

Kayser, H. (1884b), *Sitz. Konigl. Preuss. Akad. Wiss. Berlin* 1119.

Kayser, H. (1885), "Uber Blitzphotografien," *Ann. Physik.* **25** 131. (NBL)

Keil, K. (1950), *Handworterbuch der Meteorology*, p. 306, F. Knapp, Frankfurt.

Kekez, M. N., and Savic, P. (1977), "The glow (spark) discharge electrolysis-combustion concept of ball lightning," in *Proc. 13th International Conf. Phenomena in Ionized Gases, Part I.* p. 395.

Kelvin, M. (1895), *Proc. Glasgow Phil. Soc.* **26** 233.

Kenvan, G. (1902), "The tragedy of Pelee," *The Outlook, New York* **71** 115.*

Kerkhoff, L. (1913), "Ein Beitrag zum kapitel Kugelblitz," *Zeit. Angew. Meteorol.* **30** 286.

Kerner, B. S., and Osipov, V. V. (1987), "Ball lightning in the mixture of neutral gases," *Doklady Akad. Nauk. SSSR* **292(1)** 82.

Keul, A. G. (1974), "Feuerkugel uber dem Voralpengebiet," *Wetter und Leben* **2(6)** 48.*

Keul, A. G. (1977), "Kugelbilitz in Wien," *Wetter und Leben* **2(9)** 188.*

Keul, A. G. (1980), "Kugelblitze," *Wetter und Leben* **32** 167.

Keul, A. G. (1981), "Ball lightning reports," *Naturwiss.* **68(3)** 134.

Keul, A. G. (1992), "Possible ball lightning colour photograph from Sankt Gallenkirch, Vorarlberg, Austria," *J. Met.* **17(167)** 73.

Keul, A. G. (1993a), "Ball lightning—a case for physics and psychology" in A. G. Keul (ed.) *Progress in Ball Lightning Research: Proc. VIZOTUM*, p. 36, The Vizotum Project, Salzburg, Austria.

Keul, A. G. (1993b), "What is a ball lightning photograph?," *J. Met.* **18** 257.

Keul, A. G. (ed.) (1993c), *Progress in Ball Lightning Research: Proceedings of the Interdisciplinary Congress VIZOTUM*, The Vizotum Project, Salzburg, Austria.

Keul, A. G. (1994), "Ball lightning colour photograph from Senning, Lower Austria," *J. Met.* **19(185)** 10.

Keul, A. G. (1996), "Ball lightning photographs—testing the limits," *J. Met.* **21(207)** 82.

Keul, A. G. (1997a), "Ball lightning car collision near Traunstein, Bavaria," *J. Met.* **22(222)** 284.

Keul, A. G. (1997b), "The southern Bavaria ball lightning car collision," in Y. H. Ohtsuki and H. Ofuruton (eds.) *Proceedings: 5th International Symposium on Ball Lightning (ISBL97)*, p. 27.

Keul, A. G., and Schwarzenbacher, K. (1989), "Phenomenological and psychological analysis of 150 Austrian ball lightning reports," in Y. H. Ohtsuki (ed.) *Science of Ball Lightning (Fireball)*, p. 58, World Scientific, Singapore.

Keul, A. G., Gugenbauer, A., and Diendorfer, G. (1993), "A ball lightning trace case at Perg, Upper Austria" (in Russian), *J. Met.* **18(182)** 287.

Khalatow, S., and Khalatow, D. (1907), "Eclair en boule," *Bull. Soc. Astron. Fr.* **21** 452.

Khazen, A. M. (1977a), "Sharovaia molniia-statsionarnoe sostoianie podvod energii usloviia vozniknoveniia," *Doklady Akad. Nauk. SSSR* **235** 288.

Khazen, A. M. (1977b), "Ball lightning: stationary state energy supply, conditions of occurrence," *Sov. Phys. Doklady* **22(7)** 371.

Kidd, T. D. (1994), "Report of ball lightning in northern England," *J. Met.* **19(189)** 170.

Kikuchi, H. (1989), "Electrical reconnection, critical ionization velocity, ponderomotive force, and their application to triggered and ball lightning," in Y. H. Ohtsuki (ed.) *Science of Ball Lightning (Fireball)*, p. 288, World Scientific, Singapore.

Kikuchi, H. (1995), "Ball lightning," in H. Volland (ed.) *Handbook of Atmospheric Electrodynamics*, Vol. I, p. 167, CRC Press, Boca Raton, Florida.

Kikuchi, H. (1996), "A new model of ball lightning," in *IEEE Conference Record—Abstracts. 1996 IEEE International Conference on Plasma Science*, p. 110. Cat. No. 96CH35939. Institute of Electrical and Electronics Engineers, New York.

Kikuchi, H. (1997a), "Relevance of EHD, EHD-chemical (electro-chemical), and MHD models to natural and artificial ball lightning," in Y. H. Ohtsuki and H. Ofuruton (eds.) *Proceedings: 5th International Symposium on Ball Lightning (ISBL97)*, p. 162.

Kikuchi, H. (1997b), "EHD and MHD models of fireballs and their relevance to natural and artifical ball lightning," in *IEEE Conference Record—Abstracts. 1997 IEEE International Conference on Plasma Science*, p. 319. Cat. No. 97CH36085. Institute of Electrical and Electronics Engineers, New York.

Kilburn-Scott, M. (1923), "Globular lightning," *Nature* **112** 760.

Kilinski, K. von (1958), *Lehrbuch der Luftelektrizitat*, Akademische Verlagsgesellschaft, Geest and Portig, K-G, Leipzig.

Kimpara, A. (1965), "Electromagnetic energy radiated from lightning," in S. C. Coroniti (ed.) *Problems of Atmospheric and Space Electricity*, p. 352, Elsevier, Amsterdam.

King, E. (1773), "Account of the effects of lightning at Steeple Ashton, and Holt in the County of Wilts. on the 20th June 1772," *Phil. Trans. Roy. Soc. London* **63** 231.

King, M. B. (1976a), "Energy source implication of a helicon toroid model for ball lightning," *VFRC-QPR* **18** 109.

King, M. B. (1976b), "An experiment to create ball lightning," *VFRC-QPR* **22** 70.

King, M. B. (1992), "Progress and results in zero-point energy research," in *Proceedings of the 27th Intersociety Energy Conversion Engineering Conference*, p. 4.297, Society of Automotive Engineers, Warrendale, Pennsylvania.

King, R. (1972), "Ball lightning," *Weather* **27** 302.

Kingston, H. R. (1931), "Another case of ball lightning," *Roy. Astron. Soc. Canada J.* **25** 420.

Kircherus, A. (1682), "O onderaardse weereld," *Amsterdam* **1** 279.

Kirko, D. L., Samonchev, P. V., Martynov, A. A., Savjelov, A. S., Dmitrov, S. K., Zhdanov, S. K., Smirnov, V. M., and Korotaev, K. N. (1993), "Localized luminous formations in liquid nitrogen as a ball lightning model," in A. G. Keul (ed.) *Progress in Ball Lightning Research: Proc. VIZOTUM*, p. 41, The Vizotum Project, Salzburg, Austria.

Klass, P. J. (1961), "Plasma theory may explain many UFO's," *Aviation Week Space Technol.* **75(23)** Aug 22, p. 52.

Klass, P. J. (1966), "Many UFO's are identified as plasmas," *Aviation Week Space Technol.* **85(14)** Oct. 3, p. 54.

Klass, P. J. (1968), *UFO's Identified,* Random House, New York.

Klein, J. (1897), "Rathelselhafte Blitzerscheinungen," *Zeit. Angew. Meteorol.* **14** 166.*

Kleiner, H. (1904a), "Kugelblitz in Mullersdorf, 2 April 1904" (summary), *Fortschr. Phys.* **60(3)** 311.

Kleiner, H. (1904b), "Kugelblitz in Mullersdorf, 2 April 1904," *Zeit. Angew. Meteorol.* **21** 118.*

Klimov, A. I., and Mishin, G. I. (1993), "Anomalous wave and gas dynamic properties of long-lived high-energy plasmoids," *Pis'ma v Zhurnal Tekh. Fiz.* **19(13–14)** 19.

Klimov, A. I., Gridin, A. Yu., Avramenko, R. F., and Nikolaeva, V. I. (1993), "Experimental investigation of compact high-energy plasmoids," in A. G. Keul (ed.) *Progress in ball lightning research: Proc. VIZOTUM*, p. 44, The Vizotum Project, Salzburg, Austria.

Klose, J. (1950), "Beschreibung eines Kugelblitzes," *Zeit. fur Meteorol.* **4** 91.

Knoche, W. (1909), "Uber elektrische Entladungen in der Kordillere," *Meteorol. Zeit.* **26** 355.

Knoche, W. (1942), *Ann. Soc. Cient. Argentina* **134** 237.

Koch, T. (1890), "Meteorologische Notizen und Korrespondenzen," *Zeit. Angew. Meteorol.* **7** 165.

Koechlin, J. (1924), *La Nature* **2628**.

Kogan-Beletskii, G. I. (1957), "K voprosu o priroda sharovoi molnii," *Priroda* **4** 71.

Kogan-Beletskii, G. I. (1959), "On the question of the nature of ball lightning," paper AFCRL-TN-59-795, U.S. Air Force Cambridge Research Laboratory.

Kogan-Beletskii, G. I. (1961), "The nature of ball lightning," in D. J. Ritchie (ed.) *Ball Lightning*, Consultants' Bureau, New York.

Kohl, T. (1882), *Gaea* **18** 569.

Kohl, T. (1884), "Uber den Kugelblitz" (summary), *Fortschr. Phys.* **40(3)** 663.

Kohl, T. (1885), "Kugelblitz" (summary), *Fortschr. Phys.* **41(3)** 651.

Kohlberg, O. A. (1901a), "Kugelblitz" (summary), *Fortschr. Phys.* **57(3)** 260.

Kohlberg, O. A. (1901b), "Kugelblitz," *Zeit. Angew. Meteorol.* **18** 24.*

Kohldmasov, A. (1972), "Ball lightning in liquids," *Tekh. Molod.* **8** 24.*

Kollert, H. (1889), *Elektro. Zeit.* **10**.

Kolobkov, N. V. (1951), "Grozy I shkvaly," *Gos. Izdat. Tekh.-Teor. Lit. Moscow* 47.*

Koloc, P. (1977), "A new model for ball lightning," unpublished paper, Neophysics Research Ltd., College Park, Maryland.

Koloc, P. M. (1989), "The PLASMAK (TM) configuration and ball lightning," in Y. H. Ohtsuki (ed.) *Science of Ball Lightning (Fireball)*, p. 289, World Scientific, Singapore.

Koloc, P. M. (1997), "Fusion implications of free-floating PLASMAK (TM) magnetoplasmoids," in *Current Trends in International Fusion Research. Proc. First Internat. Symp. on Evaluation of Current Trends in Fusion Research*, p. 475, Plenum Press, New York.

Kolosovskii, O. A. (1981), "Observation of ball lightning track on the window glass," *Zh. Tekh. Fiz.* **51(4)** 856.

Kondo, N., Ofuruton, H., and Ohtsuki, Y. H. (1996), "Numerical investigation of electromagnetic field with fireball formation," in *IEEE Conference Record—Abstracts. 1996 IEEE International Conference on Plasma Science*, p. 153. Cat. No. 96CH35939. Institute of Electrical and Electronics Engineers, New York.

Kondo, N., Ofuruton, H., Kamogawa, M., Aoki, M., and Ohtsuki, Y. H. (1996), "Measurement of electromagnetic field near a fireball," in *IEEE Conference Record—Abstracts. 1996 IEEE International Conference on Plasma Science*, p. 152. Cat. No. 96CH35939. Institute of Electrical and Electronics Engineers, New York.

Kondo, N., Kamogawa, M., Ofuruton, H., and Ohtsuki, Y. H. (1997), "Numerical investigation of the electromagnetic field with a plasma fireball," in Y. H. Ohtsuki and H. Ofuruton (eds.) *Proceedings: 5th International Symposium on Ball Lightning (ISBL97)*, p. 230.

Konieczny, J. (1963), "New weapon X-ball shaped thunder (globular fire-balls)," *Przeglad Lotniczy* **2** 72.*

Koppen, W. (1888a), "Bermerkenswerthe Blitzschlage," *Meteorol. Zeit.* **5** 158.

Koppen, W. (1888b), "Bermerkenswerthe Blitzschlage," *Meteorol. Zeit.* **5** 202.

Koppen, W. (1926), "Kugelblitz und Fohn in der Krin," *Meteorol. Zeit.* **43** 33.

Korshunov, V. K. (1990), "Drift motion of the magnetic monopole of Poljakov-'t Hooft in the air and the ball-lighting phenomenon (in memory of Paul Dirac)" (in Russian), *Mod. Phys. Lett. A* **5(21)** 1629.

Kosarev, E. L., and Serezhkin, Yu. G. (1974a), "Narrow-band decimeter radiation from lightning," *Sov. Phys. Tech. Phys.* **19(2)** 229.

Kosarev, E. L., and Serezhkin, Yu. G. (1974b), "Narrow-band decimeter radiation from lightning" (in Russian), *Zh. Tekh. Fiz.* **44** 364.

Kosarev, E. L., Vaganov, A. B., Zakirov, B. S., Luganskii, L. B., Narusbek, E. A., and Samosyuk, V. N. (1968), "Results of an expedition to study radio emission in the decimeter range accompanying linear lightning" (in Russian), *Zh. Tekh. Fiz.* **38(11)** 1831.

Kosarev, E. L., Vaganov, A. B., Zakirov, B. S., Luganskii, L. B., Narusbek, E. A., and Samosyuk, V. N. (1969), "Results of an expedition to study radio emission in the decimeter range accompanying linear lightning," *Sov. Phys. Tech. Phys.* **14** 1477.

Kozlov, B. N. (1975), "Principles of the relaxation theory of ball lightning" (in Russian), *Dokl. Akad. Nauk., SSSR* **221(4)** 802.

Kozlov, B. N. (1976), "Principles of the relaxation theory of ball lightning," *Sov. Phys. Doklady* **20(4)** 261.

Kozlov, B. N. (1978a), "O maksimal'nom energovydelenii sharovoi molnii," *Doklady Akad. Nauk. SSSR* **238** 61.

Kozlov, B. N. (1978b), "Teoriia sharovoi molnii," *Fiz. Plazmy* **4** 159.*

Kozlov, B. N. (1978c), "A theory of ball lightning," *Sov. J. Plasma Phys.* **4** 91.*

Kozlov, B. N. (1978d), "Maximum energy liberated by ball lightning," *Sov. Phys. Doklady* **23(1)** 41.

Kraemer, H., "Die entstehung eines Gewitters," in Weltall und Menscheit, insert bild IV, opposite page XII, Metzger and Wittig, Leipzig.

Krebs, W. (1912), "Les éclairs en boule et la question de la protection contre la foudre," *Ciel et Terre* **33** 257.

Krider, E. P., Weidman, C. D., and Noggle, R. C. (1977), "The electric field produced by lightning stepped leaders," *J. Geophys. Res.* **82(6)** 951. (NBL)

Krirsky, L. (1958), "Zare po blescich u vzdalene vourky," *Meteorologicke Zpravy (Prague)* **11(16)** 167.

Krummel, H. (1966), "Uber das Auftreten von Kugelblitzen bei dem Gewitter am 21 Juli 1965 in Marquardt bei Potsdam," *Zeit. Meteorol.* **19** 112.

Kruskal, M., and Schwarzschild, M. (1954), "Some instabilities of a completely ionized plasma," *Proc. Roy. Soc. (London)* **A223** 348.

Kudryavyev, V. V. (1955), *M. V. Lomondsov,* Izdatel'stvo, Moscow.

Kuhn, E. (1951), "Ein Kugelblitz auf einer Moment-Aufnahme?," *Naturwiss.* **38** 518.

Kulakov, A. V., and Rumyantsev, A. A. (1991a), "Ball lightning as a quantum condensate" (in Russian), *Doklady Akad. Nauk. SSSR* **320(4–6)** 1103.

Kulakov, A. V., and Rumyantsev, A. A. (1991b), "Ball lightning as a quantum condensate" (English translation), *Sov. Phys. Doklady* **36(10)** 701.

Kunin, W., and Furow, L. W. (1993), "Investigation of metastable toroidal plasma vortices," in A. G. Keul (ed.) *Progress in Ball Lightning Research: Proc. VIZOTUM*, p. 50, The Vizotum Project, Salzburg, Austria.

Kunin, V. N., Pleshivtsev, V. S., and Furov, L. V. (1997a), "Experimental studies of the nature of ball lightning" (in Russian), *Teplofizika Vysokikh Temperatur* **35(6)** 866.

Kunin, V. N., Pleshivtsev, V. S., and Furov, L. V. (1997b), "Experimental studies of the nature of ball lightning" (English translation), *High Temperature* **35(6)** 853.

Kunin, V. N., Pleshivtsev, V. S., and Furov, L. V. (1997c), "Experimental laboratory research of the ball lightning nature," in Y. H. Ohtsuki and H. Ofuruton (eds.) *Proceedings: 5th International Symposium on Ball Lightning (ISBL97)*, p. 225.

Kuschewitz, H. (1905), "Kugelblitz," *Elektro. Zeit.* **26(35)** 829.

Lacroix, A. (1904), *La Montagne Pelee et ses Eruptions,* Chapter 10, Masson et Cie, Paris.*

Ladikov, Y. P. (1960), "Magneto-vortex rings," *Izv. Akad. Nauk. SSSR* **4** 7.

Ladikov, Y. P. (1961), "Magneto-vortex rings" (in English), in D. J. Ritchie (ed.) *Ball Lightning,* Consultants' Bureau, New York.

Lagrange, E., (1910), *Ciel et Terre*, **31** 385.

Lalande, M. de (1852), "Fails observés à la station de Beuzeville pendant l'orage du 17 mai," *C. R. Hebd. Séances Acad. Sci.* **35** 24.

Lally, V. E. (1969), "Balloons—types, flight profiles and visibility," in E. U. Condon (ed.) *Scientific Study of Unidentified Flying Objects*, Section IV, Chapter 8, p. 755, Bantam Books, New York. (NBL)

Lalung, A. (1899a), "Correspondance-curieux cas de foudre en boule," *Bull. Soc. Astron. Fr.* **13** 471.

Lalung, A. (1899b), "La foudre globulaire aux Bureaux du Cable Francais," *Bull. Soc. Astron. Fr.* **13** 500.

Lancaster, A. (1878), *Ann. Obser. Roy. Bruxelles* 279.

Lancaster, A. (1888), "Les Orages," *Ciel et Terre* **9** 193.

Lancaster, A. (1905), "Coup de foudre," *Ciel et Terre* **26** 241.

Lane, F. W. (1945), *The Elements Rage,* Country Life Ltd., London.

Langmuir, I. (1907), "Lightning phenomena," *Mon. Weather Rev.* **35** 357.

Lardry, M. (1929), "Deux cas de foudre globulaire," *La Météorologie* **5** 278.

Laroche, P. (1982), "Neutralized electrical charges location during triggered lightning flashes," in *Proceedings: International Aerospace Conference on Lightning and Static Electricity*, Public Relations Section, Culham Laboratory, Abingdon, Oxon., UK. (NBL)

Larsson, A. (1998), *Klotblixten—ett fenomen som fortfarande gäckar vetenskapen,* Institutionen för högspänningsforskning, Uppsala University, Uppsala, Sweden.

Laskowski, B. R. (1927), "Ball lightning," *Tycos* **17** 16.*

Latham, J. (1964), "The electrification of snowstorms and sandstorms," *Roy. Meteorol. Soc. Q. J.* **90** 91. (NBL)

Latham, J., and Stromberg, I. M. (1977), "Point-discharge," in R. H. Golde (ed.) *Lightning*, Vol. I, p. 4, Academic Press, New York. (NBL)

Laugier, A. (1883a), "Effets produits par un coup de foudre à Rambouillet," *C. R. Hebd. Séances Acad. Sci.* **97** 1160.

Laugier, A. (1883b), "Effets produits par un coup de foudre à Rambouillet" (summary), *Fortschr. Phys.* **39(3)** 523.

Lazarides, M. (1971), "Lightning balls," *Nature* **231** 134.

Le Vine, D. M. (1980), "Sources of the strongest RF radiation from lightning," *J. Geophys. Res.* **85** 4091. (NBL)

Leconte, L. (1891), "Observation de foudre en boule," *La Météorologie* **39** 353.

Leduc, S. (1899a), "Wandernde Kugelformig Funken," *Beibl. Ann. Phys.* **23** 680.

Leduc, S. (1899b), "Etincelle globulaire ambulante," *C. R. Hebd. Séances Acad. Sci.* **129** 37.

Leduc, S. (1899c), *L'Eclairage Electrique* **21** 142.

Leduc, S. (1901), *L'Eclairage Electrique* **29** 54.

Leduc, S. (1902), "Etincelle globulaire ambulante," *Bull. Soc. Astron. Fr.* **16** 546.

Lee, W. R. (1977), "Lightning injuries and death," in R. H. Golde (ed.) *Lightning*, Vol. II, p. 16, Academic Press, New York. (NBL)

Lemonnier, M. (1891a), "Foudre en boule," *Le Cosmos* **juin** 282.*

Lemonnier, M. (1891b), "Foudre en boule," *La Météorologie* **39** 200.

Leonov, R. A. (1965a), *Zagadka Sharovyoy Molnii,"* Nauka, Moscow.*

Leonov, R. A. (1965b), "The Riddle of Ball Lightning," *TT-66-33253, JPRS: 36823,* U.S. Dept. Commerce, Washington, DC.

Leonov, R. A. (1965c), "The Enigma of Ball Lightning," Izd. Nauka, Moscow, Transl. TT: 66-33253, Clearinghouse Fed. Sci. Tech. Inform., U.S. Dept. of Commerce, Springfield, Virginia.

Leonov, R. A. (1966), "The Enigma of Ball Lightning," (report on 'Zagadka sharavyoy molnii', Moskva, IZD-VO, Nauka, 1965), *Foreign Science Bull.* **2(5)** 47.

Leonov, S. B., Pankova, M. B., and Tomilov, Yu. M. (1997), "Absorbing halo of hydrocarbonic plasma fire-ball," in Y. H. Ohtsuki and H. Ofuruton (eds.) *Proceedings: 5th International Symposium on Ball Lightning (ISBL97)*, p. 218.

Lepel, F. von (1889), "Uber die feuchten Funkenrohren und die Gewitterblitz," *Meteorol. Zeit.* **6** 216.

Lepel, F. von (1890a), *Elektro. Zeit.* **10** 487.

Lepel, F. von (1890b), "Uber wandernde funken" (summary), *Fortschr. Phys.* **46(3)** 420.

Lepel, F. von (1890c), "Uber wandernde funken," *Meteorol. Zeit.* **7** 297.

Lepel, F. von (1892), "Uber wandende funken," *La Météorologie* **40** 165.

Less, E. (1901), "Kugelblitz über Wolken," *Meteorol. Zeit.* **18** 39.

Lethbridge, M. (1981), "Cosmic rays and thunderstorm frequency," *Geophys. Res. Lett.* **8** 521. (NBL)

Letzmann, H. (1938), "Kugelblitz in Estland," *Zeit. Angew. Meteorol.* **55** 341.*

Leveille, H. (1891), "La foudre en boule," *Le Cosmos* dec, p. 4.*

Leveille, H. (1892), "La foudre en boule," *La Météorologie* **40** 64.

Lewis, B., and Elbe, G. von (1961), *Combustion, Flames and Explosions of Gases,* Academic Press, New York. (NBL)

Lewis, H. W. (1963), "Ball lightning," *Scientific American* **208** 107.

Lewis, H. W. (1980), "Ball lightning" in *Light from the Sky, Scientific American* 7.

Lewis, M. P. (1988), "Unusual lightning event. Ball lightning?," *Weather* **43(7)** 272.

Lilienfeld, P. (1970), "Ball lightning," *Nature* **226** 253.

Lindberg, L. (1960), "Experiments with plasma rings," *Nature* **185** 452.

Lindberg, L., Witalis, E., and Jacobsen, C. T. (1960), *Nature* **185**, 452. (NBL)

Lindberg, L., and Jacobsen, C. (1961), "On the amplification of poloidal magnetic flux in a plasma," *Astrophys. J.* **133** 1043. (NBL)

Lindemann, G. (1951), "Verletzungen durch Kugelblitze," *Kosmos* **47** 380.

Lindholm, F. (1927), "Eine kugelblitzartige Erscheinung," *Meteorol. Zeit.* **44** 391.

Lining, J. (1754), "Extract of a letter from John Lining, M. D. of Charlestown, in South Carolina, to Charles Pinckney, Esq; in London: with his answers to several queries sent to him concerning his experiments of electricity with a kite," *Phil. Trans. Roy. Soc. London* **48** 757.

Lippman, A. (1911), *Aeronautik* **1** 160.

List, K. (1905), "Elektrische Erscheinungen bei den Volkanausbruchen in Mittelamerik," *Meteorol. Zeit.* **22** 139.

Little, P. F. (1982), "Transmission of line model for assessing lightning hazards to aircraft," in *Proceedings: International Aerospace Conference on Lightning and Static Electricity*, Public Relations Section, Culham Laboratory, Abingdon, Oxon., UK. (NBL)

Livens, G. H. (1918), *The Theory of Electricity*, Cambridge University Press, Cambridge, England. (NBL)

Lockyer, W. J. S. (1907a), "Beaded or pearl lightning with a moving camera," *Knowledge Sci. News* **4** 121.*

Lockyer, W. J. S. (1907b), "Beaded or pearl lightning with a moving camera," *Knowledge Sci. News* **4** 145.*

Lockyer, W. J. S. (1908a), "Beaded or pearl lightning flashes; ribbon or band lightning," *Knowledge Sci. News* **5** 53.*

Lockyer, W. J. S. (1908b), "Beaded or pearl lightning flashes; ribbon or band lightning," *Knowledge Sci. News* **5** 117.*

Lockyer, W. J. S. (1908c), *Knowledge Sci. News* **5** 151.

Lodge, H. (1880), *Elektro. Zeit.* **1** 430.

Lodge, O. (1892a), "Globular lightning," *Electrician* **29** 603.

Lodge, O. (1892b), "Lightning conductors and lightning guards," *London* 138.*

Lodge, O. (1901), *L'Eclairage Electrique* **26** 58.

Lodge, O. (1905), "Lightning research," *Electrician* **55** 174.

Loeb, L. B. (1965), *Electrical Coronas*, University of California Press, Berkeley, California. (NBL)

Loeb, L. B. (1966), "Comments on a paper by W. H., Andersen, 'Energy source for ball lightning'," *J. Geophys. Res.* **71** 676.

Loeb, L. B., and Kip, A. F. (1939), "Electrical discharges in air at atmospheric pressure. The nature of the positive and negative point-to-plane coronas and the mechanism of spark propagation.," *J. Appl. Phys.* **10** 142. (NBL)

Logan, F. A. (1946), "What are fireballs?," *Electr. Rev. London* **138** 381.

Lomonosov, M. (1753), *Oratio de Metonis VI Electrica Ortis*, Royal Academy of Sciences, Petersburg.*

Long, M. (1978), "A ball lightning report from Kent," *J. Met.* **3(31)** 210.

Lowke, J. J. (1995), "Theory of ball lightning as an electric discharge," in *Proceedings: 11th International Conference on Gas Discharges and their Applications*. Institute of Electrical Engineers, Tokyo, Japan.

Lowke, J. J. (1996), "A theory of ball lightning as an electric discharge," *J. Phys. D (Appl. Phys.)* **29(5)** 1237.

Lowke, J. J., Uman, M. A., and Liebermann, R. W. (1969), "Toward a theory of ball lightning," *J. Geophys. Res.* **74** 6887.

Lugdunt, M. (1651), *B. A. Magni. Ratisb. Fp. Operum* **2** 44.*

Luizet, D. (1897), "Electricité atmospherique," *Bull. Soc. Astron. Fr.* **11** 298.

Luizet, D. (1907), "Observation d'un éclair en chapelet," *C. R. Hebd. Séances Acad. Sci.* **145** 780.

Luizet, D. (1908), "Perlschnurblitz," *Meteorol. Zeit.* **25** 43.

Lunev, V. I. (1992a), "Luminous balls in Siberia and the Far East: phenomenology, experiment, hypotheses" (in Russian), *Izvestiya Vysshikh Uchebnykh Zavedenii, Fizika* **35(3)** 65.

Lunev, V. I. (1992b), "Luminous balls in Siberia and the Far East: phenomenology, experiment, hypotheses" (English translation), *Sov. Phys. J.* **35(3)** 256.

Lynn, W. T. (1869), "Correspondence," *Proc. Meteorol. Soc. Lond.* **4** 374.

Lyons, W. A. (1994), "Characteristics of luminous structures in the stratosphere above thunderstorms as imaged by low-light video," *Geophys. Res. Letts.* **21** 875. (NBL)

Mack, K. (1901), "Blitz mit deutlich Erkennbarer Fortpflanzungsrichtung," *Meteorol. Zeit.* **18** 427.

MacLean, M. (1890), *Proc. Glasgow Phil. Soc.* **21** 22.

MacLear, J. P. (1904), "Ball lightning," *Meteorol. Mag.* **39** 187.

Maiorov, S. A., Tkachev, A. N., and Yakovlenko, S. I. (1991a), "Nonergodic Coulomb plasma and a possible explanation of the nature of ball lightning" (in Russian), *Sbornik Kratkie Soobshcheniya po Fizike, AN SSSR, Fizicheskii Institut im. P.N. Lebedeva* **10** 22.

Maiorov, S. A., Tkachev, A. N., and Yakovlenko, S. I. (1991b), "Nonergodic Coulomb plasma and a possible explanation of the nature of ball lightning" (English translation), *Sov. Phys.—Lebedev Institute Reports* **10** 19.

Maiorov, S. A., Tkachev, A. N., and Yakovlenko, S. I. (1992a), "Certain properties of an electrically insulated Coulomb plasma. Numerical simulation of microfields and thermodynamic properties. The problem of ball lightning" (in Russian), *Izvestiya Vysshikh Uchebnykh Zavedenii, Fizika* **35(2)** 10.

Maiorov, S. A., Tkachev, A. N., and Yakovlenko, S. I. (1992b), "Certain properties of an electrically insulated Coulomb plasma. Numerical simulation of microfields and thermodynamic properties. The problem of ball lightning" (English translation), *Sov. Phys. J.* **35(2)** 108.

Maiorov, S. A., Tkachev, A. N., and Yakovlenko, S. I. (1993a), "Recombination of a Coulomb plasma and nonbinary interaction effects" (in Russian), *Izvestiya Vysshikh Uchebnykh Zavedenii, Fizika* **36(1)** 68.

Maiorov, S. A., Tkachev, A. N., and Yakovlenko, S. I. (1993b), "Recombination of a Coulomb plasma and nonbinary interaction effects" (English translation), *Russian Physics J.* **36(1)** 55.

Maiorov, S. A., Tkachev, A. N., and Yakovlenko, S. I. (1994a), "Metastable state of supercooled plasma" (in Russian), *Usp. Fiz. Nauk.* **164(3)** 297.

Maiorov, S. A., Tkachev, A. N., and Yakovlenko, S. I. (1994b), "Metastable state of supercooled plasma" (English translation), *Physica Scripta* **51(4)** 498.

Maksimenko, V. (1964), "Vystrel sharovoiy molniey," *Voyennyye Znaniya* **7** 37.

Mal, R. K., and Sengupta, P. R. (1993), "Radial motion of a non-homogeneous solid spherical body in a magnetic field (Earth)," *Proc. Indian National Science Acad., Part A* **59(2)** 157.

Malan, D. J. (1961), "Unusual types of lightning," *Ann. Geophys.* **17** 388.

Malan, D. J. (1963), *Physics of Lightning*, p. 6, England University Press, London.

Malone, T. M. (1951), *Compendium of Meteorology*, p. 141, American Meteorological Society, Boston.

Malsch, W. (1956a), "Beobachtung einer schwachen Trombe und eines Kugelblitzes," *Meteorol. Rund.* **9(7, 8)** 150.

Malsch, W. (1956b), "Beobachtung einer schwachen Trombe und eines Kugelblitzes," *Meteorol. Rund.* **9(9, 10)** 138.

Malter, L., Webster, W. M., and Johnson, E. O. (1951), "Studies of externally heated hot cathode arcs," *RCA Review* **12** 415. (NBL)

Mancini, E. (1892), "Sopra un fenomeno analogo al fulmine globulare provocato per induzione," *Atti. D. Real. Accad. Lincei* **1** 308.*

Mancini, E. (1893a), "Ein Kugelblitzformige Erscheinung, die durch Induktion entstanden ist," *Jahrbuch der Astronomie und Geophysik.* **4** 345.

Mancini, E. (1893b), "Kugelblitz," *Météorologie* **43** 322.

Mancini, E. (1893c), "Kugelblitz," *Météorologie* **10** 157.

Mancini, E. (1893d), "Uber ein dem Kugelblitz ahnliches Pharomen, das durch Induction entstanden ist," *Naturwiss. Rund.* **8** 62.*

Mankin, R. W. (1979), "Insects as unidentified flying objects: Author's reply to comments; 2," *Appl. Opt.* **18(16)** 2725.

Mantel, G. (1885), "Bermerkenswerthe Blitzschlage," *Meteorol. Zeit.* **2** 18.

Manykin, E. A., Ojovan, M. I., and Poluectov, P. P. (1993), "Ball lightning is very much like Rydberg matter, isn't it?" in A. G. Keul (ed.) *Progress in Ball Lightning Research: Proc. VIZOTUM*, p. 54, The Vizotum Project, Salzburg, Austria.

Manykin, E. A., Ozhovan, M. I., and Poluektov, P. P. (1982), "On the nature of ball lightning," *Zh. Tekh. Fiz.* **52(7)** 5.

Marcel, M. du (1854), "Reproduction artificielle des éclairs," *Météorologie* **2** 57.

Marcet, W. (1888), "On atmospheric electricity," *Roy. Meteorol. Soc. Q. J.* **14** 197.

Marchant, E. W. (1930), *Nature* **125** 128.

Marcuse, A. (????), "Erkphysik—die Atmosphaere," in *Weltall und Menschheit*, p. 450, Metzer and Wittig, Leipzig.

Margs, G. (1956), "Kugelblitz," *Meteorol. Rund.* **9** 72.

Markson, R. (1978), "Solar modulation of atmospheric electrification and possible implications for the Sun-weather relationship," *Nature* **273** 103. (NBL)

Markson, R., and Muir, M. (1980), "Solar wind control of the earth's electric field," *Science* **208** 979. (NBL)

Marsh, O. C. (1895), "Globular lightning," *Nature* **53** 152.

Marsh, O. C. (1896a), "Note on globular lightning," *Amer. J. Sci. Arts* **151** 13.

Marsh, O. C. (1896b), "Globular lightning" (summary), *Fortschr. Phys.* **52(3)** 301.

Marsh, O. C. (1896c), "Kugelblitz," *Meteorol. Zeit.* **13** 189.

Marsh, O. C. (1896d), "Globular lightning," *Naturwiss. Rund.* **11** 235.*

Marsh, O. C. (1899), "Kugelblitz," *Météorologie* **47** 171.

Marsh, O. C. (1902), *Zeit. Angew. Meteorol.* **19** 239.

Martin, T. C. (1894), "The inventions, researches and writings of Nikola Tesla," in *The Electrical Engineer*, pp. 167, 171, van Nostrand, New York.

Martin, T. H. (1866), "La foudre, l'électricité et la magnetisme chez les anciens," *Paris* 178, 187, 288.*

Martynov, U. L. (1970), "Okislenie azota I sharovaya molniya," unpublished paper, USSR.

Marum, D. van (1795), "Tweede vervolg der proefneemingen gedaan met Teeyler's electric machine," *Haarlem.**

Marum, D. van (1800), *Phil. Mag.* **8** 313.

Mascart, E. (1876), *Trait. Elect. Statique* **2** 563.

Mascart, J. (1910), "Communications verbales—l'observations de feux St.-Elme," *Bull. Soc. Astron. Fr.* **24** 524. (NBL)

Mascart, M. (1890), "Sur l'existence du tonnerre en boule," *C. R. Hebd. Séances Acad. Sci.* **111** 496.

Mason, B. J. (1962), *Clouds, Rain and Rainmaking,* Chapter 7, p. 112, Cambridge University Press, Cambridge, England. (NBL)

Mason, C. (1742), "Fire-ball seen in the air, and a great explosion heard, Dec. 11, 1741," *Phil. Trans. Roy. Soc. London* **42** 1.

Mason, D. (1964), "Lightning strikes on aircraft—II, St. Elmo's fire," *Weather* **19** 254.

Massey, R. S., and Holden, D. N. (1995), "Phenomenology of transionospheric pulse pairs," *Radio Science* **30(5)** 1645. (NBL)

Mathias, E. (1908), "La matière fulminante," *Paris* **7** 113.*

Mathias, E. (1916a), "Sur trois observations d'éclairs en boule faites au sommet du Puy de Dome," *Ann. Physique* **5** 365.

Mathias, E. (1916b), "Sur trois observations d'éclairs en boule faites au sommet du Puy de Dome," *C. R. Hebd. Séances Acad. Sci.* **162** 642.

Mathias, E. (1916c), "Ball lightning on Puy de Dome," *Mon. Weather Rev.* **44** 516.

Mathias, E. (1924a), "Sur les formes terminales des éclairs fulgurants," *C. R. Hebd. Séances Acad. Sci.* **179** 136.

Mathias, E. (1924b), "Sur le bruit de l'éclair," *C. R. Hebd. Séances Acad. Sci.* **179** 372.

Mathias, E. (1924c), "Retour sur la théorie de l'éclair fulgurant," *C. R. Hebd. Séances Acad. Sci.* **179** 550.

Mathias, E. (1924d), Traite d'Électricité Atmospherique et Tellurique, p. 296, University of France Press, Paris.

Mathias, E. (1925a), "Contribution à l'étude de la matière fulminante: I. Les heterogeneites thermiques," *C. R. Hebd. Séances Acad. Sci.* **181** 1038.

Mathias, E. (1925b), "Contribution à l'étude de la matière fulminante: La couleur. Heterogeneites chimiques.," *C. R. Hebd. Séances Acad. Sci.* **181** 1111.

Mathias, E. (1926a), "Contribution à l'étude de la matière fulminante: Les globes noirs et blancs sans lumière propre," *C. R. Hebd. Séances Acad. Sci.* **182** 32.

Mathias, E. (1926b), "Contribution à l'étude de la matière fulminante: L'energie par centimetre cube et par gramme au moment de l'explosion," *C. R. Hebd. Séances Acad. Sci.* **182** 194.

Mathias, E. (1927a), "La foudre, ses differentes formes, la matière fulminante," *Ann. Post. Teleg. Telep.* **16** 965.*

Mathias, E. (1927b), "Contribution à l'étude de la matière fulminante: Decomposition spontanée et sans bruit sensible de certains éclairs sphériques de grand diametre," *C. R. Hebd. Séances Acad. Sci.* **184** 312.

Mathias, E. (1927c), "Contribution à l'étude de la matière fulminante: exemples de decomposition spontanée," *C. R. Hebd. Séances Acad. Sci.* **184** 565.

Mathias, E. (1927d), "Contribution à l'étude de la matière fulminante: celle-ci est-elle chaude où froide?" *C. R. Hebd. Séances Acad. Sci.* **184** 653.

Mathias, E. (1928a), *Ann. Post. Teleg. Telep.* **Juillet**.*

Mathias, E. (1928b), "Sur une curieuse photographie d'éclair obtenue dans la region du Luc des Quatre-Cantons," *C. R. Hebd. Séances Acad. Sci.* **186** 1334.

Mathias, E. (1929a), "Contribution à l'étude de la matière fulminante: son explosion par choc," *C. R. Hebd. Séances Acad. Sci.* **188** 1208.

Mathias, E. (1929b), "Contribution à l'étude de la matière fulminante: la tension superficielle. Partage d'un globe en plusieurs autrès sans l'influence d'un choc où d'un rebondissement," *C. R. Hebd. Séances Acad. Sci.* **189** 512.

Mathias, E. (1929c), "Contribution à l'étude de la matière fulminante: abaissement de sa tension superficielle par des impuretés," *C. R. Hebd. Séances Acad. Sci.* **189** 607.

Mathias, E. (1929d), "Contribution à l'étude de la matière fulminante: les formes serpentines," *C. R. Hebd. Séances Acad. Sci.* **189** 813.

Mathias, E. (1929e), "Contribution à l'étude de la matière fulminante: les globes excavateurs," *C. R. Hebd. Séances Acad. Sci.* **189** 1049.

Mathias, E. (1930a), "La foudre globulaire," *Bull. Mens. Soc. Fr. Electr.* **10** 22.*

Mathias, E. (1930b), "La foudre et ses differentes formes," *Bull. Soc. Astron. Fr.* **44** 416.

Mathias, E. (1930c), "Monographie de l'éclair fulgurant. Complements relatifs à ses formes terminales. Remarques sur la pression electrostatique des foudres sphériques.," *C. R. Hebd. Séances Acad. Sci.* **191** 393.

Mathias, E. (1930d), "De la matière fulminante," *C. R. Hebd. Séances Acad. Sci.* **191** 1420.

Mathias, E. (1931a), "Sur l'existence où la non-existence du filet des éclairs en chapelet," *C. R. Hebd. Séances Acad. Sci.* **192** 653.

Mathias, E. (1931b), "Sur l'éclair en chapelet avec grains," *C. R. Hebd. Séances Acad. Sci.* **193** 1140.

Mathias, E. (1931c), "Sur les éclairs en chapelet avec traits," *C. R. Hebd. Séances Acad. Sci.* **193** 1375.

Mathias, E. (1932a), "Variation de la tension superficielle de la matière fulminante en fonction de la temperature et du poids moleculaire," *C. R. Hebd. Séances Acad. Sci.* **194** 413.

Mathias, E. (1932b), "Les éclairs globulaires et ascendants dans les montagnes et les plateaux elevés," *C. R. Hebd. Séances Acad. Sci.* **194** 2257.

Mathias, E. (1932c), *Soc. Fran. Elect. Bull.* **2** 729.

Mathias, E. (1933a), "Les éclairs en chapelet avec grains," *Bull. Soc. Astron. Fr.* **47** 82.

Mathias, E. (1933b), "Electricité atmospherique," *C. R. Hebd. Séances Acad. Sci.* **196** 654.

Mathias, E. (1933c), "Sur la realité du reste d'éclaire spherique," *C. R. Hebd. Séances Acad. Sci.* **197** 962.

Mathias, E. (1934a), "Sur les foudres globulaires bleues," *C. R. Hebd. Séances Acad. Sci.* **199** 505.

Mathias, E. (1934b), "La notion d'impureté dans les foudres globulaires bleues," *C. R. Hebd. Séances Acad. Sci.* **199** 1083.

Mathias, E. (1934c), "Au sujet des foudres bleues," *Ciel et Terre* **50** 300.

Mathias, E. (1934d), *Mem. Meteorol. Natur.* **No. 24.**

Mathias, E. (1935a), "La théorie de Dauzere sur la conductibilité de l'air dans les regions exposés à la foudre," *C. R. Hebd. Séances Acad. Sci.* **201** 314.

Mathias, E. (1935b), "La foudre et la forme globulaire," *C. R. Hebd. Séances Acad. Sci.* **201** 535.

Mathias, E. (1935c), "La foudre et sa forme globulaire," Memorial de l'Office National Meteorologique de France, Paris, No. 24.*

Mathouillot, H. (1908), "Deux photographies de foudre globulaire," *C. R. Hebd. Séances Acad. Sci.* **146** 603.

Matopa, H. M. (1962), "Ball lightning," *Priroda* No. 1–6 71.

Matsumoto, T. (1997), "Ball lightning during underwater spark discharges and the Matsumae easthquakes," in Y. H. Ohtsuki and H. Ofuruton (eds.) *Proceedings: 5th International Symposium on Ball Lightning (ISBL97)*, p. 193.

Matthias, B. T., and Buchsbaum, S. J. (1962), "Pinched lightning," *Nature* **194** 327.

Matts, E. (1964), "A fireball?," *Weather* **19** 228.

Maudemain, A. (1919), "Eclairs en boule," *Bull. Soc. Astron. Fr.* **33** 284.

Maurain, C. (1948), "Eclairs en boule," in *La Foudre*, p. 100, Librairie Armand Colin, Paris, France.

Mavrogordato, M. (1886a), "Bolides où foudre en boule," *L'Astronomie* **5** 309.*

Mavrogordato, M. (1886b), "Bolides où foudre en boule," *Météorologie* **34** 310.

McCann, R., and Koloc, P. (1977), "Energy of ball lightning," *Bull. Amer. Phys. Soc.* **22(9)** 1176.

McClelland, J. A. (1919), "Globular lightning," *Nature* **103** 284.

McClenahan, D. H., and Plumer, J. A. (1982), "Protection of advanced composites against the direct effects of lightning," in *Proceedings: International Aerospace Conference on Lightning and Static Electricity*, Public Relations Section, Culham Laboratory, Abingdon, Oxon., UK. (NBL)

McCrosky, R. E. (1971), "Phenomenology of a bead (?) lightning event," unpublished report, Smithsonian Institution Astrophysical Observatory, Cambridge, Massachusetts.

McCrosky, R. E., and Boeschenstein, H. (1965), "The Prairie Meteorite Network," Special Report No. 173, Smithsonian Institution Astrophysical Observatory, Cambridge, Massachusetts. (NBL)

McDonough, T. (1887), "Globular lightning," *Nature* **11** 62.

McEachron, K. B. (1941), *Electr. Rev.* **60** 885.

McIntosh, D. H. (1973), "Lightning damage," *Weather* **28** 160.

McMillan, W. G. (1889), "On the phenomenon of the lightning discharge as illustrated by the striking of a house in Cossipore, Calcutta," *Nature* **40** 295.

McNally, J. R. (1961), "Ball lightning—a survey," *Bull. Amer. Phys. Soc.* **6** 202.

McNally, J. R. (1966), "Preliminary Report on Ball Lightning," Report No. ORNL-3938, Oak Ridge National Laboratory, Oak Ridge, Tennessee.

Meaden, G. T. (1973), "Four storm stories from Steeple Aston: 1670, 1772 and 1973," *Weather* **28** 520.

Meaden, G. T. (1976), "Remarkable occurrence of unusually-shaped ball lightning, 15 August 1975," *J. Met.* **1(10)** 310.*

Meaden, G. T. (ed.) (1985), "First Conference on Tornadoes, Waterspouts, Wind-Devils and Severe Storm Phenomena Oxford, UK Tornado & Storm Res Organ 29 June 1985," *J. Met.* **10(100)**

Meaden, G. T. (1990a), "Preliminary analysis of the video recording of a rotating ball-of-light phenomenon which happened at Ashford, Kent, in England on 10 September 1989," *J. Met.* **15(148)** 128.

Meaden, G. T. (1990b), "Low, overhead lightning in Aylesbury, England, followed by an incidence of ball lightning inside a house, 20 December 1989," *J. Met.* **15(148)** 143.

Meaden, G. T. (1990c), "Ball lightning bibliography from the Journal of Meteorology 1975–1990: issue numbers 1–147," *J. Met.* **15(148)** 172.

Meaden, G. T. (1990d), "Spontaneous combustion—an eye-witness case from Hungary," *J. Met.* **15(152)** 320.

Meaden, G. T. (1995), "The ball of light which was reported by a witness as passing through her face," *J. Met.* **20** 311.

Medawar, P. B. (1969), *The Art of the Soluble,* p. 11, Pelican Books, Harmondsworth, England. (NBL)

Meek, J. M., and Craggs, J. D. (1953), *Electrical Breakdown of Gases,* p. 242, Oxford University Press, Oxford, England.

Meek, J. M., and Perry, F. R. (1946), "The lightning discharge," *Phys. Soc. London, Rpt. Prog. Phys.* **10** 314.

Meese, A. D., and Evans, E. H. (1962), "Charge transfer in the lightning stroke as determined by the magnetograph," *J. Franklin Inst.* **273** 375. (NBL)

Meinardus, W. (1894), "Uber das Wetterleuchten," *Meteorol. Zeit.* **12** 14.

Meissner, A. (1930), "Uber Kugelblitz," *Meteorol. Zeit.* **47** 17.

Mendenhall, T. C. (1890), "On globular lightning," *Amer. J. Meteorol.* **6** 437.

Menin, H. (1881), *Les Mondes* **56** 387.

Menshutkin, B. N. (1911), *Mikhailo Vasil'evich Lomonosov,* Zizneopisanie, Izdanke.

Menshutkin, B. N. (1952), *Russia's Lomonosov,* Princeton University Press, Princeton, New Jersey.

Menzel, D. H., and Ford, L. G. (1965), *The World of Flying Saucers,* Doubleday, New York.

Mercanton, P. L. (1945), "De quelques éclairs rares," *Lausanne Bull. Soc. Vaud. Sci. Natur.* **63** 153.*

Merhaut, O. (1944), "Eine bemerkenswerthe Blitzaufnahme," *Naturwiss.* **32** 212.

Mesenyashin, A. I. (1991), "Electrostatic and bubble nature of ball lightning," *Appl. Phys. Letters* **58(23)** 2713.

Mesenyashin, A. I. (1993), "Ball lightning as an electrically charged shell of water molecules" in A. G. Keul (ed.) *Progress in Ball Lightning Research: Proc. VIZOTUM,* p. 57, The Vizotum Project, Salzburg, Austria.

Meskill, D. (1990), "Horizontal lightning bolt and a 'ball of light' in a house south Limerick, Ireland, 23 February 1990," *J. Met.* **15(148)** 153.

Meskill, D. (1996), "Heavy thunderstorm hits Charleville, 15 September 1995—and ball lightning too," *J. Met.* **21(206)** 58.

Mettetal, M. (1895a), "Observation d'un phénomène electrique," *C. R. Hebd. Séances Acad. Sci.* **121** 596.

Mettetal, M. (1895b), *La Nature* **23** 407.

Mettetal, M. (1895c), "Kugelblitz," *Naturwiss. Rund.* **11** 64.*

Mettetal, M. (1896), "Kugelblitz" (summary), *Fortschr. Phys.* **52(3)** 309.

Meunier, A. (1852a), "Double cas de foudre en boule observé dans un très-court espace de temps," *C. R. Hebd. Séances Acad. Sci.* **35** 195.

Meunier, A. (1852b), "Double cas de foudre en boule observé dans un très-court espace de temps," *Cosmos* **1** 351.*

Meunier, S. (1886), "Substance singulière recueillie à la suite d'un metéore rapporté à la foudre," *C. R. Hebd. Séances Acad. Sci.* **103** 837.

Meunier, S. (1887), *L'Astronomie* **6** 104.

Meunier, S. (1888), "Bemerkenswerthe Blitzschlage," *Météorologie* **37** 167.

Meusburger, D. (1904), "Kugelblitz," *Meteorol. Zeit.* **21** 283.

Mey, A. (1927), "Kugelblitze," *Meteorol. Zeit.* **44** 188.

Meyer, J. (1897), "Kugelblitz am 22 mai 1897 in Hamburg beobachtet," *Ann. Hydro. Marit. Meteorol.* **25** 257.

Michimoto, K. (1997), "Ball lightning observation on 15th July 1996 at Kawagoe City, Japan," in Y. H. Ohtsuki and H. Ofuruton (eds.) *Proceedings: 5th International Symposium on Ball Lightning (ISBL97)*, p. 12.

Milham, W. I. (1918), *Meteorology*, p. 470, Macmillan, New York.

Mills, A. A. (1971), "Ball lightning and thermoluminescence," *Nature* **233** 131.

Mills, A. A. (1980), "Will-o'-the-Wisp," *Chem. Brit.* **16(2)** 69. (NBL)

Mimno, H. R. (1970), "Ball lightning," *IEEE Trans. Aero. Elect. Sys.* **6** 846, 254.

Minchin, G. M. (1895a), "Personal injury from a fire-ball," *Nature* **53** 5.

Minchin, G. M. (1895b), "A fire ball," *Scientific American* **73** 374.

Minin, V. F., and Baibulatov, F. K. (1969), "The nature of pearl lightning" (in Russian), *Doklady Akad. Nauk. SSSR* **118(4)** 795.

Minin, V. F., and Baibulatov, F. K. (1970), "The nature of pearl lightning," *Sov. Phys. Doklady* **14(10)** 979.

Minnaert, M. G. J. (1954), *The Nature of Light and Color in the Open Air* (translated by H. M. Kremer-Priest; revised by K. E. B. Jay), Dover, New York. (NBL)

Minnaert, M. G. J. (1993), *Light and Color in the Outdoors* (translated and revised by L. Seymour), Springer-Verlag, New York. (NBL)

Mohn, H. (1908), "Kugelblitz," *Meteorol. Zeit.* **25** 314.

Mohr, F. B. (1960), "A truly remarkable fly," *Science* **151** 634.

Moigno, F. (1859), *Le Cosmos* **14** 672.

Moller, F. (1949), "Der geheimnisvolle Kugelblitz," *Umschau* **49** 458.

Moller, F. (1951a), *Hamel en Dampkring* **49** 158.

Moller, F. (1951b), "Was ist ein Kugelblitz?," *Kosmos* **47** 86.

Moncel, T. du (1853), "Nouveau système d'inflammation à distance de substances inflammables par le courant d'une pile de Daniell et des conducteurs très fins," *C. R. Hebd. Séances Acad. Sci.* **37** 953.

Moncel, T. du (1854a), "Note sur les éclairs en boule," *C. R. Hebd. Séances Acad. Sci.* **38** 408.

Moncel, T. du (1854b), "Theorie des éclairs," *Mem. Soc. Sci. Nat. Cherbourg* **2** 49.*

Moncel, T. du (1854c), *Mem. Soc. Sci. Nat. Cherbourg* **2** 85.

Moncel, T. du (1854d), "Reproduction artificielles des éclairs," *Météorologie* **2** 57.

Moncel, T. du (1857), "Notice historique et théorique sur le tonnerre et les éclairs," *Paris* 54.*

Moncel, T. du (1881), "Eclairs en boule," *Bull. Soc. Encourag.* **7** Nov.*

Moncel, T. du (1882a), "Eclairs en boule" (summary), *Fortschr. Phys.* **38(3)** 556.

Moncel, T. du (1882b), "Eclairs en boule," *La Lumière Electrique* **6** 145.*

Moncel, T. du (1882c), *La Lumière Electrique* **6** 169.

Monnot, M. (1948), "Foudre en boule," *Bull. Soc. Astron. Fr.* **62** 323.

Moon, F. D. (1925), "Ball lightning," *Marine Observer* **2** 129.

Moore, C. B., and Vonnegut, B. (1977), "The thundercloud," in R. H. Golde (ed.) *Lightning*, Vol. I, p. 3, Academic Press, New York. (NBL)

Moore, M. (1977), "Machine makes lightning," *Brevard Today*, Brevard County Newspaper, Cocoa Beach, Fla. July 18.

Moorsel, J. van (1889), "Globular lightning," *Electrician* **22** 520.

Morgan, J. H. W. (1922), "Ball lightning," *Meteorol. Mag.* **57** 46.

Morris, W. (1936), "A thunderstorm mystery," *London Daily Mail* Nov 5.

Morrison, P. (1972), "The nature of scientific evidence: a summary," in C. Sagan and T. Page (eds.) *UFO's—A Scientific Debate*, Chapter 15, p. 276, Cornell University Press, Ithaca, New York. (NBL)

Morse, R. L. (1991), "Plasmas," in *Encyclopedia of Physics*, 2nd ed., R. G. Lerner and G. L. Trigg (eds.), VCH Publishers, New York. (NBL)

Mortley, W. S. (1973), "Ball lightning enigma," *New Scientist* **57** 42.

Moskovits, M. (ed.) (1995), *Science and Society, the John C. Polanyi Nobel Laureate's Lectures*, Anansi Press, Concord, Ontario, Canada. (NBL)

Mott, F. T. (1880), "Curious electrical phenomenon," *Nature* **22** 193.

Moureaux, T. (1887a), "Ancien exemple de foudre en boule," *La Nature Bull.* No. 562.*

Moureaux, T. (1887b), "Ancien exemple de foudre en boule," *Météorologie* **35** 221.

Moureaux, T. (1892a), "Foudre globulaire," *La Nature* **juin** 33.*

Moureaux, T. (1892b), "Foudre globulaire," *Météorologie* **40** 215.

Moussette, C. (1884), "Theorie mécanique de la foudre," *C. R. Hebd. Séances Acad. Sci.* **107** 435.

Moussette, C. (1886a), "La foudre en spirale," *C. R. Hebd. Séances Acad. Sci.* **103** 30.

Moussette, C. (1886b), "La foudre en spirale" (summary), *Fortschr. Phys.* **42(3)** 654.

Moussette, C. (1886c), "La foudre en spirale," *La Lumière Electrique* **21** 123.*

Moussette, C. (1886d), "Orages du 12 mai 1886, la foudre en spirale," *Météorologie* **34** 298.

Moussette, C. (1886e), "La foudre en spirale," *Rev. Electr.* **2** 109.*

Moussette, C. (1887), *La Lumière Electrique* **25** 534.

Muhleisen, R. (1972), "Der Kugelblitz—ein Plasmazustand?," *Kosmos* **68** 159.

Muirhead, H. (1883), "Globe lightning," *Meteorol. Mag.* **18** 57.

Muirhead, H. (1883), "Globe lightning," *Electrician* **12** 50.

Mukharev, L. A. (1986), "The nature of ball lightning," *Sov. J. Commun. Technol. Electron.* **30(9)** 77.

Muldrew, D. B. (1990), "The physical nature of ball lightning," *Geophysical Res. Lett.* **17(12)** 2277.

Muldrew, D. B. (1996), "Solid-core model of ball lightning," in *IEEE Conference Record—Abstracts. 1996 IEEE International Conference on Plasma Science*, p. 151. Cat. No. 96CH35939. Institute of Electrical and Electronics Engineers, New York.

Muller, M. (1888), *La Nature* **16** 415.

Müller-Hillebrand, D. (1955), *Tekn. Zeit.* **85** 541.

Müller-Hillebrand, D. (1957), "Rückwärtiger Blitzeinschlag in Häuser und Energieumsatz im eingeengten Blitzkanal" ETZ-A 78, p. 548, cited in R. H. Golde (1973). (NBL)

Müller-Hillebrand, D. (1963), "Zur Frage des Kugelblitzes," *Elektrie* **17** 211.

Müller-Hillebrand, D. (1965a), "Lightning protection," in S. C. Coroniti (ed.) *Problems of Atmospheric and Space Electricity*, p. 407, Elsevier, Amsterdam.

Müller-Hillebrand, D. (1965b), "Ball lightning," in S. C. Coroniti (ed.) *Problems of Atmospheric and Space Electricity*, p. 455, Elsevier, Amsterdam.

Muratov, M. (1960), *Lomonosov*, Izdatel'stvo, Moscow.

Murchie, G. (1954), *Song of the Sky*, p. 218, Houghton, Mifflin, Boston, Massachusetts.

Muschenbroek, P. van (1769), *Cours de Physique Experimentale et Mathematique*, Vol. 3, p. 405, Sigaud de la Fond, Leyden.*

Musya, K. (1931), "On the luminous phenomenon that attended the Idu earthquake, Nov 26 1930," *Tokyo Univ. Earthquake Res. Bull.* **9** 177.

Naber, H. (1897a), "Kugelblitz" (summary), *Fortschr. Phys.* **53(3)** 300.

Naber, H. (1897b), "Kugelblitz," *Zeit. Angew. Meteorol.* **14** 144.*

Nartmann, D. (1889), *L'Astronomie* **8** 90.

Nasse, E. (1872a), "Un exemple de foudre globulaire observé à Brives, le vendredi 17 mai," *C. R. Hebd. Séances Acad. Sci.* **74** 1384.

Nasse, E. (1872b), "Foudre globulaire à Brives, le vendredi 17 mai" (summary), *Fortschr. Phys.* **28** 852.

Nasse, E. (1872c), "Foudre globulaire à Brives, le vendredi 17 mai," *Mondes* **28** 280.*

Nasser, E. (1971), *Fundamentals of Gaseous Ionization and Plasma Electronics*, Wiley Interscience, New York. (NBL)

Nauer, H. (1953), "Modellversuche zum Kugelblitz," *Zeit. Angew. Phys.* **5(12)** 441.

Nauer, H. (1956), "Wie entsteht ein Kugelblitz?," *Umschau* **56** 75.

Neda, T., Ofuruton, H., and Ohtsuki, Y-H. (1989), "A calculation of electric field of ball lightning," in Y. H. Ohtsuki (ed.) *Science of Ball Lightning (Fireball)*, p. 260, World Scientific, Singapore.

Needham, B. (1993), "Bolt from the blue," *New Scientist (Australian edition)* **28** Aug, p. 4.

Neesen, F. (1900), "Gewitter und Blitzableiter," *Himmel und Erde* **13** 151.*

Neesen, F. (1901), *Himmel und Erde* **13** 145.

Neish, W. J. P. (1990), "Lightning phenomena in the Drakensberg Mountains of Natal, RSA," *J. Met.* **15(154)** 377.

Nernst, H. (1909), *Theor. Chem.* **6** 679.

Neugebauer, T. (1937), "Zu dem Problem des Kugelblitzes," *Zeit. Phys.* **106(7, 8)** 474.

Neugebauer, T. (1975), "The theory of ball lightning," *Fiz. Szle (Hungary)* **25(2)** 49.*

Neugebauer, T. (1977), "Zu der quantenmechanischen Theorie des Kugelblitzes," *Acta Physica* **42(1)** 29.*

Neuntenfl, J. (1951), "Wintergewitter St. Elmsfeuer und Kugelblitz," *Wetter und Leben* **3** 49.

Newman, M. M. (1960), "Thunderstorm electrical discharges intercepted by aircraft and related ball lightning and spherics phenomena," *J. Geophys. Res.* **65** 1966.

Newman, M. M. (1965), "Use of triggered lightning to study the discharge process in the channel," in S. C. Coroniti (ed.) *Problems of Atmospheric and Space Electricity*, p. 482, Elsevier, Amsterdam. (NBL)

Newman, M. M. (1968), "Lightning discharge simulation and triggered lightning," in S. C. Coroniti and J. Hughes (eds.) *Planetary Electrodynamics*, Vol. 2, p. 213, Gordon and Breach, New York. (NBL)

Newman, M. M., and Robb, J. D. (1977), "Protection of aircraft," in R. H. Golde (ed.) *Lightning*, Vol. II, p. 21, Academic Press, New York. (NBL)

Newman, M. M., Stahman, J. R., Robb, J. D., Lewis, E. A., Martin, S. G., and Zinn, S. V. (1967), "Triggered lightning strokes at very close range," *J. Geophys. Res.* **72** 4761. (NBL)

Newman, W., and Kemppainen, A. O. (1946), "High-voltage installation of the precipitation-static project," *Proc. IRE* **34** 247.

Nguyen Manh Duc (1987), "Lightning observations at Tam Dao," Papers on atmospherical electricity: Publications of the Institute of Geophysics, Polish Academy of Sciences D-26, p. 75.

Nichols, J. R. (1928), *Bells through the Ages*, p. 277, London, Chapman and Hall.

Nicholson, M. (1774), "An account of a storm of lightning observed on the 1st of March 1774 near Wakefield in Yorkshire," *Phil. Trans. Roy. Soc. London* **64** 350.

Nickel, K. L. E. (1989a), "A fluid-dynamical model for ball lightning and bead lightning," in Y. H. Ohtsuki (ed.) *Science of Ball Lightning (Fireball)*, p. 135, World Scientific, Singapore.

Nickel, K. L. E. (1989b), "The lifetime of Hill's vortex," in Y. H. Ohtsuki (ed.) *Science of Ball Lightning (Fireball)*, p. 177, World Scientific, Singapore.

Niese, H. (1948), *Start* **16** 7.

Niessl, G. von (1898a), "Kugelblitzbeobachtung" (summary), *Fortschr. Phys.* **54(3)** 313.

Niessl, G. von (1898b), "Kugelblitzbeobachtung," *Verh. Naturf. Ver. Brunn* **36** 41.*

Niessl, G. von (1901a), "Elektrische Lichtphanomene an den Leitungsdrahten" (summary), *Fortschr. Phys.* **57(3)** 260.

Niessl, G. von (1901b), "Elektrische Lichtphanomene an den Leitungsdrahten," *Verh. Naturf. Ver. Brunn* **40**.*

Nippoldt, M. (1916), "Kugelblitz mit nachfolgenden Blitzstrahl," *Meteorol. Zeit.* **33** 475.

Nixon, J. (1750), "A letter from the Rev. Mr. John Nixon, FRS, to the President serving to accompany a letter from Mr. William Smith to Mr. Nixon, giving a very particular account of the earthquake felt on Sep. 30, 1750," *Phil. Trans. Roy. Soc. London* **46** 727.

Noad, H. M. (1855), *Manual of Electricity*, Vol. I p. 204.

Nonhebel, G., Colvin, J., Patterson, H. S., and Whytlaw-Gray, R. (1927), "The coagulation of smokes and the theory of Smoluchowski," *Roy. Soc. Lond. Proc.* **116** 540.

Norinder, H. (1923), *Tekn. Tidskrift (Stockholm)* **33** 190.*

Norinder, H. (1939), "Om blixtens natur," *Kungliga Ventenskapsocietetens Arsbok* **94** 39.

Norinder, H. (1949), "On the measurement and nature of atmospherics produced by electrical discharges in snow squalls and from other sources," *Tellus* **1(2)** 1. (NBL)

Norinder, H. (1965), "Comments," in S. C. Coroniti (ed.) *Problems of Atmospheric and Space Electricity*, p. 455, Elsevier, Amsterdam.

Norinder, H., and Siksna, R. (1953), "Experiments concerning electrification of snow," *Arkiv foer Geofysik* **2** 59.

Norton, A. P. (1978), *Norton's Star Atlas* 17th ed., G. E. Satterthwaite (ed.), Gall and Inglis, Edinburgh, Scotland. (NBL)

O'Neill, J. J. (1971), *Prodigal Genius, the Life of Nikola Tesla,* Ives Washburn, David McKay, New York.

Obermayer, A. von (1888), "Uber die bei Beschreibung von Elmsfeuer notwendigen angaben," *Meteorol. Zeit.* **5** 224. (NBL)

Obermayer, A. von (1892), "St. Elmsfeuer beobachtungen," *Jahrbuch der Astronomie und Geophysik* **3** 325. (NBL)

Ofuruton, H., and Ohtsuki, Y-H. (1989), "Experimental research on ball lightning," in Y. H. Ohtsuki (ed.) *Science of Ball Lightning (Fireball)*, p. 310, World Scientific, Singapore.

Ofuruton, H., and Ohtsuki, Y-H. (1990), "Experimental research on ball lightning," *Nuovo Cimento* **13C(4)** 761.

Ofuruton, H., Kondo, N., Kamogawa, M., Aoki, M., and Ohtsuki, Y. H. (1997a), "Experimental conditions of laboratory ball lightning produced by microwave and discharge," in Y. H. Ohtsuki and H. Ofuruton (eds.) *Proceedings: 5th International Symposium on Ball Lightning (ISBL97)*, p. 215.

Ofuruton, H., Ohtsuki, Y H., Kondo, N., Kamogawa, M., Kato, M., and Takahashi, T. (1997b), "Nature of ball lightning in Japan," in Y. H. Ohtsuki and H. Ofuruton (eds.) *Proceedings: 5th International Symposium on Ball Lightning (ISBL97)*, p. 17.

Ofuruton, H., Ohtsuki, Y-H., Neda, T., Kondo, N., Kamogawa, M., Miyajima, T., Shiina, Y., Hashimoto, K., and Sasaki, K. (1997c), "Charged aerosol model of ball lightning," in Y. H. Ohtsuki and H. Ofuruton (eds.) *Proceedings: 5th International Symposium on Ball Lightning (ISBL97)*, p. 67.

Ohnishi, T., and Suganuma, J. I. (1997), "Numerical simulation of ball lightning as a standing lightning," in Y. H. Ohtsuki and H. Ofuruton (eds.) *Proceedings: 5th International Symposium on Ball Lightning (ISBL97)*, p. 125.

Ohtsuki, Y. H. (ed.) (1989), *Science of Ball Lightning (Fireball)*, World Scientific, Singapore.

Ohtsuki, Y. H. (1991), "Creation of the plasma vortex," in *International Conference on the Circles Effect*, p. 84, Tornado and Storm Research Organisation, Cambridge, United Kingdom.

Ohtsuki, Y. H., and Ofuruton, H. (1987), "Nature of ball lightning in Japan," *Nuovo Cimento* **10C(5)** 577.

Ohtsuki, Y. H., and Ofuruton, H. (1989), "Nature of ball lightning in Japan," in Y. H. Ohtsuki (ed.) *Science of Ball Lightning (Fireball)*, p. 31, World Scientific, Singapore.

Ohtsuki, Y. H., and Ofuruton, H. (1990), "Observations in Japan of ball lightning (fireballs)," *J. Met.* **15(148)** 154.

Ohtsuki, Y. H., and Ofuruton, H. (1991), "Plasma fireballs formed by microwave interference in air," *Nature* **350(6314)** 139.

Ohtsuki, Y. H., and Ofuruton, H. (eds.) (1997), *Proceedings: 5th International Symposium on Ball Lightning (ISBL97)*.

Ol'khovatov, A. Yu. (1997), "Explosions at Sasovon, Russia, as examples of explosions of ball lightning-like formations endogenic origin," in Y. H. Ohtsuki and H. Ofuruton (eds.) *Proceedings: 5th International Symposium on Ball Lightning (ISBL97)*, p. 20.

Olivier, C. P. (1925), *Meteors,* Williams and Wilkins, Baltimore, Maryland. (NBL)

Olivier, C. P. (1965), "Meteor" in *Encyclopedia Americana*, 1965 ed., Vol. 18, p. 713. (NBL)

Ollendorf, H. (1933), *Ark. Elektr.* **27** 169.

Olsson, B. A. (1982), "Lightning testing of aircraft fuel tank details," in *Proceedings: International Aerospace Conference on Lightning and Static Electricity*, Public Relations Section, Culham Laboratory, Abingdon, Oxon., UK.

Oltramare, G. (1879), "Explication du bolide de Geneve du 7 juin 1879," *C. R. Hebd. Séances Acad. Sci.* **88** 1319.

Orekhov, B. K. (1989), "Ball lightning," PERA-OTIS 89/9749, Moscow Dept. of Trade Industry, Overseas Technical Information Service.

Oriel, A. (1901), "Foudre en boule," *Bull. Soc. Astron. Fr.* **15** 303.

Orville, R. E. (1966), "High speed, time resolved spectrum of a lightning stroke," *Science* **157** 451. (NBL)

Orville, R. E. (1972), "Lightning between clouds and ground," *Weatherwise* **25** 108. (NBL)

Orville, R. E. (1977), "Lightning spectroscopy," in R. H. Golde (ed.) *Lightning*, Vol. 1, p. 8, Academic Press, New York. (NBL)

Osborne, J. (1978), "Ball lightning," *J. Met.* **3(31)** 215.

Ostapenko, V. I., and Tolpygo, K. B. (1984), "The plasma theory of ball lightning," *Ukrainskii Fizicheskii Zh.* **29(2)** 210.

Oswalt, P. (1915), "Foudre en boule," *Bull. Soc. Astron. Fr.* **29** 437.

Otto, M. (1901), "L'éclair en boule," *La Nature* **29** 361.*

Owen, R. (1886), "Ball, or globe lightning," *Amer. J. Meteorol.* **3** 383.

Owens, A. L. (1924), "Ball lightning," *Marine Observer* **1** 117.

Painter, B. F. (1920), "Reproducing ball lightning," *QST* Dec, p. 59.

Palazzo, L. (1903), "Kugelblitz," *Meteorol. Zeit.* **20** 188.

Palffymuhoray, P. (1984), "Ball lightning," *New Scientist* **101(1399)** 43.

Palmer, J. (1752), "An account of the effects of lightning at South Molton in Devonshire," *Phil. Trans. Roy. Soc. London* **47** 330.

Paneth, F. A. (1956), "The frequency of meteorite fall throughout the ages," *Vistas in Astronomy* **2** 1681.

Paradise, W. (1773), "Account of the effects of lightning at Steeple Ashton and Holt, in the county of Wilts., on 20th June 1772," *Phil. Trans. Roy. Soc. London* **63** 240.

Parent, E. (1873a), "Sur les effets produits par la foudre, à Troyes, le 26 juillet 1873; observations de nombreux globes de feu," *C. R. Hebd. Séances Acad. Sci.* **77** 370.

Parent, E. (1873b), "Sur les effets produits par la foudre, à Troyes, le 26 juillet 1873; observations de nombreux globes de feu" (summary), *Fortschr. Phys.* **29** 1002.

Parent, E. (1873c), "Sur les effets produits par la foudre, à Troyes, le 26 juillet 1873; observations de nombreux globes de feu," *Mondes* **31** 716.*

Parisot, M. (1821), "Sur quelques phénomènes de Météorologie observés à Epinol, dans le Département des Vosges, en août 1821," *Ann. Chim. Phys.* **19** 86.

Parry, A. (1958a), "Soviet affairs," *Missiles Rockets* **3(4)**, April, p. 136.

Parry, A. (1958b), "Soviet affairs," *Missiles Rockets* **3(6)**, June, p. 159.

Pascal, P. V. H. (1956), "Nouveau traité de chimie minérale," Masson, Paris, **10** 368; **13** 266. (NBL)

Patterson, H. S., Whytlaw-Gray, R., and Cawood, W. (1929a), "The process of coagulation in smokes," *Roy. Soc. Lond. Proc.* **124** 502. (NBL)

Patterson, H. S., Whytlaw-Gray, R., and Cawood, W. (1929b), "The structure and electrification of smoke particles," *Roy. Soc. Lond. Proc.* **124** 523. (NBL)

Paw-u, K. T. (1979), "Insects as unidentified flying objects: comment," *Appl. Opt.* **18(16)** 2723.

Pcholko, I. G. (1963), *Aviatsionnaya Meteorologiya*, p. 239, Meteorologicheskoy, Izdatel'stvo, Leningrad.

Pcholko, I. G. (1966), "Aviation meteorology" (translation of Pcholko 1963), *Foreign Science Bull.* **2(4)** Appendix I 50.

Peake, N. B. (1962), "Anti-bubble ball lightning," *New Scientist* **13** 334.

Pelletier, L. (1934), "Les formes étranges de la foudre," *Rev. Ciel. Bourges*.*

Pellissier, G. (1894), "Les éclairs en chapelet," *La Nature* **22** 323.*

Pelzl, E. (1953), "Anleitung fur die Beobachter an den Wetterbeobachtungstellen des meteorologischen und hydrologischen dienstes der Deutschen Demokratischen Republik" Veroffentlichungen des Meteorologischen und Hydrologischen dienstes der Deutschen Demokratischen Republik, Nr. 7, p. 136, Akademie Verlag, Berlin.

Peppler, W. (1919), "Eine merkwurdige kugelblitzartige elektrische Entladung," *Zeit. Angew. Meteorol.* **36** 29.

Perala, R. A., and Rudolph, T. H. (1982), "The interaction of electromagnetic fields with aircraft during a lightning event," in *Proceedings: International Aerospace Conference on Lightning and Static Electricity*, Public Relations Section, Culham Laboratory, Abingdon, Oxon., UK.

Peroux, E. (1907), "Eclair en boule," *Bull. Soc. Astron. Fr.* **21** 453.

Perrin, P. (1873), "Etude sur les éclairs," *Paris* 77.*

Peterbaude, H. (1888), *Prog. Gymn. Hirschberg* 6.

Petersen, H. (1954a), "Ball lightning," *Geophysica* **5** 49.

Petersen, H. (1954b), "Ball lightning," *Weather* **9** 73.

Petersen, H. (1954c), "Photography and ball lightning," *Weather* **9** 321.

Petrie, W. M. F. (1881), "Slow lightning," *Nature* **24** 284.

Petrz'ilka, V. A. (1967), *Ball Lightning*, Institute of Plasma Physics, Czechoslovak Academy of Science, Prague.*

Petrz'ilka, V. A. (1969), "A possible explanation of ball lightning motions," *Ceskoslov. Casopis Fysiku* **19** 160.

Pfeil, H. (1886a), "Die hirschberger Kugelblitze" (summary), *Fortschr. Phys.* **42(3)** 670.

Pfeil, H. (1886b), "Die hirschberger Kugelblitze," *Gaea* 5.*

Pflegel, R. (1956), "Beobachtung eines Kugelblitzes," *Zeit. Meteorol.* **10(11)** 347.

Phillips, R. (1863), On Atmospheric Electricity, p. 45, Hardwicke, Piccadilly, London.*

Phillips, R. (1889), "Globular and other forms of lightning," *Nature* **41** 58.

Piddocke, B. G. (1974), "Ball lightning," *Weather* **29** 314.

Pierce, E. T. (1977a), "Atmospherics and radio noise," in R. H. Golde (ed.) *Lightning*, Vol. I, p. 10, Academic Press, New York.

Pierce, E. T. (1977b), "Lightning location and warning systems," in R. H. Golde (ed.) *Lightning*, Vol. II, p. 15, Academic Press, New York. (NBL)

Pierce, E. T., Nadile, R. M., and McKinnon, P. J. (1960), "An investigation of negative point-plane corona and its relation to ball lightning," Report AFCRL-TR-60-354 U.S. Air Force Cambridge Research Lab.

Pierce, S., and Carlstead, E. M. (1966), "Observation of a curious electrical phenomenon in Hawaii," *Mon. Weather Rev.* **94** 272.

Pike, W. S. (1982), "A remarkable ball lightning incident in Berkshire, 3 July 1982," *J. Met.* **7(72)** 253.

Pilaye, M. de (1843), "Foudre en boule," *C. R. Hebd. Séances Acad. Sci.* **17** 630.

Pilleux, L. (1895a), "Sur le aurores boréales et les éclairs en boule," *Météorologie* **43** 249.

Pilleux, L. (1895b), "Sur le aurores boréales et les éclairs en boule," *Le Cosmos* **31** 420.*

Pilyugin, N. N., Talipov, R. F., and Khlebnikov, V. S. (1997a), "Supersonic flow with gasdynamic and physicochemical nonuniformities past bodies" (in Russian), *Teplofizika Vysokikh Temperatur* **35(2)** 322.

Pilyugin, N. N., Talipov, R. F., and Khlebnikov, V. S. (1997b), "Supersonic flow with gasdynamic and physicochemical nonuniformities past bodies" (English translation), *High Temperature* **35(2)** 319.

Pippard, B. (1982), "Ball of fire?," *Nature* **298(5886)** 702.

Piren, M. C. de (1911), "Foudre en boule," *Bull. Soc. Astron. Fr.* **25** 284.

Piscicellitaeggi, C. (1988), "Ball lightning," *New Scientist* **117(1599)** 69.

Pittock, A. B. (1977), "A comment on 'On the problem of ball lightning' by W. I. Arabadji," *J. Geophys. Res.* **82(24)** 3499.

Planté, G. (1860), *Bibl. Univ. Geneve.* **7** 332.

Planté, G. (1868), "Les courants secondaires et leurs applications," *Ann. Chim. Phys.* **5** 5.

Planté, G. (1872), "Sur l'emploi des courants secondaires pour accumuler ou transformer les effets de la pile voltaique," *C. R. Hebd. Séances Acad. Sci.* **74** 592.

Planté, G. (1873), "Suite de récherches sur les courants secondaires, et leurs applications," *C. R. Hebd. Séances Acad. Sci.* **77** 466.

Planté, G. (1875a), "Récherches sur les phénomènes produits dans les liquides par des courants electriques de haute tension," *C. R. Hebd. Séances Acad. Sci.* **80** 1133.

Planté, G. (1875b), "Récherches sur les phénomènes produits par des courants electriques de haute tension, et sur leurs analogies avec les phénomènes naturels," *C. R. Hebd. Séances Acad. Sci.* **81** 185.

Planté, G. (1875c), "Sur la formation de la grèle," *C. R. Hebd. Séances Acad. Sci.* **81** 616.

Planté, G. (1875d), "Sur la formation de la grèle," *C. R. Hebd. Séances Acad. Sci.* **81** 618.

Planté, G. (1875e), "Uber eine den Kugelblitzen anaboge Erscheinung" (summary), *Fortschr. Phys.* **31** 913.

Planté, G. (1875f), "Uber eine den Kugelblitzen anaboge Erscheinung," *Zeit. Oster. Ges. Meteorol.* **10** 303.*

Planté, G. (1876a), "Sur les trombes," *C. R. Hebd. Séances Acad. Sci.* **82** 220.

Planté, G. (1876b), "Sur la formation de la grèle," *C. R. Hebd. Séances Acad. Sci.* **83** 314.

Planté, G. (1876c), "Sur la foudre globulaire," *C. R. Hebd. Séances Acad. Sci.* **83** 321.

Planté, G. (1876d), "Eclairs en chapelet," *C. R. Hebd. Séances Acad. Sci.* **83** 484.

Planté, G. (1876e), *Elektr. Ersch.* **1** 25.

Planté, G. (1876f), "Rosenkranzformige Blitz" (summary), *Fortschr. Phys.* **32** 1216.

Planté, G. (1876g), "Sur la foudre globulaire" (summary), *Fortschr. Phys.* **32** 1216.

Planté, G. (1876h), "Sur la foudre globulaire," *Mondes* **40** 623.*

Planté, G. (1876i), "Rosenkranzformige Blitz," *Mondes* **40** 762.*

Planté, G. (1876j), "Rosenkranzformige Blitz," *Naturforsch.* **8** 424.*

Planté, G. (1877a), "Lumière electrosilicique," *C. R. Hebd. Séances Acad. Sci.* **84** 914.

Planté, G. (1877b), "Machine rhéostatique," *C. R. Hebd. Séances Acad. Sci.* **85** 294. (NBL)

Planté, G. (1877c), "Suite de récherches sur les effets produits par des courants electriques de haute tension, et sur leurs analogies avec les phénomènes naturels," *C. R. Hebd. Séances Acad. Sci.* **85** 619.

Planté, G. (1877d), "Machine rhéostatique," *C. R. Hebd. Séances Acad. Sci.* **85** 795. (NBL)

Planté, G. (1878a), "Wandernde elektrische Funken," *Beibl. Ann. Phys.* **2** 620.

Planté, G. (1878b), "Sur les effets de la machine rhéostatique," *C. R. Hebd. Séances Acad. Sci.* **86** 761. (NBL)

Planté, G. (1878c), "Etincella electrique ambulante," *C. R. Hebd. Séances Acad. Sci.* **87** 325.

Planté, G. (1879a), "Récherches sur les phénomènes produits par des courants electriques de haute tension, et sur leurs analogies avec les phénomènes naturels," *C. R. Hebd. Séances Acad. Sci.* **88** 442.

Planté, G. (1879b), "Récherches sur l'Electricité," *Paris* 200.*

Planté, G. (1884a), "Foudre en boule," *Ciel et Terre* **5** 365.

Planté, G. (1884b), "Sur la foudre globulaire," *C. R. Hebd. Séances Acad. Sci.* **99** 273.

Planté, G. (1884c), "On globe lightning," *Electrician* **13** 366.

Planté, G. (1884d), "Sur la foudre globulaire" (summary), *Fortschr. Phys.* **40(3)** 650.

Planté, G. (1884e), "Sur la foudre globulaire," *La Lumière Electrique* **13** 286.*

Planté, G. (1884f), "Sur la foudre globulaire," *La Nature* **12(2)** 195.*

Planté, G. (1884g), "Sur la foudre globulaire," *Météorologie* **32** 314.

Planté, G. (1884h), "Sur la foudre globulaire," *Naturforsch.* **16** 369.*
Planté, G. (1884i), "Sur la foudre globulaire," *Rev. Sci.* **34** 601.*
Planté, G. (1885a), "On globe lightning," *Electrician* **14** 433.
Planté, G. (1885b), *Zeit. Elektro.* **3** 12.
Planté, G. (1887), "Récherches sur l'Electricité," *Paris*.*
Planté, G. (1888), "Les phénomènes Electriques de l'Atmosphère," *Paris* 60.*
Planté, G. (1890), "Reproduction du phénomène de la foudre en boule," *Ciel et Terre* **11** 424.
Planté, G. (1891), "Electrical Phenomena in the Atmosphere," *Paris*.*
Planté, G. (1901), *La Nature* **29** 562.
Pockels, F. (1893), "Merkwurdige Blitze," *Meteorol. Zeit.* **10** 73.
Pockels, F. (1898), "Ein Versuch, die bei Blitzschlagen erreichte maximale Stromstarke zu schatzen," *Meteorol. Zeit.* **15** 41.
Poey, M. (1855), "Des caractères physiques des éclairs en boules et de leurs affinité avec l'état sphéroidal de la matière," *C. R. Hebd. Séances Acad. Sci.* **40** 1183.
Poey, M. (1856a), "Sur l'origine et la nature des éclairs sans tonnerre et des tonnerres sans éclairs," *C. R. Hebd. Séances Acad. Sci.* **43** 985.
Poey, M. (1856b), "Sur l'origine et la nature des éclairs sans tonnerre et des tonnerres sans éclairs," *Cosmos* **9** 536.*
Poiterin, H. (1859a), "Effets de la foudre en boule," *Cosmos* **15** 58.*
Poiterin, H. (1859b), "Effets de la foudre en boule" (summary), *Fortschr. Phys.* **15** 62.
Pollock, R. E. (1898), "Not ball lightning," *Mon. Weather Rev.* **26** 363.
Pollog, C. H. (1930), "Ein Kugelblitz," *Meteorol. Zeit.* **47** 79.
Polychronakis, M. C. ven (1908), "Remarquable coup de foudre en boule," *Bull. Soc. Astron. Fr.* **22** 479.
Ponnamperuma, C., and Weller, F. (1964), "Differences in the character of C-to-C hydrocarbons from gaseous methane in low frequency electric discharges," *Nature* **203** 272. (NBL)
Popov, Y. A. (1959), "Razryad molnii na fyuzelyaz samoleta," *Priroda* **48(12)** 111.
Potts, L. M. (1910), "Ball lightning," *Science* **31** 144.
Pouillet, M. (1852), "Coup de foudre extraordinaire," *C. R. Hebd. Séances Acad. Sci.* **35** 400.
Poulpiquet, H. (1889), "Kugelblitz," *Jahrbuch Naturwiss.* **5** 254.
Poulpiquet, H. (1889a), "La foudre globulaire," *La Nature* **17** 39.*
Poulpiquet, H. (1889b), "La foudre globulaire," *Météorologie* **37** 72.
Poulpiquet, H. (1889c), "Kugelblitz," *Zeit. Angew. Meteorol.* **6** 69.
Poulpiquet, H. (1889d), *Zeit. Angew. Meteorol.* **6** 179.
Poulpiquet, H. (1890), "Globular lightning," *Electrician* **22** 390.
Poulter, R. M. (1935), "Lightning photographs," *Meteorol. Mag.* **70** 289.
Poulter, R. M. (1954a), "Photography and ball lightning," *Weather* **9** 121.
Poulter, R. M. (1954b), "Ball lightning," *Weather* **9** 321.
Poumier, M. (1886a), "La foudre en boule," *L'Astronomie* **5** 348.*
Poumier, M. (1886b), "La foudre en boule," *Météorologie* **34** 343.
Powell, J. R., and Domish, R. (1969), "Ball lightning and experimental near-spherical flames in DC fields," *Bull. Amer. Phys. Soc.* **14(11)** 1026.
Powell, J. R., and Finkelstein, D. (1969), "Structure of ball lightning," in H. E. Landsberg and J. van Mieghem (eds.) *Advances in Geophysics*, Vol. 13, p. 141, Academic Press, New York.
Powell, J. R., and Finkelstein, D. (1970), "Ball lightning," *Amer. Scientist* **58** 262.
Powell, J. R., Zucker, M. S., Manwaring, J. F., and Finkelstein, D. (1965), "Laboratory production of self-sustained atmospheric luminosities," (unpublished) Report 10625, Brookhaven National Laboratory, Upton, New York.
Powell, J. R., Finkelstein, D., Zucker, M. S., and Manwaring, J. F. (1966), "Laboratory production of self-sustained atmospheric luminosities," paper presented at American Physical Society, 8th Annual Meeting, Division of Plasma Physics.

Powell, J. R., Zucker, M. S., Manwaring, J. F., and Finkelstein, D. (1967), "Laboratory production of self-sustained atmospheric luminosities" Abstract 2C-2, *Bull. Amer. Phys. Soc.* **12** 751.

Pozwolski, A. E. (1977), "The ball lightning," *Indian J. Meteorol. Hydrol. Geophys.* **28(1)** 74.

Pratt, C. E. (1961), "Ball lightning," *Marine Observer* **31** 127.

Preaubert, E. (1904), "Note sur un éclair à propagation lente," *Météorologie* **52** 270.

Preece, W. H. (1872a), "On lightning and lightning conductors," *J. Soc. Teleg. Engr.* **1** 333.

Preece, W. H. (1872b), "Lightning and lightning conductors—discussions," *J. Soc. Teleg. Engr.* **1** 362.

Prentice, S. A. (1977), "Frequency of lightning discharges" in R. H. Golde (ed.) *Lightning*, Vol. I, p. 14, Academic Press, New York. (NBL)

Price-Williams, D. R. (1972), "Psychology and epistemology of UFO interpretations," in C. Sagan and T. Page (eds.) *UFO's—A Scientific Debate*, Chapter 10, p. 224, Cornell University Press, Ithaca, New York. (NBL)

Pringle, C. S. (1904), "Ball lightning," *Meteorol. Mag.* **39** 153.

Prinz, H. (1977), "Lightning in history," in R. H. Golde (ed.) *Lightning*, Vol. I, p. 1, Academic Press, New York. (NBL)

Prinz, W. (1888), "Etude de la structure des éclairs par la photographie," *Bull. Acad. Roy. Belgique* **III, 16** 244.*

Prinz, W. (1905), "Observations d'éclairs," *Ciel et Terre* **26** 249.

Pritt, A. T., Coombe, R. D., Pilipovich, D., Wagner, R. I., Benard, D., and Dymek, C. (1977), "Chemical generation of a population between the spin-orbit states of atomic iodine," *Appl. Phys. Lett.* **31(11)** 745. (NBL)

Prochnow, O. (1928), *Erdball und Weltall*, Hugo Bermuhler Verlag, Berlin-Lichterfelde.*

Prochnow, O. (1930a), "Zur Blitzforschung," *Physik. Zeit.* **31** 335.

Prochnow, O. (1930b), "Ein Perlenschnurblitz," *Zeit. Angew. Meteorol.* **47** 316.

Prohaska, K. (1892), "Ein merkwurdiges St. Elmsfeuer," *Jahrbuch der Astronomie und Gephysik* **3** 326. (NBL)

Prohaska, K. (1893), "Elmsfeuer am Schafberg," *Meteorol. Zeit.* **10** 222. (NBL)

Prohaska, K. (1900), "Blitzschaden in Steiermark und Karnten im Jahre 1899," *Meteorol. Zeit.* **17** 331.

Prohaska, K. (1901), "Feuerkugel in Pollou, Steiermark," *Meteorol. Zeit.* **18** 94.

Prohaska, K. (1903), "Uber blitzschaden und bemerkenswerthe Entladungsformen der atmosphärischen Elektrizitat im Jahr 1901," *Meteorol. Zeit.* **20** 315.

Prokh, L. Z. (1972), "Sharovaya molniya vo vremya zimney grozy v Kiev," *Meteorol. I. Gidrol.* **5** 103.*

Protasevich, E. T. (1993), "Gas discharge in moist air as a means of ball lightning simulation" in A. G. Keul (ed.) *Progress in Ball Lightning Research: Proc. VIZOTUM*, p. 65, The Vizotum Project, Salzburg, Austria.

Protasevich, E. T. (1995a), "Air discharge pecularities at non-uniform humidity and pressure 1-20 torr," in *Proceedings: 11th International Conference on Gas Discharges and their Applications*, Vol. 2, p. 52, Institute of Electrical Engineers, Japan, Tokyo.

Protasevich, E. T. (1995b), "Experiment simulating the ionisation of humid air in a lightning channel and its relevance to ball and bead lightning," *J. Met.* **20(196)** 41.

Protasevich, E. T., and Skavinskii, V. P. (1993), "Analysis of luminous, ball-lightning-type plasma in the period 1914–1916" in A. G. Keul (ed.) *Progress in Ball Lightning Research: Proc. VIZOTUM*, p. 61, The Vizotum Project, Salzburg, Austria.

Pudovkin, A. K. (1996a), "Ball lightning in Novosibirsk Academgorodok" (in Russian), *Usp. Fiz. Nauk.* **166(11)** 1253.

Pudovkin, A. K. (1996b), "Ball lightning in Novosibirsk Academgorodok" (English translation), *Physics—Usp.* **39(11)** 1189.

Pühringer, A. (1965), "Comments," in S. C. Coroniti (ed.) *Problems of Atmospheric and Space Electricity*, p. 460, Elsevier, Amsterdam.

Pühringer, A. (1967), "Beobachtung eines Kugelblitzes im Salzkammergut und ein Erklarungsversuch," *Wetter und Leben* **19** 57.

Pühringer, A. (1993a), "Pearl lightning above Enzenreith, Lower Austria, on August 2, 1988," in A. G. Keul (ed.) *Progress in Ball Lightning Research: Proc. VIZOTUM*, p. 68, The Vizotum Project, Salzburg, Austria.

Pühringer, A. (1993b), "The physical conditions for ball lightning formation" in A. G. Keul (ed.) *Progress in Ball Lightning Research: Proc. VIZOTUM*, p. 72, The Vizotum Project, Salzburg, Austria.

Raab, O. (1940), *Bildmessg. und Luftbildwes.* **15** 57.

Raasche, L. (1896), *Korr. Natur. Ver. Rigi.* **39** 70.

Racine, E. (1909), "Curieux coup de foudre," *Bull. Soc. Astron. Fr.* **23** 461.

Rahir, M. E. (1901), "Curieux coup de foudre," *Ciel et Terre* **22** 155.

Raillard, M. (1856a), "Sur les éclairs sans tonnerre et les tonnerres sans éclairs," *C. R. Hebd. Séances Acad. Sci.* **43** 816.

Raillard, M. (1856b), "Sur les éclairs sans tonnerre et les tonnerres sans éclairs" (summary), *Fortschr. Phys.* **12** 589.

Raiser, R., and Lortz, D. (1995), "Ball lightning as an example of a magnetohydrodynamic equilibrium," *Phys. Rev. E* **52(3)B** 3034.

Ranada, A. F., and Trueba, J. L. (1996), "Ball lightning an electromagnetic knot?," *Nature* **383(6595)** 32.

Randjelovic, J. (1969), "Ball lightning and electromagnetic radiation from linear lightning discharges," *Foreign Science Bull.* **5(2)** 65.

Ratcliffe, J. A. (1960), *Physics of the Upper Atmosphere*, Academic Press, New York. (NBL)

Ravetz, J. (1971), *Scientific Knowledge and its Social Problems*, Oxford University Press, Oxford, England. (NBL)

Rayle, W. D. (1966), "Ball lightning characteristics," NASA Tech. Note, NASA-TN-D-3188, National Aeronautics and Space Administration, Washington, DC.

Raymond, G. (1887a), "Un coup de foudre remarquable," *La Nature* **mars** 227.*

Raymond, G. (1887b), "Un coup de foudre remarquable," *Météorologie* **35** 221.

Rees, J. A. (ed.) (1973), *Electrical Breakdown in Gases*, Macmillan Press, Basingstoke, England. (NBL)

Rees, M. (1980), "Ball lightning in the bedroom," *Weather* **35** 342.

Reichenbach, E. (1889), "Meteorologische Notizen und Korrespondenzen," *Zeit. Angew. Meteorol.* **6** 119.

Reid, W. (1838), "An attempt to develop the law of storms," *London* 402.*

Reimann, D. (1886a), "Beobachtung von Kugelblitzen zu Hirschberg in Schleisien" (summary), *Fortschr. Phys.* **42(3)** 654.

Reimann, D. (1886b), "Beobachtung von Kugelblitzen zu Hirschberg in Schleisien," *Meteorol. Zeit.* **3** 510.

Reimann, D. (1887), "Weitere Berichte über Gewittererscheinungen im Schlesische Gebirge," *Meteorol. Zeit.* **4** 164.

Reimann, D. (1896), "Kugelblitz," *Zeit. Angew. Meteorol.* **13** 25.*

Reimann, D. (1897a), "Kugelblitz," *Zeit. Angew. Meteorol.* **14** 167.*

Reimann, D. (1897b), "Kugelblitz," *Zeit. Angew. Meteorol.* **14** 263.*

Reimann, D. (1905), "Ein vom Wind Bewegter Blitz," *Meteorol. Zeit.* **22** 360.

Reiners, F. (1882a), "Aus den Reisenberichten des Kapt R. Reiners von der Deutschen Bark 'Aeolus', 1, Kugelblitz," *Ann. Hydro. Marit. Meteorol.* **10** 582.

Reiners, F. (1882b), "Kugelblitz" (summary), *Fortschr. Phys.* **38(3)** 544.

Reisch-Perkallen, H. (1896), "Eigenthumliche elektrische Erscheinung," *Zeit. Angew. Meteorol.* **13** 143.*

Remscheid, M. (1907), *Umschau* **18** 576.*

Rendorf, O. (1934), *Atmosfernoye Elektrichestvo*, Leningrad.*

Renou, E. (1876), "Sur quelques particulaires de la foudre," *C. R. Hebd. Séances Acad. Sci.* **83** 1002.

Reynolds, W. C. (1923), "Globular lightning," *Nature* **112** 903.

Reynolds, W. C. (1930), "Globular lightning," *Nature* **125** 413.

Reynolds, W. C. (1931), "Charged aerosols and ball lightning," *Nature* **128** 584.

Rhighi, M. (1896a), "La foudre globulaire," *Le Cosmos* **juin** 353.

Rhighi, M. (1896b), "La foudre globulaire," *Météorologie* **44** 116.

Rhine, M. W. (1969), "Psychological aspects of UFO reports," in E. U. Condon (ed.) *Scientific Study of Unidentified Flying Objects*, Section VI, Chapter 3, p. 590, Bantam Books, New York. (NBL)

Rice-Evans, P. C. (1982), "Ball lightning in the laboratory," *Nature* **299(5886)** 774.

Rice-Evans, P. C., and Franco, I. J. (1980), "Observations of a localised oscillatory mode of an optical emission from an air plasma," *J. Phys. D. Appl. Phys.*, **13** 1079. (NBL)

Richter, P. (1912), "Gewitter," *Zeit. Angew. Meteorol.* **29** 233.*

Richter, P. (1913), "Kugelblitz," *Zeit. Angew. Meteorol.* **30** 21.

Richter, P. (1914a), "Kugelblitz" (summary), *Fortschr. Phys.* **70(3)** 230.

Richter, P. (1914b), "Kugelblitz," *Zeit. Angew. Meteorol.* **31** 215.

Ridpath, I. (1976), "Woman burned by ball lightning," *New Scientist* **70** 128.

Riecke, E. (1895), *Gotting. Nachr.* 418.

Riggenbach-Burckhardt, A. (1897), "Perlschnur-blitz," *Meteorol. Zeit.* **14** 62.

Righi, A. (1891a), "Sulle scintille costituite da masse luminose in moto," *Atti. D. Rom. Accad. Lincei, Rend.* **IV, 7** 330.*

Righi, A. (1891b), *La Lumière Electrique* **13** 47.

Righi, A. (1891c), *La Lumière Electrique* **13** 501.

Righi, A. (1891d), *La Lumière Electrique* **13** 604.

Righi, A. (1891e), "Sopra una specie di scintille elettriche nelle quali la luminosita si propaga gradatamente da un elettrodo all altro," *Mem. R. Accad. Sci. Ist. Bologna* **1** 315.*

Righi, A. (1891f), "Ricerche sperimentali intorno a certe scintille elettriche costuite da masse luminose in moto," *Mem. R. Accad. Sci. Ist. Bologna* **V, 1** 679.*

Righi, A. (1892), *Mem. R. Accad. Sci. Ist. Bologna* **2** 379.

Righi, A. (1895), *Mem. R. Accad. Sci. Ist. Bologna* **5** 445.

Righi, A. (1896a), *Eclairage Electrique*.

Righi, A. (1896b), "Nachahung von Kugelblitzen" (summary), *Fortschr. Phys.* **52(3)** 300.

Righi, A. (1896c), "Nachahung von Kugelblitzen," *Prometheus* **7** 687.*

Righi, A. (1896d), "Nachahung von Kugelblitzen," *Rev. Sci.* **6** Juin.*

Rigollet, C. (1905), "Météorologie-l'éclair en boule," *Bull. Soc. Astron. Fr.* **19** 267.

Rihanek, M., Ladislav, M., and Postranecky, J. (1957), *Naklad. Ceskos. Adad. Ved.* 90, 463.

Ritchie, D. J. (1959a), "Red lightning—are the Soviets using ball lightning as an anti-missile weapon?" Bendix Aviation Corp., Research Lab. Div. Data Sheet.*

Ritchie, D. J. (1959b), "Reds may use lightning as weapon," *Missiles and Rockets* **24** Aug, p. 13.

Ritchie, D. J. (1961), *Ball Lightning—A Collection of Soviet Research in English Translation,"* Consultants' Bureau, New York.

Ritchie, D. J. (1963), "Ball lightning in nature and in the laboratory," *IEEJ, London* **9** 202.

Ritchie, D. J., and Southward, B. G. (eds.) (1964), *The Nature of Ball Lightning,* Plenum Press, New York.

Rive, A. de la (1858), "Traite de l'electricité théorique et appliqué," *Paris* **3** 197.

Robert, A. (1909), "Note sur un phénomène lumineux," *Météorologie* **57** 296.

Roche, M. (1904), "Observations sur la foudre en boule tombe à Autun le 16 juillet," *C. R. Hebd. Séances Acad. Sci.* **139** 465.

Roche, M. (1905), "Wirkungen eines Kugelblitzes," *Jahrbuch der Astronomie und Geophysik* **16** 346.

Rock, I. (1975), *An Introduction to Perception*, Macmillan, New York. (NBL)

Rodewald, M. (1954), "Kugelblitz Beobachtungen," *Zeit. Meteorol.* **8** 27.

Rodionov, B. U. (1997), "The earth as a ball lightning," in Y. H. Ohtsuki and H. Ofuruton (eds.) *Proceedings: 5th International Symposium on Ball Lightning (ISBL97)*, p. 152.

Rodney, P. F., and Tompkins, D. R. (1975), "A theory of ball lightning," *Bull. Amer. Phys. Soc.* **20** 659.

Rood, O. N. (1869), "On the nature and duration of the discharge of a Leyden jar connected with an induction coil," *Amer. J. Sci.* **48** 153.

Rood, O. N. (1873), "Observations on the duration and multiple character of flashes of lightning," *Amer. J. Sci. Arts* **105** 163.

Rose, D. J., and Clark, M. (1961), *Plasmas and Controlled Fusion,* MIT Press, Cambridge, Massachusetts. (NBL)

Rose, H., and Rose, S. (1969), *Science and Society,* Pelican Books, Harmonsdworth, England. (NBL)

Rossman, M. (1927), "St. Elmsfeuer-beobachtungen," *Zeit. Angew. Meteorol.* **44** 262. (NBL)

Rossmann, F. (1939a), "Blitzbeobachtungen: wolken-und Erdblitz, Linien- und Perlschnur- Blitz," *Meteorol. Zeit.* **56** 194.

Rossmann, F. (1939b), "Blitz und Hagel uber die elektrische Natur des Gewitters," *Meteorol. Zeit.* **56** 372.

Rossmann, F. (1949), "Uber den Kugelblitz," *Wetter und Klima* **2** 75.

Rotch, L. (1903a), "Météorologie-éclair en boule dans la Tour Eiffel," *Bull. Soc. Astron. Fr.* **17** 483.

Rotch, L. (1903b), "Eclair en boule observé à Paris," *Ciel et Terre* **24** 544.

Rotch, L. (1903c), "Eclair en boule observé à Paris," *Météorologie* **51** 205.

Roth, F. (1889), "Beobachtung des Blitzes durch einen von demselben getroffenen," *Meteorol. Zeit.* **6** 231.

Roth, J. R. (1990), "Ball lightning as a route to fusion energy," in *Proc. IEEE 13th Symposium on Fusion Engineering*, Vol. 2, p. 1407, Cat. No.89CH2820-9. Institute of Electrical and Electronics Engineers, New York.

Roth, J. R. (1993), "Ball lightning: What nature is trying to tell the plasma research community," in *IEEE Conference Record—Abstracts. 1993 IEEE International Conference on Plasma Science,* Cat. No.93CH3334-0. Institute of Electrical and Electronics Engineers, New York.

Roth, J. R. (1995), "Ball lightning: What nature is trying to tell the plasma research community," *Fusion Technol.* **27(3)** 255.

Roth, J. R. (1996), "A mechanism for a sustained ball lightning glow discharge plasma at one atmosphere pressure," in *IEEE Conference Record—Abstracts. 1996 IEEE International Conference on Plasma Science*, p. 151, Cat. No. 96CH35939. Institute of Electrical and Electronics Engineers, New York.

Roth, J. R. (1997), "Ball lightning: what nature is trying to tell the fusion community," in *Current Trends in International Fusion Research. Proc. First Internat. Symp. on Evaluation of Current Trends in Fusion Research*, p. 459, Plenum Press, New York.

Rothman, M. A. (1988), *A Physicist's Guide to Skepticism,* Prometheus Books, Amherst, New York. (NBL)

Rousselot, U. (1904), "Foudre en boule extraordinaire," *Bull. Soc. Astron. Fr.* **18** 510.

Rowe, M. W. (1984), "Another unusual ball lightning incident," *J. Met.* **9(89)** 135.

Rowe, M. W. (1987), "A possible case of ball lightning at Danbury, Essex, in 1402," *J. Met.* **12(120)** 202.

Rowe, M. W. (1990), "Twelve more reports of ball lightning observed indoors," *J. Met.* **15(148)** 159.

Rowe, M. W., and Meaden, G. T. (1990), "A case of ball lightning inside a bedroom in Greater Manchester," *J. Met.* **15(148)** 157.

Roy, A. (1972), "Qu'est donc la foudre en boule?" *Constellation, Paris* **221** 141.

Rozema, M. (1951), "Foudre en boule," *Bull. Soc. Astron. Fr.* **65** 55.

Rozzi, N. (1911), "Foudre en boule," *Bull. Soc. Astron. Fr.* **25** 260.

Rudder, S. (1779), *The New History of Gloucestershire,* Cirencester, England.

Rudski, D. (1905), "Gewitter und Kugelblitz in Krakau," *Meteorol. Zeit.* **22** 284.

Rumeker, G. A. L. (1899), "Blitzphotographie," *Meteorol. Zeit.* **16** 422.

Rusck, S. (1977), "Protection of distribution systems," in R. H. Golde (ed.) *Lightning*, Vol. II, p. 23, Academic Press, New York. (NBL)

Russell, A. (1930), "Ball lightning," *Nature* **126** 809.

Russell, A. E. (1905), "Ball lightning," *Mon. Weather Rev.* **33** 409.

Russell, B. (1945), *The History of Western Philosophy*, p. 472, Simon and Schuster, New York. (NBL)

Russell, S. C. (1904a), "Foudre en boule," *Bull. Soc. Astron. Fr.* **18** 509.

Russell, S. C. (1904b), "Ball lightning," *Meteorol. Mag.* **39** 153.

Rustan, P. L., Kuhlman, B. P., DuBro, G. A., Reazer, M. J., Risley, M. D., and Serrano, A. V. (1982), "Correlated airborne and ground measurements of lightning," in *Proceedings: International Aerospace Conference on Lightning and Static Electricity*, Public Relations Section, Culham Laboratory, Abingdon, Oxon., UK. (NBL)

Rutgers, G. A. W. (1958), "Ein Kugelblitz?," *Phys. Blatter* **14** 417.

Rutherford, E. C. (1925), "The thunderstorm of July 22nd 1924 (at Woking)," *Meteorol. Mag.* **59** 153.

Ryan, G. M. (1895a), "Foudre globulaire," *Le Cosmos* **32** 191.*

Ryan, G. M. (1895b), "Foudre globulaire," *Météorologie* **43** 303.

Ryan, G. M. (1895c), "Note," *Nature* **52** 394.

Ryan, G. M. (1897a), "Les caprices de la foudre. Foudre en boule," *Bull. Soc. Astron. Fr.* **11** 300.

Ryan, G. M. (1897b), *Le Cosmos* **46** 95.

Sacc, M. (1861), "Sur un coup de foudre qui a frappé le telegraphe electrique entre Montelimar et Lyon," *C. R. Hebd. Séances Acad. Sci.* **53** 646.

Sagan, C., and Page, T. (eds.) (1972), *UFO's, a Scientific Debate*, Cornell University Press, Ithaca, New York.

Salanave, L. E. (1965), "Discussions" in S. C. Coroniti (ed.) *Problems of Atmospheric and Space Electricity*, p. 464, Elsevier, Amsterdam.

Sall, S. A. (1992a), "On the nature of radiation emission from ball lightning" (in Russian), *Zh. Tekh. Fiz.* **62(6)** 205.

Sall, S. A. (1992b), "On the nature of radiation emission from ball lightning" (English translation), *Sov. Phys. Tech. Phys.* **37(6)** 685.

Sall, S. A. (1997), "Relativistic effects in ball lightning" in Y. H. Ohtsuki and H. Ofuruton (eds.) *Proceedings: 5th International Symposium on Ball Lightning (ISBL97)*, p. 142.

San, M. G. de (1987), "Formation of ball lightning," *Speculations Sci. Technol.* **10(3)** 241.

San, M. G. de (1990a), "High insulating properties of the Kapitza filaments in ball lightning," *Speculations Sci. and Technol.* **13(1)** 18.

San, M. G. de (1990b), "Deuterium's depletion on Earth related to Jupiter's red spot?," *Speculations Sci. Technol.* **13(2)** 93.

Sanduloviciu, M. (1993), "On the ball lightning phenomenology" in *Proceedings of the International Workshop on Plasma Physics*, p. 235, World Scientific, Singapore.

Sanduloviciu, M., and Lozneanu, E. (1997), "Key processes of self-organization at the origin of ball lightning," in Y. H. Ohtsuki and H. Ofuruton (eds.) *Proceedings: 5th International Symposium on Ball Lightning (ISBL97)*, p. 120.

Sanduloviciu, M., Lozneanu, E., and Leu, G. (1997), "Ball lightning like structures formed under controllable laboratory conditions" in Y. H. Ohtsuki and H. Ofuruton (eds.) *Proceedings: 5th International Symposium on Ball Lightning (ISBL97)*, p. 170.

Saraoja, E. K. (1977), "Lightning earths," in R. H. Golde (ed.) *Lightning*, Vol. II, p. 18, Academic Press, New York. (NBL)

Sauter, F. (1890), "Uber Kugelblitz, I. Theil: Theorie der Kugelblitze" (summary), *Fortschr. Phys.* **46(3)** 423.

Sauter, F. (1892a), "Uber Kugelblitz, II. Theil: Beispiele von Kugelblitzen" (summary), *Fortschr. Phys.* **48(3)** 367.

Sauter, F. (1892b), "Beispiele der Kugelblitze," *Programme des Kgl. Real-Gymnasiums in Ulm zu Schlusse des Schuljahres* **1891/92 Ulm** 20.*

Sauter, F. (1892c), "Uber Kugelblitz, II. Theil: Beispiele von Kugelblitzen," *Prometheus* **5** 481.*

Sauter, F. (1893), "Der Kugelblitz," *Jahrbuch der Astronomie und Geophysik* **4** 314.

Sauter, F. (1895), "Uber Kugelblitze," *Meteorol. Zeit.* **12** 241.

Sauter, F. (1896), "Uber Kugelblitz," *Sammlung Gemeinverstandlicher Wissenschaftlicher Vortrage* Ser. 2, 9 121.

Sauter, F. (1897a), "Uber Kugelblitz" (summary), *Fortschr. Phys.* **53(3)** 313.

Sauter, F. (1897b), "Uber Kugelblitz," *Naturwiss. Rund.* **12** 217.*

Sauter, F. (1898), "Uber Kugelblitz," *Météorologie* **46** 101.

Scheminzky, F., and Wolf, F. (1948), "Photographie eines Perlschnurblitzes," *Sitzung. Akad. Wiss. Wien, Math.-Naturwiss. Klasse* **IIA, 156(1/2)** 1.

Scheurer, F. (1931), "Observations de foudre globulaire," *La Météorologie* **7** 249.*

Schick, T., and Vaughn, L. (1995). How to Think about Weird Things: Critical Thinking for a New Age, Mayfield Publishing Company. (NBL)

Schiodt, F. von (1893a), "Kugelquet" (summary), *Fortschr. Phys.* **49(3)** 384.

Schiodt, F. von (1893b), "Keglelynet," *Tidsskrift for Physik og Chemi* **III, 2** 242.*

Schloer, H. (1953), *Natur und Kultur* **45** 242.

Schlohbohm, K. A. (1961), "Blitzschlag und Kugelblitz," *Meteorol. Rund.* **14** 93.

Schmauss, A. (1909), "Perlschnurblitz," *Phys. Zeit.* **10** 968.

Schmauss, A. (1910), "Perlschnurblitz," *Meteorol. Zeit.* **27** 83.

Schmauss, A. (1918), "Randbemerkungen I," *Meteorol. Zeit.* **35** 181.

Schmauss, A. (1940), "Kugelblitz?," *Zeit. Angew. Meteorol.* **57** 238.*

Schmidt, E. E. (1860), *Lehrbuch Meteorol.* 781.*

Schmidt, K. E. F. (1905), *Elektro. Zeit.* **26** 903.

Schmidt-Boecking, H., Droerner, R., Jagutzi, O., Mervel, V., Spielberger, L., Stoehlker, T., Schneider, D., and Schenkel, T. (1997), "On the formation of 'quasi stable' hollow atoms towards a high power electric energy storage," in Y. H. Ohtsuki and H. Ofuruton (eds.) *Proceedings: 5th International Symposium on Ball Lightning (ISBL97)*, p. 133.

Schmitt, A. (1993), "Myth and folklore—ETI and ball lightning" in A. G. Keul (ed.) *Progress in Ball Lightning Research: Proc. VIZOTUM*, p. 77, The Vizotum Project, Salzburg, Austria.

Schneider, E. (1894), "Tonnerre en boule," *C. R. Hebd. Séances Acad. Sci.* **119** 506.

Schneidermann, K. (1934), "Blitzaufnahmen von Linien- und Kugel-blitz," *Feuer-Wehr-Verbands-Zeitung* **44(1)** 1.

Schneidermuhl, M. (1921), "Beobachtung von Kugelblitz," *Meteorol. Zeit.* **38** 183.

Schonert, H. (1965), "Auftreten eines Kugelblitzes in Heringsdorf," *Zeit. Meteorol.* **17** 61.

Schonland, B. F. J. (1950), *The Flight of Thunderbolts,* Oxford University Press, Oxford, England.

Schonland, B. F. J. (1956), "The lightning discharge," in *Handbuch der Physik,* Vol. 22, p. 576, Berlin.

Schonland, B. F. J. (1962a), "Lightning and the long electric spark," *Adv. Sci.* **19** 306.

Schonland, B. F. J. (1962b), "Lightning and the long electric spark," *Nature* **195** 880.

Schoute, C. (1926), "The term fireball," *Meteorol. Mag.* **61** 238.

Schove, D. J. (1951), "Hail in history, AD 1630–1680," *Weather* **6** 17.

Schrade, M. von (1962), "Auftreten eines Kugelblitzes in Heringsdorf," *Zeit. Meteorol.* **17** 61.

Schramen, M. (1890), "Foudre en boule," *Ciel et Terre* **11** 348.

Schreiber, D. (1901), "Beitrage zur Hageltheorie," *Meteorol. Zeit.* **18** 58.

Schroth, H. (1940), *Zeit. Angew. Meteorol.* **57** 306.

Schroth, H. (1942), *Zeit. Angew. Meteorol.* **59** 200.

Schubert, C. W., Jr., and Richmond, R. D. (1982), "Rocket-triggered lightning investigation," in *Proceedings: International Aerospace Conference on Lightning and Static Electricity*, Public Relations Section, Culham Laboratory, Abingdon, Oxon., UK.

Schultze, H. S., and Reitz, H. (1907), "Kugelblitz," *Prometheus* **18** 416.*

Schumann, G. A. (1848), *Prog. Gymn. Quedlinberg* 28.

Schuster, A. (1896a), "L'électricité atmospherique," *Ciel et Terre* **17** 318.

Schuster, A. (1896b), "L'électricité atmospherique," *Ciel et Terre* **17** 441.

Schuster, A. (1896c), "L'électricité atmospherique," *Ciel et Terre* **17** 462.

Schuster, A. (1896d), "Atmospheric electricity" (summary), *Fortschr. Phys.* **52(3)** 286.

Schuster, A. (1896e), "Atmospharische elektrizitat," *Meteorol. Zeit.* **13** 215.

Schuster, A. (1896f), "Atmospheric electricity," *Nature* **53** 207.

Schuster, A. (1896g), "Atmospheric electricity," *Naturwiss. Rund.* **11** 133.*

Schuster, A. (1896h), "Atmospheric electricity," *Naturwiss. Rund.* **11** 145.*

Schwarzenbacher, K. (1993), "Ball lightning research needs testable hypotheses" in A. G. Keul (ed.) *Progress in Ball Lightning Research: Proc. VIZOTUM*, p. 81, The Vizotum Project, Salzburg, Austria.

Schwegler, H. (1951), "Uber der Kugelblitz," *Naturwiss. Rund.* **4** 169.

Scott, E. K. (1923a), Letter to*Yorkshire Post* Oct 20.

Scott, E. K. (1923b), "Globular lightning," *Nature* **112** 760.

Scott, G. W., and Crewson, W. F. J. (1982), "Northrop's lightning laboratory and test techniques on composites and radomes for an advanced fighter aircraft," in *Proceedings: International Aerospace Conference on Lightning and Static Electricity*, Public Relations Section, Culham Laboratory, Abingdon, Oxon., UK.

Scott, J. R. (1955), "Photography and ball lightning," *Weather* **10** 98.

Scott, M. (1878a), "Newspaper accounts from M. Michelski, 'Curious phenomenon during the gale' in London Daily Papers Oct. 12 1877," *Roy. Meteorol. Soc. Q. J.* **4** 167.

Scott, M. (1878b), "Newspaper accounts from M. Michelski, 'Singular phenomenon at Lower Cumber' in Londonderry Sentinel, June 1 1878," *Roy. Meteorol. Soc. Q. J.* **4** 167.

Seeger, A. (1981), *Die Erklarung des Kugelblitzes* (The explanation of ball lightning), Alfa Sierra Urfysik, Darmstadt, Germany.

Séguir (1852), *Comptes Rend.* **34** 871.

Seigner, R. (1966), "Perlschnurblitz," *Wetter und Leben* **18** 34.

Seliger, H. H., and Morton, R. A. (1968), "A physical approach to bioluminescence," in *Photophysiology*, Vol. IV, Chapter 13, p. 253. (NBL)

Seliger, H. H., Buck, J. B., Fastie, W. G., and McElroy, W. D. (1964), "The spectral distribution of the firefly light," *J. Gen. Physiology* **48(1)** 95. (NBL)

Sentman, D. D., and Wescott, E. M. (1996), "Red sprites and blue jets: high altitude optical effects linked to lightning," *EOS Trans. Am. Geophys. Union* **77(1)** 1. (NBL)

Sentman, D. D., Wescott, E. M., Osborne, D. L., Hampton, D. J., and Heavner, M. J. (1995a), "Preliminary results from the Sprites94 Aircraft Campaign: 1. Red sprites," *Geophys. Res. Letts.* **22** 1205. (NBL)

Sequier, M. (1852), "Globe de feu," *C. R. Hebd. Séances Acad. Sci.* **34** 871.

Sestier, F. (1866), *De la Foudre, de ses Formes et de ses Effets*, Vol. 1, pp. 7, 119, Balliere et Fils, Paris.*

Seward, C. (1996), "Compact toroid generation using long pulses," in *IEEE Conference Record—Abstracts. 1996 IEEE International Conference on Plasma Science*, p. 152. Cat. No. 96CH35939. Institute of Electrical and Electronics Engineers, New York.

Seward, C. (1997), "Compact toroidal properties compared to ball lightning," in *IEEE Conference Record—Abstracts. 1997 IEEE International Conference on Plasma Science*, p. 320. Cat. No. 97CH36085. Institute of Electrical and Electronics Engineers, New York.

Seyboth, R. (1902), "Ball lightning at sea," *Scientific American* **86** 36.

Seymour, I. P. (1978), "Ball lightning," *New Scientist* **77–84** 817.

Shafranov, V. D. (1957a), "On magnetohydrodynamical equilibrium configurations" (English translation), *Sov. Phys. JETP* **6** 545.

Shafranov, V. D. (1957b), "On magnetohydrodynamical equilibrium configurations," (in Russian), *Zh. Eksp. Teor. Fiz., SSSR* **33** 710.

Shafranov, V. D. (1961), "On magnetohydrodynamical equilibrium configurations," in D. J. Ritchie (ed.) *Ball Lightning*, Consultants' Bureau, New York.

Shagin, I. (1960), "Ball lightning photograph," *Ogonek* **20** 34.

Shapiro, A. R., and Watson, W. K. R. (1963), "Three dimensional containment of charged particles by orthogonal standing waves," *Phys. Rev.* **131** 495. (NBL)

Shapiro, M. L. (1979), "Shedding light on UFO's," *Optical Spectra* **13(4)** 13.

Shaw, N., and Austin, E. (1926), *Manual of Meteorology*, Vol. 1, p. 247, Vol. 3, p. 393, Cambridge University Press, Cambridge, England.

Shelton, J. D. (1989), "Eddy current model of ball lightning," *Naturwiss.* **76(8)** 373.

Shick, T., Jr., and Vaughn, L. (1995), *How to Think about Weird Things—Critical Thinking for a New Age*," Mayfield, Toronto. (NBL)

Shipley, J. F. (1941), "Lightning and its symbols," *Roy. Meteorol. Soc. Q. J.* **67** 135.

Shohet, J. (1971), *The Plasma State*, Academic Press, New York.

Shunaman, F. (1976), "Twelve million volts," *Radio Electronics* **47(6)** 32.

Sie, N. (1895), *Le Cosmos* **32** 191.

Sieberg, A. (1923), *Erdbebenkunde-Jena*.*

Silberg, P. A. (1961a), "On the question of ball lightning," internal report, Raytheon Co., Waltham, Massachusetts.

Silberg, P. A. (1961b), "On the question of ball lightning," *J. Appl. Phys.* **32** 30.

Silberg, P. A. (1962), "Ball lightning and plasmoids," *J. Geophys. Res.* **67** 4941.

Silberg, P. A. (1963), "Investigation of the generation and propagation of high-energy spherical plasmoids," USAF ASD-Tech. Document Report No. ASD-TDR-63-47, Aeronautical Systems Division.

Silberg, P. A. (1964), "A standing-wave discharge in air," *J. Appl. Phys.* **35** 2264. (NBL)

Silberg, P. A. (1965), "A review of ball lightning," in S. C. Coroniti (ed.) *Problems of Atmospheric and Space Electricity*, p. 436, Elsevier, Amsterdam.

Silberg, P. A. (1978), "A note on the formation of the fireball plasma," *J. Appl. Phys.* **49(3)** 1110.

Silberg, P. A. (1981a), "On the formation of ball lightning," *Nuovo Cimento* **4C** 221.

Silberg, P. A. (1981b), "The effect of damping on some nonlinear oscillators," *Nuovo Cimento* **4C** 511.

Simon, K. H. (1963), "Der Ratselhafte Kugelblitz," *Naturwiss. Rund.* **16** 450.

Simons, P. (1996), *Weird Weather*, p. 44, Little, Brown, London.

Simpson, G. C. (1916), "Nature and cause of ball lightning," *Mon. Weather Rev.* **44** 121.

Simpson, G. C. (1923), "Thunderstorms and globular lightning," *Nature* **112** 727.

Simpson, G. C. (1924), "Ball lightning," *Nature* **113** 677.

Singer, S. (1963), "The unsolved problem of ball lightning," *Nature* **198** 745.

Singer, S. (1965), "Unsolved problems of ball lightning," in S. C. Coroniti (ed.) *Problems of Atmospheric and Space Electricity*, p. 461, Elsevier, Amsterdam.

Singer, S. (1968), "Unknown things in the sky," *Astronaut. Aeronaut. J.* **6(12)** 12.

Singer, S. (1971), *The Nature of Ball Lightning*, Plenum Press, New York.

Singer, S. (1977), "Ball lightning," in R. H. Golde (ed.) *Lightning*, Vol. I, p. 409, Academic Press, New York, Chapter 12, p. 409.

Singer, S. (1991), "Ball lightning: great balls of fire," *Nature* **350(6314)** 108.

Singer, S. (1997), "The first decade of International Symposia on ball lightning," in Y. H. Ohtsuki and H. Ofuruton (eds.) *Proceedings: 5th International Symposium on Ball Lightning (ISBL97)*, p. 1.

Singer, S. (1998), "Ball of fire—recent studies on ball lightning," in *Proceedings of the Third International Symposium on Ball Lightning*, Springer-Verlag, New York.

Sinitsyn, V. I. (1977a), "Properties of a low pressure spherical discharge," *Sov. Phys. Tech. Phys.* **22(5)** 576.

Sinitsyn, V. I. (1977b), "O nekotorykh svoistvakh sfericheskogo razriada pri ponizhennom davlenii," *Zh. Tekh. Fiz.* **47** 966.

Sinkevich, O. A. (1997a), "Long-lived plasma formations and problems associated with ball lightning" (in Russian), *Teplofizika Vysokikh Temperatur* **35(4)** 651.

Sinkevich, O. A. (1997b), "Long-lived plasma formations and problems associated with ball lightning" (English translation), *High Temperature* **35(4)** 639.

Sinkevich, O. A. (1997d), "Long-lived plasma formations and problems associated with ball lightning" (in Russian), *Teplofizika Vysokikh Temperatur* **35(6)** 968.

Sinkevich, O. A. (1997e), "Long-lived plasma formations and problems associated with ball lightning" (English translation), *High Temperature* **35(6)** 955.

Sinkevich, O. A. (1997f), "On boundaries of ball lightning stability," in Y. H. Ohtsuki and H. Ofuruton (eds.), *Proceedings: 5th International Symposium on Ball Lightning (ISBL97)*, p. 63.

Skowronek, M. (1960), "Decharge filiforme emise par une pointe fine portee a haut tension positive par rapport a un plan," *C. R. Hebd. Séances Acad. Sci.* **250** 1808. (NBL)

Smirnov, B. M. (1975a), "Analysis of the nature of ball lightning" (English translation), *Sov. Phys. Usp.* **18** 636.

Smirnov, B. M. (1975b), "Ob analize prirody sharovoi molnii" (in Russian), *Usp. Fiz. Nauk.* **116** 731.

Smirnov, B. M. (1976a), "Vozniknovenie sharovoi molnii" (in Russian), *Doklady Akad. Nauk. SSSR* **226** 806.

Smirnov, B. M. (1976b), "Formation of ball lightning" (English translation), *Sov. Phys. Doklady* **21(2)** 89.

Smirnov, B. M. (1977a), "Ball-lightning model," (English translation), *Sov. Phys. Tech. Phys.* **22(4)** 488.

Smirnov, B. M. (1977a), "Ball-lightning model" (in Russian), *Zh. Tekh. Fiz.* **47** 814.

Smirnov, B. M. (1977b), "Protsessy v sharovoi molnii," *Zh. Tekh. Fiz.* **47** 814.

Smirnov, B. M. (1981), *Physics of Weakly Ionized Gases*, Mir Publishers, Moscow. (NBL)

Smirnov, B. M. (1987a), "(Ball lightning—what is it?)," *Priroda* No. 2 15.

Smirnov, B. M. (1987b), "Electrical phenomena in ball lightning" (in Russian), *Doklady Akad. Nauk. SSSR* **292(6)** 1363.

Smirnov, B. M. (1987c), "The properties and the nature of ball lightning," *Phys. Reports—Review Sec. of Phys. Lett.* **152(4)** 7.

Smirnov, B. M. (1989a), "Electrical and radiative properties of ball lightning," in Y. H. Ohtsuki (ed.) *Science of Ball Lightning (Fireball)*, p. 192, World Scientific, Singapore.

Smirnov, B. M. (1989b), "The candle flame as a model of ball lightning," in Y. H. Ohtsuki (ed.) *Science of Ball Lightning (Fireball)*, p. 220, World Scientific, Singapore.

Smirnov, B. M. (1989c), "Electrical and radiative processes in ball lightning," *Nuovo Cimento* **12C**, Ser. 1 (5) 575.

Smirnov, B. M. (1990a), "Physics of ball lightning" (in Russian), *Usp. Fiz. Nauk.* **160(4)** 1.

Smirnov, B. M. (1990b), "Physics of ball lightning" (in English), *Sov. Phys. Usp.* **33(4)** 261.

Smirnov, B. M. (1990c), "The properties of fractal clusters," *Phys. Reports* **188(1)** 1.

Smirnov, B. M. (1991a), "A tangle of fractal fibers as a new state of matter," *Usp. Fiz. Nauk.* **161(8)** 141.

Smirnov, B. M. (1991b), "Propagation of thermal waves along a fractal filament" (in Russian), *Zh. Tekh. Fiz.* **61(12)** 82.

Smirnov, B. M. (1991c), "Propagation of thermal waves along a fractal filament" (English translation), *Sov. Phys. Tech. Phys.* **36(12)** 1376.

Smirnov, B. M. (1992a), "Observational properties of ball lightning" (in Russian), *Usp. Fiz. Nauk.* **162(8)** 43.

Smirnov, B. M. (1992b), "Observational properties of ball lightning" (English translation), *Sov. Phys. Usp.* **35(8)** 650.

Smirnov, B. M. (1993a), "Observational parameters of ball lightning," *Physica Scripta* **48(5)** 638.

Smirnov, B. M. (1993b), "Observational parameters of ball lightning" in A. G. Keul (ed.) *Progress in Ball Lightning Research: Proc. VIZOTUM*, p. 85, The Vizotum Project, Salzburg, Austria.

Smirnov, B. M. (1993c), "Fractal discharge and ball lightning," in A. G. Keul (ed.) *Progress in Ball Lightning Research: Proc. VIZOTUM*, p. 90, The Vizotum Project, Salzburg, Austria.

Smirnov, B. M. (1993d), "Physics of ball lightning," *Phys. Rep.—Review Sec. Of Phys. Lett.* **224(4)** 151.

Smirnov, B. M. (1993e), "Gasdynamics of a fractal ball," *Zh. Tekh. Fiz.* **63(4)** 190.

Smirnov, B. M. (1993f), "Radiation of some fractal structures," *Int'l. J. Theoret. Phys.* **32(8)** 1453.

Smirnov, B. M. (1993g), "Radiation of some fractal structures," *Internat. J. Theoret. Phys.* **32(8)** 1453.

Smirnov, B. M. (1993h), "Physics of ball lightning," *Phys. Reports* **224(4)** 151.

Smirnov, B. M. (1993i), "Radiative processes involving fractal structures" (in Russian), *Usp. Fiz. Nauk.* 163(7) 51.

Smirnov, B. M. (1993j), "Radiative processes involving fractal structures" (English translation), *Physics-Usp.* **36(7)** 592.

Smirnov, B. M. (1996), "Comparison of histograms of data-banks,," *Physica Scripta* **54(1)** 125.

Smirnov, B. M. (1997a), "Treatment of observations of ball lightning," in Y. H. Ohtsuki and H. Ofuruton (eds.) *Proceedings: 5th International Symposium on Ball Lightning (ISBL97)*, p. 235.

Smirnov, B. M. (1997b), "Large fractal structure and processes of their generation," in Y. H. Ohtsuki and H. Ofuruton (eds.) *Proceedings: 5th International Symposium on Ball Lightning (ISBL97)*, p. 242.

Smirnov, B. M., and Strizhev, A. Ju. (1994), "Analysis of observational ball lightning by correlation Methods," *Physica Scripta* **50(5)** 606.

Smith, B. W. (1876), "Remarkable form of lightning," *Nature* **18** 302.

Smith, P. (1860), "Singular effect of lightning," *J. Franklin Inst.* **70** 54.

Smith, W. F. (1880a), "Ball lightning" (summary), *Fortschr. Phys.* **36(3)** 360.

Smith, W. F. (1880b), "Ball lightning," *Nature* **22** 267.

Smith, W. G. (1884), "Lightning," *Nature* **30** 241.

Smith-Gordon, L. (1923), "Ball lightning," *Meteorol. Mag.* **57** 336.

Snigier, P. (1976), "Fusion energy—will experiments in ball lightning provide the key?" *Electronic Design News* **20** Apr, p. 17.

Snow-Harris, W. (1843), *On the Nature of Thunderstorms,* p. 36, J. W. Parker, London.

Soubbotine, N. de (1901), "Météorologie—cas extraordinaire de foudre en boule," *Bull. Soc. Astron. Fr.* **15** 476.

Soubbotine, N. de (1902), "Cas de foudre en boule," *Bull. Soc. Astron. Fr.* **16** 117.

Soucaze, A. (1884a), "La description des effets produits par un coup de foudre à Campan, le 24 juillet dernier," *C. R. Hebd. Séances Acad. Sci.* **99** 262.

Soucaze, A. (1884b), "Effets produits par un coup de foudre à Campan, le 24 juillet" (summary), *Fortschr. Phys.* **40(3)** 656.

Sourdillon, M. (1952), "Etude à la chambre de Boys de l'éclair dans l'air et du coup de foudre à cime horizontal," *Ann. Geophysique* **8** 349.

Spitzer, L., Jr. (1962), *Physics of Fully Ionized Gases,* Interscience, New York. (NBL)

Sprague, J. T. (1875), *Electricity,* London.*

St. Privat d'Allier Research Group (1982), "Eight years of triggered lightning research at St Privat d'Allier," in *Proceedings: International Aerospace Conference on Lightning and Static Electricity,* Public Relations Section, Culham Laboratory, Abingdon, Oxon., UK.

Stade, H. (1921), "Kugelblitz," *Meteorol. Zeit.* **38** 314.

Stakhanov, I. P. (1973), "Concerning the nature of ball lightning" (in Russian), *Zh. Etf. Teor. Fiz.* **18** 193.

Stakhanov, I. P. (1974c), "Concerning the nature of ball lightning" (English translation), *JETP Lett.* **18** 114.

Stakhanov, I. P. (1974b), "Stability of ball lightning" (in Russian), *Zh. Tekh. Fiz.* **44** 1373.

Stakhanov, I. P. (1975), "Stability of ball lightning" (English translation), *Sov. Phys. Tech. Phys.* **19** 861.

Stakhanov, I. P. (1976a), "Cluster plasma and radiation from ball lightning" (English translation), *Sov. Phys. Tech. Phys.* **21(1)** 44.

Stakhanov, I. P. (1976b), "Klasternaya plazma I illuchenie sharovoi molnii" (in Russian), *Zh. Tekh. Fiz.* **46(1)** 82.

Stakhanova, I. G. (1997), "Ball lightning pictures from the I. P. Stakhanov's archive and their interpretation," in Y. H. Ohtsuki and H. Ofuruton (eds.) *Proceedings: 5th International Symposium on Ball Lightning (ISBL97)*, p. 35.

Standler, R. B. (1972), "How to report ball lightning," *Weatherwise* **25** 186.

Steadworthy, M. A. (1917), "Les mystères de la foudre," *Bull. Soc. Astron. Fr.* **31** 19, 197. (NBL)

Steinheim, M. (1853), "Coup de tonnerre en boule," *C. R. Hebd. Séances Acad. Sci.* **36** 744.

Stekol'nikov, I. S. (1943), *Fizika Molnii i Grozozasnita* (*The Physics of Thunder and Lightning Protection*), p. 145, Izd. An-SSSR, Moscow-Leningrad.*

Stekol'nikov, I. S. (1955), *Izucheniye Molnii Grozozashchity" (Studies of Lightning and Lightning Protection*), AS-USSR Press.*

Stekol'nikov, I. S. (1960), *Priroda Dlinnoy Iskry*, Moscow-Leningrad.*

Stenhoff, M. (1976), "Ball lightning," *Nature* **260** 596.

Stenhoff, M. (1981), "Ball lightning reported in Warminster—preliminary report" (unpublished).

Stenhoff, M. (1985), "Ball lightning," *J. Met.* **10(100)** 231.

Stenhoff, M. (1986), "Smethwick revisited," *Ball Lightning Newsletter* No. 6 6.

Stenhoff, M. (1988a), "Ball lightning," M.Phil. Thesis, University of London.

Stenhoff, M. (1988b), "A survey of ball lightning," *J. Met.* **13(129)** 197.

Stenhoff, M. (1988c), "Ball lightning," *New Scientist* **117(1599)** 69.

Stenhoff, M. (1990), "Comments on video tape of a possible ball lightning event at Willesborough, Ashford, Kent on 10 September 1989," *J. Met.* **15(148)** 141.

Stenhoff, M. (1992a), "Ball lightning reported in Conwy," *J. Met.* **17(173)** 308.

Stenhoff, M. (1992b), "Physical evidence for the existence of ball lightning," in M. Stenhoff (ed.) *Proceedings: Fourth TORRO Conference: Ball Lightning*, p. 41, Tornado and Storm Research Organisation (TORRO), Oxford Brookes University, Richmond, England.

Stenhoff, M. (ed.) (1992c), *Proceedings: Fourth TORRO Conference: Ball Lightning*, Tornado and Storm Research Organisation (TORRO), Oxford Brookes University, Oxford, United Kingdom.

Stenhoff, M. and Wooding, E. R. (1977), "Some observations of ball lightning" (unpublished report),.

Stepanov, S. I. (1990a), "On the energy of ball lightning" (in Russian), *Zh. Tekh. Fiz.* **60(2)** 211.

Stepanov, S. I. (1990b), "On the energy of ball lightning" (in English), *Sov. Phys. Tech. Phys.* **35(2)** 267.

Stepanov, S. I. (1993), "How can an electrochemical model solve the mysteries of ball lightning?" in A. G. Keul (ed.) *Progress in Ball Lightning Research: Proc. VIZOTUM*, p. 94, The Vizotum Project, Salzburg, Austria.

Stepanov, S. I. (1997), "On energy of ball lightning," in Y. H. Ohtsuki and H. Ofuruton (eds.) *Proceedings: 5th International Symposium on Ball Lightning (ISBL97)*, p. 61.

Stepanov, S. I., Sall, S. A., and Arutjunan, A. V. (1997), "Electric machine in ball lightning," in Y. H. Ohtsuki and H. Ofuruton (eds.) *Proceedings: 5th International Symposium on Ball Lightning (ISBL97)*, p. 183.

Steyaert, J. J. (1993), "Energetics of ball lightning," in A. G. Keul (ed.) *Progress in ball lightning research: Proc. VIZOTUM*, p. 98, The Vizotum Project, Salzburg, Austria.

Stohr, A. (1912a), "Kugelblitz" (summary), *Fortschr. Phys.* **68(3)** 311.

Stohr, A. (1912b), "Kugelblitz," *Zeit. Angew. Meteorol.* **29** 188.*

Stokes, G. G. (1900), "Letters to the editor," *Nature* **61** 343.

Stolterfoth, M. H. (1899), "Eclair en reban boucle," *Bull. Soc. Astron. Fr.* **13** 45.

Stotte, I. R. (1993), "Fireball over Limassol," *Marine Observer* **63(321)** 132.

Strand, E. (1993), "Project Hessdalen, instrumentation of possible ball lightning," in A. G. Keul (ed.) *Progress in Ball Lightning Research: Proc. VIZOTUM*, p. 102, The Vizotum Project, Salzburg, Austria.

Stranz, H. (1940), *Zeit. Angew. Meteorol.* **57** 306.

Stranz, H. (1942), "Ein frontgewitter mit Kugelblitz," *Zeit. Angew. Meteorol.* **59** 200.*

Strickland, J. (1845), *English Mech.* **2** 125.

Stromeyer, C. E. (1899), "Remarkable lightning flashes," *Nature* **60** 520.

Stuck, E. (1897), "Kugelblitz," *Zeit. Angew. Meteorol.* **14** 214.*

Stunder, J. (1885a), "Elektrische Erscheinungen am Santis 28 Juni 1885 wahrend eines heftigen Gewitters," *Jahrb. Schweizer Alpen-Club.**

Stunder, J. (1885b), *Ann. Schweizer Meteorol. Cent. Aust.* **22**, Appendix 3.*

Stunder, J. (1886a), "Elektrische Erscheinungen am Santis 28 Juni 1885 wahrend eines heftigen Gewitters" (summary), *Fortschr. Phys.* **42(3)** 673.

Stunder, J. (1886b), "Elektrische Erscheinungen am Santis 28 Juni 1885 wahrend eines heftigen Gewitters," *Oesters. Tour. Ztg.* 211.*

Stunder, P. (1888), "Bemerkenswerthe Blitzschlage," *Meteorol. Zeit.* **5** 158.

Stunder, P. (1889), "Bemerkenswerthe Blitzschlage," *Météorologie* **37** 167.

Subramaniam, A. H. (1962), "Ball lightning at Madras on 20 October 1962," *Ind. J. Meteorol. Geophys.* **14** 358.

Sullivan, W. (1975), "Ball lightning studies," *Weatherwise* **28** 220.

Sunderland, P. G. (1988), "Ball lightning in Yorkshire, May 1985," *Weather* **43(9)** 343.

Suring, R. (1914), "Merkwurdige Blitzwirkung," *Meteorol. Zeit.* **31** 90.

Sutton, M. (1983), "Ball lightning—All in the mind," *Nature* **301(5900)** 460.

Sviatsky, D. (1901), "Météorologie—les feux St.-Elme en Russe," *Bull. Soc. Astron. Fr.* **15** 476. (NBL)

Swain, W. (1933), "Another case of ball lightning," *Roy. Astron. Soc. Canada J.* **27** 254.

Swart, C. D. (1887), "Rare electrical phenomenon at sea," *Monthly Weather Rev.* **15** 84.*

Swart, C. D. (1888), "Rare electrical phenomenon at sea," *Amer. J. Meteorol.* **4** 98.

Sweeney, M. (1992), "The severe thunderstorms and ball lightning in Eire on 20 March 1992," *J. Met.* **17(169)** 167.

Swinhoe, P. (1978), "Ball lightning," *J. Met.* **3(31)** 216.

Symons, G. (1872), *J. Soc. Tel. Engr.* **1** 376.

Symons, M. (1878), "Discussions on waterspouts and globular lightning," *Roy. Meteorol. Soc. Q. J.* **4** 165.

Szpor, S. (1977), "Eclair globulaire, éclair en chapelet, nouvelle theorie," *Acta Geophys. Pol.* **25(4)** 327.

Tait, C. (1895), "Fireballs at sea," *Scientific American* **73** 405.

Tait, J. B. (1952), "Bolbliksem te Warnsveld op 28 November 1951" (Ball lightning at Warnsveld on 28 Nov 1951), *Hamel en Dampkring* **50(718)** 134.*

Tait, P. G. (1880a), "Thunderstorms," *Nature* **22** 339.

Tait, P. G. (1880b), "Thunderstorms," *Nature* **22** 364.

Tait, P. G. (1880c), "Thunderstorms," *Nature* **22** 408.

Tait, P. G. (1880d), "Thunderstorms," *Nature* **22** 436.

Tait, P. G. (1884), "Thunderstorms," *Electrician* **13** 114.

Tait, P. G. (1911), "Thunderstorms," in *Life and Scientific Work of Peter Guthrie Tait*, p. 312, by C. G. Knott, Cambridge University Press, Cambridge, England (supplement).

Taker, L. J. (1960), *Flying Saucers and the USAF*, van Nostrand, Princeton, New Jersey.

Talman, C. F. (1932), "Ball lightning," *Amer. Mercury* **26** 69.

Tamura, Y. (1954), "An analysis of electric field after lightning discharge," *J. Geomag. Geoelect.* **6** 34.

Tanaka, H., and Tanaka, M. (1997a), "Is ball lightning 'Anderson localization'?" *Appl. Phys. Lett.* **71(26)** 3793.

Tanaka, H., and Tanaka, M. (1997b), "Is ball lightning 'Anderson localization'?" in Y. H. Ohtsuki and H. Ofuruton (eds.) *Proceedings: 5th International Symposium on Ball Lightning (ISBL97)*, p. 137.

Tanaka, H., Ohtsuki, Y. H., and Kamogawa, M. (1997), "The interaction between bulk-plasmons and photons via surface roughness," in Y. H. Ohtsuki and H. Ofuruton (eds.) *Proceedings: 5th International Symposium on Ball Lightning (ISBL97)*, p. 91.

Tanenbaum, B. S. (1967), *Plasma Physics*, Chapter 2, McGraw-Hill, New York. (NBL)

Tantor, T. L. (1913), "Globular discharge of lightning," *Roy. Astron. Soc. Canada J.* **12** 530.

Taranenko, Y. N., and Roussel-Dupré, R. (1995), "Upward discharges and gamma-ray flashes: a manifestation of runaway air breakdown," 1995 Annual Cedar Meeting, June 25–30, Boulder, Colorado.

Tardy, M. (1891a), "Eclairs en boule," *Le Cosmos* **août** 73.*

Tardy, M. (1891b), "Eclairs en boule," *Météorologie* **39** 285.

Tardy, M. (1896), "Eclairs en boule," *Météorologie* **44** 66.

Tarnovskii, A. S. (1990a), "Possible mechanism for enhanced energy release in a ball-lightning explosion" (in Russian), *Zh. Tekh. Fiz.* **60(3)** 200.

Tarnovskii, A. S. (1990b), "Possible mechanism for enhanced energy release in a ball-lightning explosion" (English translation), *Sov. Phys. Tech. Phys.* **35(3)** 395.

Tastes, M. de (1884), "L'orage du 1er fevrier 1884, a Tours," *Météorologie* **32** 105.

Tastes, M. de (1885a), "Das Gewitter," *Amer. J. Meteorol.* **2** 142.

Tastes, M. de (1885b), "Das Gewitter vom 1 Februar 1884 in Tours," *Meteorol. Zeit.* **2** 114.

Taylor, A. R. (1977), "Protection of trees," in R. H. Golde (ed.) *Lightning*, Vol. II, p. 26, Academic Press, New York. (NBL)

Taylor, J. E. (1954), "Photography and ball lightning," *Weather* **9** 321.

Taylor, W. D. (1884), "Ball lightning in South Wales," *Meteorol. Record* **4** 4.

Teich, M. (1955), "Eine Kugelblitzbeobachtung," *Zeit. Meteorol.* **9** 379.

Teletov, G. S. (1966a), "Ball lightning" from 'Priroda' No. 9, 1966, p. 84, *Foreign Science Bull.* **2(12)** 53.

Teletov, G. S. (1966b), "Ball lightning" in JPRS-39032, U.S. Dept. Commerce, Washington, DC.

Teletov, G. S. (1966c), "Sharovaya molniia," *Priroda* **9** 84.

Tepper, M. (1958), "Tornadoes," *Scientific American* May. (NBL)

Ter Haar, D. (1989), "An electrostatic-chemical model of ball lightning," *Physica Scripta* **39(6)** 735.

Terada, T. (1931), "On luminous phenomena accompanying earthquakes," *Tokyo Univ. Earth-quake Res. Bull.* **9** 225.

Tesla, N. (1891), *Electrical Engineer (NY)* **7** 549. (NBL)

Tesla, N. (1904), "The transmission of electrical energy without wires," *Electrical World Engineer* **35** 429.

Tesla, N. (1956), in *Nikola Tesla, Lectures—Patents—Articles*, p. A-152, Nolit Publishing House, Kultura Press, Belgrade, Yugoslavia.

Tesla, N. (1978), *Nikola Tesla, Colorado Springs Notes, 1899–1900*, Nikola Tesla Museum, Nolit, Belgrade, Yugoslavia.

Tessan, M. de (1859a), "Sur la foudre en boule," *C. R. Hebd. Séances Acad. Sci.* **49** 189.

Tessan, M. de (1859b), "Sur la foudre en boule" (summary), *Fortschr. Phys.* **15** 62.

Thate, J. B. (1952), "Bolbliksem te Warnsveld op 28 November 1951," *Hemel en Dampkring* **50** 134.*

Theron, H. (1897), "Les caprices de la foudre, foudre en boule," *Bull. Soc. Astron. Fr.* **11** 299.

Theunert, S. (1991), "Kugelblitzerscheinung," *Mitteilungen, Deutsche Meteorologische Gesellschaft* **1/91** 11.

Thompson, D. (1945), "Explosion of a fireball at Kamaron Island, Red Sea," *Roy. Meteorol. Soc. Q. J.* **71** 39.
Thompson, E. (1909), "Atmospheric electricity," *Science* **30** 857.
Thompson, E. (1910), "Atmospheric electricity" (summary), *Fortschr. Phys.* **66(3)** 334.
Thompson, J. J. (1891), "On the discharge of electricity through exhausted tubes without electrodes," *Phil. Mag.* **32** 320, 444. (NBL)
Thompson, W. (1888), "Discussion on lightning conductors," *Brit. Assn. Adv. Sci.* **58** 604.
Thompson, W., Lord Kelvin (1888). Report of the 58th Meeting of the British Association for the Advancement of Science, BAAS, London.
Thornton, W. M. (1911a), "On thunderbolts" (summary), *Fortschr. Phys.* **67(3)** 342.
Thornton, W. M. (1911b), "Sur la foudre globulaire," *J. Phys. Rad. Ser.* **5, 1** 488.
Thornton, W. M. (1911c), "On thunderbolts," *Naturwiss. Rund.* **26** 409.*
Thornton, W. M. (1911d), "On thunderbolts," *Phil. Mag.* **6(21)** 630.
Thornton, W. M. (1912a), "Uber Kugelblitz" (summary), *Fortschr. Phys.* **68(3)** 317.
Thornton, W. M. (1912b), "Uber Kugelblitz," *Meteorol. Zeit.* **29** 39.
Thornton, W. M. (1913), *Le Cosmos* **18** 142.
Thost, E. (1931), "Ein Kugelblitz," *Meteorol. Zeit.* **48** 269.
Tikhomnrov, E. I. (1932), "Sharovaya molniya I ee novaya teoriya," *Klimat i Pogoda* No. 7 109.*
Tissander, G. (1884), "Eclair en chapelet observé au-dessus de Paris," in *L'Ocean Aerien*, p. 151, G. Masson, Paris.*
Tissander, G. (1887a), "Un foudre en mer," *La Nature* mars 254.*
Tissander, G. (1887b), "Un foudre en mer," *Météorologie* **35** 221.
Tissander, G. (1891), "Sur le cyclone de la Martinique du 18 août 1891," *C. R. Hebd. Séances Acad. Sci.* **113** 421.
Toepler, M. (1897), "Geschichtete Entladung in freier Luft," *Ann. Physik* **63** 109.
Toepler, M. (1898a), *Abh. Naturwis. Ges. Isis, Dresden* **1** 10.
Toepler, M. (1898b), "Uber die Eigenschaften des Gesichteten Buschellichtbogens" in freier Luft, *Ann. Physik* **66** 660. (NBL)
Toepler, M. (1898c), "Uber gleitende Entladung langs einer Glasoberflachen," *Ann. Physik* **66** 1061. (NBL)
Toepler, M. (1899), "Verhalten des Buschellichtbogens in Magnetfelde," *Ann. Physik* **69** 680. (NBL)
Toepler, M. (1900a), "Le foudre en boule," *Ciel et Terre* **21** 401.
Toepler, M. (1900b), "Uber die Abhangigkeit des Charakters Elektrischer Dauerentladung in atmospharischer Luft von der dem Entladungsraume kontinuielich zugefuhrten Elektrizitätsmenge nebst einem anhange zur Kenntnis der Kugelblitze," *Ann. Physik* **IV 307** 560.
Toepler, M. (1900c), "L'éclair en boule," *Le Cosmos* sep 287.*
Toepler, M. (1900d), "Zur Kenntniss der Kugelblitz," *Meteorol. Zeit.* **17** 543.
Toepler, M. (1901a), "Ein fluss von Diaphragmen auf elektrische Dauerentladung durch Luft von Atmospharendruck," *Ann. Physik* **IV 6** 339.
Toepler, M. (1901b), "Kugelblitz," *Jahrbuch der Astronomie und Geophysik* **12** 409.
Toepler, M. (1901c), *L'Eclairage Electrique* **26** 58.
Toepler, M. (1901d), "L'éclair en boule," *Météorologie* **49** 53.
Toepler, M. (1901e), "Fragen zur Erforschung der Kugelblitz," *Meteorol. Zeit.* **18** 533.
Toepler, M. (1902), "Grenzpotentialdifferenzen den elektrischen Entladung in Luft von Atmospharendruck," *Ann. Physik* **7** 477. (NBL)
Toepler, M. (1903), "Uber Funkenlagen und Anfangsspannungen in Luft von Atmospherendruck," *Ann. Physik* **10** 730.
Toepler, M. (1916), *Abh. Naturwis. Ges. Isis, Dresden* **8** 13.
Toepler, M. (1917a), "Zur Kenntniss der Gesetze der Bildung von Leuchtmassen (Perlen) bei Perlschnurblitzen," *Meteorol. Zeit.* **34** 225.

Toepler, M. (1917b), *Verbands mitt. D. Dresdner Elektrotechn. Vereins U. D. Dresdner Bezirksvereins Dtsch. Ing.* **22** Feb.*

Toepler, M. (1926a), *Mitt. Hermsdorf-Schomberg Isolatorem* **25** 18.

Toepler, M. (1926b), *Mitt. Hermsdorf-Schomberg Isolatorem* **25** 734.

Toepler, M. (1929), *Zeit. Tech. Phys.* **10** 113.

Toepler, M. (1944), "Bisher unbekannte form des Elmsfeuers," *Naturwiss.* **1** 365.*

Toepler, M. (1954), "Blitz," *Naturwiss. Rund.* **7** 326.

Toepler, M. (1959), "Kugelblitz, St. Elmsfeuer," *Phys. Blatter* **15** 190.

Toepler, M. (1960), "Kugelblitz und Elmsfeuer?," *Phys. Blatter* **16** 348.

Tomlinson, C. (1859), *The Thunderstorm,* p. 151, Society for Promoting Christian Knowledge.*

Tomlinson, C. (1888a), "On some effects of lightning," *Phil. Mag.* **26** 114.

Tomlinson, C. (1888b), "Remarks on the weathering of rocks and certain electrical phenomena, suggested by some statements in a popular novel," *Phil. Mag.* **26** 475.

Tomlinson, C. (1888c), "On the weathering of rocks and certain electrical phenomena," *Phil. Mag.* **26** 484.

Tomlinson, C. (1889), "On some effects of lightning," *Nature* **40** 366.

Tompkins, D. R., and Rodney, P. F. (1977), "Photographic observations of ball lightning," unpublished report, Terrene Corp., Refugio, Texas.

Tompkins, D. R., and Rodney, P. F. (1980), "Possible photographic observations of ball lightning," *Nuovo Cimento* **3C** 200.

Tompkins, D. R., Rodney, P. F., and Gooding, R. (1975), "Photographic observations of ball lightning," *Bull. Amer. Phys. Soc.* **20** 659.

Tompkins, D. R., Rodney, P. F., and Gooding, R. (1976), "Photographic observations of ball and bead lightning," unpublished report, University of Wyoming, Dept. of Physics and Astronomy, Laramie, Wyoming.

Tonks, L. (1960), "Electromagnetic standing waves and ball lightning," *Nature* **187** 1013.

Totch, L. (1903), "Eclair en boule observé à Paris," *Ciel et Terre* **24** 544.

Touchet, E. (1899), "Les orages d'août et Septembre 1899," *Bull. Soc. Astron. Fr.* **13** 417.

Touchet, E. (1931), "Eclair en chapelet," *Bull. Soc. Astron. Fr.* **45** 84.

Touchet, E. (1932), "Observation d'un éclair globulaire (?)," *Bull. Soc. Astron. Fr.* **46** 490.

Townsend, G. F. (1895a), *Electr. Rev.* **26** 297.

Townsend, G. F. (1895b), "A lightning photograph," *Scientific American* **72** 379.

Townsend, J. S. (1900), "The conductivity produced in gases by the motion of negatively charged ions," *Nature* **63** 340. (NBL)

Townsend, J. S. (1901), "The conductivity produced in gases by the motion of negatively charged ions," *Phil. Mag.* **1** 198. (NBL)

Townsend, J. S. (1910), *The Theory of Ionization of Gases by Collisions,* Constable and Co., London. (NBL)

Tramborg, C. (1904), "Kugelblitz auf See," *Ann. Hydro. Marit. Meteorol.* **32** 390.

Trapp, E. (1949), "Kugelblitz Erscheinung in Wien," *Wetter und Leben* **1** 274.

Trecul, A. (1875), "Un effet de foudre pendant l'orage du 18 août," *C. R. Hebd. Séances Acad. Sci.* **83** 478.

Trecul, A. (1881a), "Cas remarquable de tonnerre en boule; éclairs diffus voisins de la surface du sol," *C. R. Hebd. Séances Acad. Sci.* **92** 775.

Trecul, A. (1881b), "Cas remarquable de tonnerre en boule; éclairs diffus voisins de la surface du sol" (summary), *Fortschr. Phys.* **37(3)** 534.

Trecul, A. (1881c), "Cas remarquable de tonnerre en boule; éclairs diffus voisins de la surface du sol," *Mondes* **54** 534.*

Trier, O. (1960), "Kugelblitz-Aufnahmen," *Phys. Blatter* **16** 349.

Tripe, J. W. (1875), "Remarks about ball lightning," in "Remarks on some practical points connected with the construction of lightning conductors" by R. J. Mann, *Roy. Meteorol. Soc. Q. J.* **2** 431.

Tripe, J. W. (1887a), "Ball lightning seen during a thunderstorm, on July 11 1874," *Electrician* **19** 179.

Tripe, J. W. (1887b), "Ball lightning seen during a thunderstorm, on July 11 1874," *Meteorol. Mag.* **22** 88.

Tripe, J. W. (1887c), "Ball lightning seen during a thunderstorm, on July 11 1874," *Roy. Meteorol. Soc. Q. J.* **13** 301.

Tripe, J. W. (1888), "Ball lightning seen during a thunderstorm, on July 11 1874," *Amer. J. Meteorol.* **4** 149.

Trost, T. F., and Pitts, F. L. (1982), "Analysis of electromagnetic fields on an F-106B aircraft during lightning," in *Proceedings: International Aerospace Conference on Lightning and Static Electricity*, Public Relations Section, Culham Laboratory, Abingdon, Oxon., UK.

Trost, T. F., and Turner, C. D. (1982), "Transient electromagnetic fields on a delta-wing aircraft model with injected currents," in *Proceedings: International Aerospace Conference on Lightning and Static Electricity*, Public Relations Section, Culham Laboratory, Abingdon, Oxon., UK.

Trowbridge, J. (1907a), "Experimental ball lightning," *Astr. J.* May.*

Trowbridge, J. (1907b), "Ball lightning," *Scientific American* **96** 489.

Trowbridge, J. (1908), "Lightning and powerful electric discharges," *Mon. Weather Rev.* **36** 92.

Trowbridge, J. (1910a), "Experimental ball lightning" (summary), *Fortschr. Phys.* **66(3)** 358.

Trowbridge, J. (1910b), "Experimental ball lightning," *Knowledge Sci. News* **7** 282.

Trubnikov, B. A. (1972a), "The nature of ball lightning" (English translation), *Geophys.-Doklady Akad. Nauk. SSSR* **203** 13.

Trubnikov, B. A. (1972b), "The nature of ball lightning" (in Russian), *Doklady Akad. Nauk. SSSR (Geofiz. Ser.)* **203** 1296.

Tschirwinskij, P. N. (1936), "Zur Theorie der Kugelblitz," *Meteorol. I. Hydrol.-Moscow* **3** 78.*

Tschirwinskij, P. N. (1939), "Zur Theorie der Kugelblitz," *Ann. Hydro. Marit. Meteorol.* **67** 109.*

Tuck, J. L. (1971), "Ball lightning—a status summary to 1971," unpublished report, LA-4847-MS, Los Alamos National Laboratory, Los Alamos, New Mexico.

Tumakaev, G. K. (1996a), "The nature of the formation of toroidal plasma structures in the thunderstorm activity of the atmosphere" (in Russian), *Zh. Tekh. Fiz.* **66(8)** 33.

Tumakaev, G. K. (1996b), "The nature of the formation of toroidal plasma structures in the thunderstorm activity of the atmosphere" (English translation), *Sov. Phys. Tech. Phys.* **41(8)** 764.

Turman, B. N. (1977), "Detection of lightning superbolts," *J. Geophys. Res.* **82** 2566. (NBL)

Turman, B. N., and Edgar, B. C. (1982), "Global lightning distributions at dawn and dusk," *J. Geophys. Res.* **82** 1191. (NBL)

Turman, B. N., Edgar, B. C., and Friesen, L. N. (1978), "Global lightning distribution at dawn and dusk for Aug–Sep 1977," paper presented at the American Geophysical Unim spring meeting, abstract in EOS-Trans. AGU **59(4)** 285. (NBL)

Turner, D. J. (1994), "The structure and stability of ball lightning," *Phil. Trans. Roy. Soc. Lond. Series A* **347(1682)** 83.

Turner, D. J. (1996), "Some characteristic properties of ball lightning," *J. Met.* **21(212)** 325.

Turner, D. J. (1997a), "The interaction of ball lightning with glass window panes," *J. Met.* **22(216)** 52.

Turner, D. J. (1997b), "Reply to Mr. Campbell's letter" (Campbell 1997), *J. Met.* **22(220)** 219.

Turner, D. J. (1998), "Ball lightning and other meteorological phenomena," *Phys. Rep.—Review Section of Phys. Letters* **293(1)** 2.

Turner, P. (1962), "St. Elmo's fire" from *Selections from the History of the World, The Natural History of G. Plinus Secundus*, p. 34, Southern Illinois University Press, Carbondale, Illinois. (NBL)

Turpain, M. A. (1911), "Curieux effets d'un coup de foudre sur une antenne receptrice d'ondes electriques," *J. Phys. Rad. Ser.* **5,1** 372.

Tverskoy, P. N. (1949), *Atmosfernoye Elektrichestvo*, Moscow-Leningrad.*

Tweney, C. F., and Shirshov, C. F. (eds) (1936), *Hutchinson's Technical and Scientific Encyclopedia,* Vol. I, p. 193, Macmillan, New York.

Uman, M. A. (1962), "Bead lightning and the pinch effect," *J. Atmos. Terres. Phys.* **24** 43.

Uman, M. A. (1964), "The diameter of lightning," *J. Geophys. Res.* **69** 583. (NBL)

Uman, M. A. (1967), "Decaying lightning channels, bead lightning and ball lightning," scientific paper L7-9E4-HIVOL-P7, Westinghouse Research Labs., Pittsburgh, Pennsylvania.

Uman, M. A. (1968a), "Some comments on ball lightning," *J. Atmos. Terres. Phys.* **30** 1245.

Uman, M. A. (1968b), "Decaying lightning channels, bead lightning and ball lightning," in S. C. Coroniti and J. Hughes (eds.) *Planetary Electrodynamics*, Vol. 2, p. 199, Gordon and Breach, New York.

Uman, M. A. (1969), *Lightning,* p. 240, Advanced Physics Monograph Series, McGraw-Hill, New York.

Uman, M. A. (1971a), *Understanding Lightning,* Bek Technical Publications, Carnegie, Pennsylvania.

Uman, M. A. (1971b), "What is ball lightning?," *Science Digest* **70** (October) 21.

Uman, M. A. (1974), "The Earth and its atmosphere as a 'leaky, spherical capacitor'," *Amer. J. Phys.* **42** 1033. (NBL)

Uman, M. A. (1982), "Ball lightning wild and wonderful," *Nature* **300(5893)** 578.

Uman, M. A. (1984), *Lightning,* Appendix C, p. 243, Dover, New York.

Uman, M. A. (1986), *All About Lightning,* p. 123, Dover, New York.

Uman, M. A. (1987), *The Lightning Discharge,* p. 23, Academic Press, Orlando, Florida.

Uman, M. A., and Helstrom, C. W. (1966), "A theory of ball lightning," *J. Geophys. Res.* **71** 1975.

Uman, M. A., and Orville, R. E. (1965), "The opacity of lightning," *J. Geophys. Res.* **70** 5491. (NBL)

Uman, M. A., and Voshall, R. E. (1968), "Time interval between lightning strokes and the initiation of dart leaders," *J. Geophys. Res.* **73** 497. (NBL)

Urbanitzky, A. von (1886), *Blitz und Blitz-Schutzvorrichtungen,* Elektrotechnische Bibliothek.*

Urbanitzky, A. von (1888), *Die Elektrizitat des Himmels und der Erde,* p. 78, Vienna-Berlin.*

Vandeperre, G. F. (1924), *Elektrotech.* **11** 271.

Vannesson, G. (1886), *L'Astronomie* **5** 431.*

Varley, F. (1871), "Some experiments on the discharge of electricity through rarefied media and the atmosphere," *Roy. Soc. Lond. Proc.* **19** 236.

Varley, F. (1872), *J. Soc. Tel. Engr.* **1** 368.

Vatai, E. (1989), "Ball lightning as observational evidence of the existence of lepto-nucleons and their clusters," *Nuovo Cimento* **101A(6)** 905.

Vaughan, O. H., and Vonnegut, B. (1976), "Luminous electrical phenomena in Huntsville, Alabama, tornadoes," unpublished NASA/MSFC paper. (NBL)

Vaughan, O. H., Jr. and Vonnegut, B. (1989), "Recent observations of lightning discharges from the top of a thundercloud into the clear air above," *J. Geophys. Res.* **94** 13179. (NBL)

Veduta, A. P. (1997), "Microwave theory of the ball lightning and the physical conception of the portable self sustained nuclear fusion reactor with the DC plasma trapping," in Y. H. Ohtsuki and H. Ofuruton (eds.) *Proceedings: 5th International Symposium on Ball Lightning (ISBL97)*, p. 207.

Verdet, M. (1890), *La Nature* **18** 203, 303.*

Vernon, M. D. (1971), *The Psychology of Perception,* Penguin Books, Harmondsworth, England. (NBL)

Vidler, G. T. (1964), "Lightning strike 23 April 1964," *Met. Mag.* **93** 254.

Viemeister, P. E. (1961), *The Lightning Book,* p. 130, Doubleday, New York.

Viezee, W. (1969), "Optical mirage," in E. U. Condon (ed.) *Scientific Study of Unidentified Flying Objects*, Section VI, Chapter 4, p. 598, Bantam Books, New York. (NBL)

Villemontee, G. G. de (1912a), "Sur un cas de foudre globulaire," *C. R. Hebd. Séances Acad. Sci.* **155** 1567.

Villemontee, G. G. de (1912b), "Sur un cas de foudre globulaire," *Météorologie* **60** 307.

Violle, J. (1901a), "Sur un éclair en boule," *C. R. Hebd. Séances Acad. Sci.* **132** 1537.

Violle, J. (1901b), "La foudre en boule," *Ciel et Terre* **22** 479.

Violle, J. (1901c), "Kugelblitz," *Naturwiss. Rund.* **16** 504.

Violle, J. (1902), "Kugelblitz Beobachtung," *Meteorol. Zeit.* **19** 335.

Virier, M. (1904), "La note," *C. R. Hebd. Séances Acad. Sci.* **139** 904.

Visser, S. W. (1957a), "Twee bliksenballen bij Laren," *Hamel en Dampkring* **55** 45.*

Visser, S. W. (1957b), *Himmel und Erde* **55** 54.

Visser, S. W. (1957c), "Some remarks on lightning flashes and ball lightning," *Meteorol. Mag.* **86** 344.

Vlasov, A. A. (1959), *Sci. Sinica* **8** 266.

Vlasov, A. A., and Yakorlev, M. A. (1975), "Interaction between ions through an intermediate system (neutral gas) and the problem of the existence of a self-maintaining swarm of particles (ball lightning)," *Vestn. Mosk. Univ. Fiz. Astron. (USSR)* **16(3)** 332.*

Vlasov, A. N. (1997), "A ball lightning is a natural nuclear reactor?" in *Proceedings: 5th International Symposium on Ball Lightning (ISBL97)*, p. 75.

Vogel, M. (1941), "Beobachtung eines Kugelblitzes," *Zeit. Angew. Meteorol.* **58** 124.*

Voitsekhovskii, B. V., and Voitsekhovskii, B. B. (1974), "Priroda sharovoy molnii" (in Russian), *Akad. Nauk. SSSR Dokl. Mat. Fiz.* **218** 77.

Voitsekhovskii, B. V., and Voitsekhovskii, B. B. (1975), "The nature of ball lightning" (English translation), *Sov. Phys. Doklady* **19** 580.

Voitsekhovskii, B. V., and Voitsekhovskii, B. B. (1976), "Elektromekhanicheski effekty v sharovoi molnii," *Akad. Nauk. SSSR, Sibirskoe Otdelenie, Izvestiia, Seriia Tekn. Nauk.* Feb.

Voitsekhovskii, B. V., and Voitsekhovskii, M. B. (1986), "Incandescent-breakdown multiple-point luminescence of ball lightning," *Doklady Akad. Nauk. SSSR* **287(2)** 331.

Voitsekhovskii, B. V., and Voitsekhovskii, M. B. (1987), "Insulated St. Elmo fire on soaring concentrator and the ball lightning," *Doklady Akad. Nauk. SSSR* **295(3)** 580.

Volmer, C. R. (1888), *Elektro. Zeit.* **9** 473.

Volmer, C. R. (1896a), "Ein Kugelblitz in Paderborn" (summary), *Fortschr. Phys.* **52(3)** 298.

Volmer, C. R. (1896b), *Zeit. Angew. Meteorol.* **13** 185.

Volmer, C. R. (1897), "Kugelblitz," *Meteorol. Zeit.* **14** 34.

Vonnegut, B. (1960), "Electrical theory of tornadoes," *J. Geophys. Res.* **65** 203. (NBL)

Vonnegut, B., and Weyer, J. R. (1965), "Luminous phenomenon in noctural tornadoes," *Science* **153** 1213. (NBL)

Vonnegut, B., Moore, C. B., and Harris, C. K. (1960), "Stabilization of a high voltage discharge by a vortex," *J. Meteorol. (now J. Atmos. Terres. Phys.)* **14** 468. (NBL)

Wagner, A. (1939), "Kugelblitz," *Meteorol. Zeit.* **56** 350.

Wagner, C. F., and McCann, G. C. (1941), *Electr. Engr.* **60** 374.

Wagner, G. A. (1971), "Optical and acoustic detection of ball lightning," *Nature* **232** 187.

Wahlgren, B. I., Olsson, B. A., and Luther, C. A. (1982), "Lightning testing of the Viggen aircraft," in *Proceedings: International Aerospace Conference on Lightning and Static Electricity*, Public Relations Section, Culham Laboratory, Abingdon, Oxon., UK.

Wainhouse, W., and Pitcairn, R. (1773), "Account of the effects of lightning at Steeple Ashton and Holt, in the County of Wilts. on the 20th June 1772, contained in several letters," *Phil. Trans. Roy. Soc. London* **63** 234.

Walker III, Sydney (1968), "Establishing observer creditability: a proposed method," *J. Astronautical Sci.* **15** 92. (NBL)

Walter, B. (1902), "Zur Erklarung der wolfschen Blitzphotographie," *Ann. Hydro. Marit. Meteorol.* **30** 579.

Walter, B. (1903a), "Uber die wolfsche Blitzphotographie," *Ann. Hydro. Marit. Meteorol.* **31** 65.

Walter, B. (1903b), "Uber die Entstehungweise der Blitzen," *Ann. Physik* **10** 393.

Walter, B. (1903c), *Jahr. Hamburg-Wiss. Ansl.* **20**.

Walter, B. (1909a), *Handwerk-Buch der Naturwissenschaften.**

Walter, B. (1909b), "Uber Kugelblitz," *Meteorol. Zeit.* **26** 217.

Walter, B. (1911), "Kugelblitz," *Meteorol. Zeit.* **28** 582.

Walter, B. (1915a), "Das Wesen der Kugelblitze" (summary), *Fortschr. Phys.* **71(3)** 250.

Walter, B. (1915b), "Das Wesen der Kugelblitze," *Prometheus* **26** 408.*

Walter, B. (1928), *Erdball und Weltall,* H. Bermuhler, Berlin-Lichterfelde.*

Walter, B. (1929), "Uber die suringjche Blitzaufnahme, sowie uber eine neue, vermeintliche Kugelblitzaufnahmen," *Phys. Zeit.* **30** 26.

Wartmann, A. (1888a), "La foudre globulaire," *La Nature* mar 226.*

Wartmann, A. (1888b), "Coup de foudre en boule," *Le Cosmos* fev 279.*

Wartmann, A. (1889a), "Uber einen Kugelblitz," *Arch. Sci. Phys. Natur.* **21(1)** 75.*

Wartmann, A. (1889b), "Uber einen Kugelblitz" (summary), *Fortschr. Phys.* **45(3)** 506.

Wartmann, A. (1889c), "Uber einen Kugelblitz," *La Nature* **8(823)** 226.*

Wartmann, A. (1889d), "Coup de foudre en boule," *Météorologie* **37** 102.

Wartmann, A. (1889e), "La foudre globulaire," *Météorologie* **37** 148.

Wartmann, A. (1889f), "Uber einen Kugelblitz," *Meteorol. Zeit.* **6** 119.

Wartmann, A. (1889g), "Kugelblitz," *Zeit. Angew. Meteorol.* **6** 90.

Wartmann, A. (1897), "Foudre en boule," *Bull. Soc. Astron. Fr.* **11** 29.

Wasse, J. (1725), "Two letters on the effects of lightning," *Phil. Trans. Roy. Soc. London* **33** 366.

Watkins, D. C., Cobine, J. D., and Vonnegut, B. (1978), "Electric discharges inside tornadoes," *Science* **199(4325)** 171. (NBL)

Watson, W. (1752a), "A letter to the Royal Society concerning the electrical experiments in England on thunder clouds," *Phil. Trans. Roy. Soc. London* **47** 565. (NBL)

Watson, W. (1752b), "An account of Mr. Benjamin Franklin's treatise, lately published, entitled, Experiments and Observations on Electricity, made at Philadelphia in America," *Phil. Trans. Roy. Soc. London* **47** 202. (NBL)

Watson, W. (1754), "An answer to Dr. Lining's query relating to the death of Professor Richman," *Phil. Trans. Roy. Soc. London* **48** 765.

Watson, W. K. R. (1960), "A theory of ball lightning formation," *Nature* **185** 449.

Watt, A. D. (1957), "Characteristics of atmospheric noise from 1 to 100 KC," *Proc. IRE* **45** 787.

Waye, B. E. (1945), "Globe lightning," *Nature* **155** 752.

Webb, J. D. C. (1981), "Ball of lightning on 7 October 1980," *J. Met.* **6(59)** 150.

Webb, S. (1900), "Letters to the Editor," *Nature* **61** 343.

Webber, H. (1907), "A case of globular lightning," *J. Roy. Astro. Soc. Canada* **1** 44

Weber, L. (1879), *Schr. Nat. Ver. Sch. Holst.* **2** 28.

Weber, L. (1880), *Schr. Nat. Ver. Sch. Holst.* **3** 107.

Weber, L. (1881a), *Schr. Nat. Ver. Sch. Holst.* **4** 48.

Weber, L. (1881b), *Schr. Nat. Ver. Sch. Holst.* **4** 82.

Weber, L. (1882), *Schr. Nat. Ver. Sch. Holst.* **4** 105.

Weber, L. (1883a), *Schr. Nat. Ver. Sch. Holst.* **5** 27.

Weber, L. (1883b), *Schr. Nat. Ver. Sch. Holst.* **5** 41.

Weber, L. (1883c), *Schr. Nat. Ver. Sch. Holst.* **5** 52.

Weber, L. (1884), "Ein merkwurdiger Blitzschlag zu Ribnitz am 30 Juli 1884," *Meteorol. Zeit.* **1** 405.

Weber, L. (1885a), "Uber den gegenwartigen Stand der Kugelblitzfrage," *Amer. J. Meteorol.* **2** 142.

Weber, L. (1885b), "Uber den gegenwartigen Stand der Kugelblitzfrage," *Electr. Zeit.* **6** 401.*

Weber, L. (1885c), "Uber den gegenwartigen Stand der Kugelblitzfrage" (summary), *Fortschr. Phys.* **41(3)** 652.

Weber, L. (1885d), "Uber den gegenwartigen Stand der Kugelblitzfrage," *Gaea* **10**.*

Weber, L. (1885e), "Uber den gegenwartigen Stand der Kugelblitzfrage," *Meteorol. Zeit.* **2** 118.

Weber, L. (1896), "Beobachtung uber Kugelblitz," *Zeit. Angew. Meteorol.* **13** 36.*

Weber, L. (1911), "Kugelblitze," *Meteorol. Zeit.* **28** 582.

Weber, L. (1915), "Kugelblitze," *Meteorol. Zeit.* **32** 22.

Weekes, W. H. (1841), "Ball lightning," *Ann. Elect. Magnet. Chem. Guard. Expr. Sci.* **7** 243.

Wegener, A. (1923), "Literaturbericht; Brand, W.: Der Kugelblitz," *Meteorol. Zeit.* **40** 381.

Welker, P. A. (1895), "Electric storm on Mount Elbert, Colorado," *Science* **2(36)** 304.

Welker, P. A. (1896), "Eine elektrische Erscheinung auf Mount Elbert," *Zeit. Angew. Meteorol.* **13** 153.*

Wells, D. R. (1962), "Observation of plasma vortex rings," *Phys. Fluids* **5** 1016. (NBL)

Wells, D. R. (1964), "Axially symmetric force-free plasmoids," *Phys. Fluids* **7** 826. (NBL)

Wells, D. R. (1966), "Injection and trapping of plasma vortex rings," *Phys. Fluids* **9** 1010. (NBL)

Wells, D. R., and Ziajka, P. (1978), "Production of fusion energy," *Int. J. Fusion Energy* **1** 3. (NBL)

Wendt, O. (1906a), "Flachenblitz oder Kugelblitz" (summary), *Fortschr. Phys.* **62(3)** 287.

Wendt, O. (1906b), "Flachenblitz oder Kugelblitz," *Zeit. Angew. Meteorol.* **23** 41.*

Wertheimer, M. (1969), "Perceptual problems," in E. U. Condon (ed.) *Scientific Study of Unidentified Flying Objects*, Section VI, Chapter 1, p. 559, Bantam Books, New York.

Wescott, E. M., Sentman, D. D., Osborne, D. L., Hampton, D. J., and Heavner, M. J. (1995b), "Preliminary results from the Sprites94 Aircraft Campaign: 2. Blue jets," *Geophys. Res. Letts.* **22** 1209. (NBL)

Westphal, W. (1931), "Beobachtung eines Kugelblitzes," *Naturwiss.* **19** 19.

Westphal, W. (1958), "Photographierte Blitz," *Phys. Blatter* **14** 573.

Westrum, R. (1993), "Ball lightning reports: percolation vs. interaction," in A. G. Keul (ed.) *Progress in Ball Lightning Research: Proc. VIZOTUM*, p. 107, The Vizotum Project, Salzburg, Austria.

Wheeler, D. (1991), "Another look at May 1891," *J. Met.* **16(162)** 267.

Whipple, F. J. W. (1921), "Ball lightning seen at St. John's Wood," *Meteorol. Zeit.* **56** 195.

White, D. A. (1994), "Ball lightning and rocket lightning," *J. Met.* **19(185)** 15.

White, E. H., Rapaport, E., Seliger, H. H., and Hopkins, T. A. (1971), "The chemi- and bio-luminescence of firefly luciferin: an efficient production of electrically excited states," *Bio. Organic Chem.* **1(1/2)** 92. (NBL)

Whitehead, E. R. (1977), "Protection of transmission lines," in R. H. Golde (ed.) *Lightning*, Vol. II, p. 22, Academic Press, New York. (NBL)

Wickson, K. M. (1982), "Ball lightning inside an aircraft" (Letter), *Weather* **37(6)** 190.

Wilkinson, J. (1931), "An example of ball lightning," *Roy. Astron. Soc. Canada J.* **25** 322.

Willett, J. C., Bailey, J. C., and Krider, E. P. (1989), "A class of unusual lightning electric field waveforms with very strong high-frequency radiation," *J. Geophys. Res.* **94(D13)** 16255. (NBL)

Wilson, C. T. R. (1956), "A theory of thundercloud electricity," *Roy. Soc. Lond. Proc.* **236** 297.

Wilson, J. (1990), "Ball lightning freshly served," *Electronics World Wireles World* **96(1655)** 749.

Wilson, J. G. (1954), "St. Elmo's fire," *Marine Observer* **24** 10. (NBL)

Wilson. M. (1865), *J. Roy. Geogr. Soc.* **35**.*

Wimbush, B. S. (1924), "Ball lightning," *Roy. Astron. Soc. Canada J.* **18** 282.

Winchester, G. (1929), "A particular lightning phenomenon," *Science* **70** 501.

Winterberg, F. (1978), "Electrostatic theory of ball lightning," *Zeit. Met., Berlin* **28** 263.

Wirtz, C. (1920), "Beobachtung einer kugelblitzartigen Erscheinung in Kiel," *Ann. Hydro. Marit. Meteorol.* **48** 234.

Wislicenus, G. (1894a), "Uber Sanct-Elmsfeuer und Kugelblitze auf Hoher See" (summary), *Fortschr. Phys.* **50(3)** 422.

Wislicenus, G. (1894b), "Uber Sanct-Elmsfeuer und Kugelblitze auf Hoher See," *Prometheus* **6** 36.*

Witalis, E. A. (1990a), "Ball lightning as a magnetized air plasma whirl structure," *J. Met.* **15(148)** 121.

Witalis, E. A. (1990b), "Hall effect magnetohydrodynamics and applications," in *IEEE International Conference on Plasma Science*, p. 168, Cat. No. 90CH2857-1. Institute of Electrical and Electronics Engineers, New York.

Witalis, E. A. (1991), "Ball lightning," *Nature* **352(6333)** 290.

Wittmann, A. (1971), "In support of a physical explanation of ball lightning," *Nature* **232** 625.

Wittmann, A. (1976), "Gibt es Kugelblitz?," *Umschau Wiss. Tech.* **76(16)** 516.

Wittmann, A. D. (1993), "More on the Neustadt multiple ball lightning case," in A. G. Keul (ed.) *Progress in Ball Lightning Research: Proc. VIZOTUM*, p. 110, The Vizotum Project, Salzburg, Austria.

Wojowsky, M. (1966), "?" from Przeglad Lotniczy, No. 12, 1966, p.12, *Foreign Science Bull.* **2(4)** Appendix II 52.

Wolf, F. (1943a), "Das Gewitter und seine Entladungsformen. I Teil: Gewitterentstehung und Linienblitz," *Naturwiss.* **31(7/8)** 73.

Wolf, F. (1943b), "Das Gewitter und seine Entladungsformen. II Teil: Kugelblitze und Perlschnurblitze," *Naturwiss.* **31** 215.

Wolf, F. (1956a), "Interessante Aufnahme eines Kugelblitzes," *Naturwiss.* **43** 415.

Wolf, F. (1956b), "Interessante Aufnahme eines Kugelblitzes," *Orion* **4** 589.

Wolf, F. (1958), "Photographierte Blitz," *Phys. Blatter* **14** 573.

Wolf, K. (1915a), "Das Wesen der Kugelblitz," *Meteorol. Zeit.* **32** 416.

Wolf, K. (1915b), "Das Wesen der Kugelblitz," *Prometheus* **26** 229.*

Wolf, K. (1915c), "Das Wesen der Kugelblitz (eine Entgegnung)," *Prometheus* **26** 586.*

Wolf, K. (1915d), "The nature of ball lightning," Scientific American Suppl. **80** 54.

Wolff, C. (1725), *Vernuftige Gedanken von den Wirkungen der Natur,* p. 471, Halle.*

Wood, R. W. (1930a), "Ball lightning," *Nature* **126** 723.

Wood, R. W. (1930b), "Plasmoidal high frequency oscillating discharges in non-conducting vacua," *Phys. Rev.* **35** 673. (NBL)

Wooding, E. R. (1963), "Ball lightning," *Nature* **199** 272.

Wooding, E. R. (1972), "Laser analogue to ball lightning," *Nature* **239** 394.

Wooding, E. R. (1976), "Ball lightning in Smethwick," *Nature* **262** 279.

Wooding, E. R. (1992), "The possibility of recording ball lightning," in M. Stenhoff (ed.) *Proceedings: Fourth TORRO Conference: Ball Lightning*, p. 1, Tornado and Storm Research Organisation (TORRO), Oxford Brookes University, Richmond, England.

Wordingham, C. H. (1888), "Globe lightning," *Electrician* **22** 56.

Workman, E. J., and Holtzer, R. E. (1937), "An anomalous lightning photograph," *Phys. Rev.* **51** 149.

Workman, E. J., Brook, M., and Kitagawa, N. (1960), "Lightning and charge storage," *J. Geophys. Res.* **65** 1513. (NBL)*

Worlock, R. M. (1960), "A proposal for laboratory studies of ball lightning and charged particle containment," unpublished, Electro-Optical Systems, Inc., Pasadena, California.

Wormell, W. T. (1953), "Symons Memorial Lecture," *Roy. Meteorol. Soc. Q. J.* **79** 474.

Wu, H. M., and Chen, Y. (1989), "Magnetohydrodynamic equilibrium of plasma ball lightning," *Phys. Fluids B (Plasma Physics)* **1(8)** 1753.

Wu, H. M., and Oakes, M. E. (1991), "Magnetohydrodynamics equilibrium of a self-confined elliptic plasma ball," *Phys. Fluids B (Plasma Physics)* **3(8)** 213.

Yaffee, M. (1972), "Air transport lightning strikes studied," *Aviation Week Space Technol.* **11** Dec, p. 34.

Yakovlenko, S. I. (1992a), "Possibility of simulating ball lightning with a laser" (in Russian), *Kvantovaya Elektronika, Moskva* **19(1)** 5.

Yakovlenko, S. I. (1992b), "Possibility of simulating ball lightning with a laser" (English translation), *Sov. J. Quantum Electronics* **22(1)** 1.

Yakovlenko, S. I. (1992c), "On the possibility of fireball modeling using a laser," *Laser Physics* **2(2)** 196.

Yamamoto, H. (1989), "A model for ball lightning," in Y. H. Ohtsuki (ed.) *Science of Ball Lightning (Fireball)*, p. 229, World Scientific, Singapore.

Yankov, V. V. (1959a), "Behavior of a conducting gaseous sphere in a quasi-stationary electromagnetic field" (in Russian), *Zh. Eksp. Teor. Fiz.* **36** 560.

Yankov, V. V. (1959b), "Dynamics of a conducting gaseous sphere in a quasi-stationary electromagnetic field" (in Russian), *Zh. Eksp. Teor. Fiz.* **37** 224.

Yankov, V. V. (1961a), "Behavior of a conducting gaseous sphere in a quasi-stationary electromagnetic field," English translation in D. J. Ritchie (ed.) *Ball Lightning*, Consultants' Bureau, New York.

Yankov, V. V. (1961b), "Dynamics of a conducting gaseous sphere in a quasi-stationary electromagnetic field," English translation in D. J. Ritchie (ed.) *Ball Lightning*, Consultants' Bureau, New York.

Yasui, K. (1993), "Plasma fireballs fed by microwaves," *Phys. Lett. A* **173(6)** 451.

Yasui, K. (1997), "Plasma fireballs fed by microwaves," in Y. H. Ohtsuki and H. Ofuruton (eds.) *Proceedings: 5th International Symposium on Ball Lightning (ISBL97)*, p. 108.

Yates, T. P. (1898), "Ball lightning," *Mon. Weather Rev.* **26** 565.

Yeliseyev, A. (1960), *Vozniknoveniye nauki ob Elektrischestve v Rossii*, Moscow.*

Young, C. J. (1934), "Lightning freaks at Peterborough," The Peterborough Citizen, England, Sep 25.

Young, G. A. (1962), "A lightning strike of an underwater explosion plume," Noltr. 61-43, Naval Surface Weapons Center, White Oak Lab., Silver Spring, Maryland.

Yur'yev, A. (1960), "The secret of ball lightning," *Komsomol'shaya Pravda* **3** 5.*

Zaitsev, A. V. (1972a), "New theory of ball lightning," *Zh. Tekh. Fiz.* **42(1)** 213.

Zaitsev, A. V. (1972b), "New theory of ball lightning" (English translation), *Sov. Phys. Tech. Phys.* **17** 173.

Zaitsev, I. V., and Zaitsev, S. V. (1991a), "Electrostatic model of ball lightning" (in Russian), *Pis'ma v Zhurnal Tekh. Fiz.* **17(7–8)** 34.

Zaitsev, I. V., and Zaitsev, S. V. (1991b), "Electrostatic model of ball lightning" (English translation), *Sov. Tech. Phys. Letters* **17(4)** 249.

Zaitsev, N. K. (1997), "Model of fireball on the basis of glow discharge in air at atmospheric pressure," in Y. H. Ohtsuki and H. Ofuruton (eds.) *Proceedings: 5th International Symposium on Ball Lightning (ISBL97)*, p. 202.

Zalesskii, Y. M. (1958), "Naelektrizovannyi sneg I sharovaya molniia zimoi," *Priroda* **10** 114.

Zapletal, G. (1970), "Kugelblitz Schlins-Jupident," *Wetter und Leben* **22** 193.

Zenger, C. V. (1889), "Les figures électriques dessinées par l'éclair," *C. R. Hebd. Séances Acad. Sci.* **109** 294.

Zheng, X-H. (1990), "Quantitative analysis for ball lightning," *Physics Lett. A* **148(8–9)** 463.

Zheng, X-H. (1992), "The long life of ball lightning," in M. Stenhoff (ed.) *Proceedings: Fourth TORRO Conference: Ball Lightning*, p. 83, Tornado and Storm Research Organisation (TORRO), Oxford Brookes University, Richmond, England.

Ziman, J. (1968), *Public Knowledge*, Cambridge University Press, Cambridge, England. (NBL)

Zimmerman, P. D. (1970), "Energy content of Covington's lightning ball," *Nature* **228** 853.

Zoege von Manteuffel, P. (1938), "Eine Blitzaufnahme," *Umschau Wiss. Tech.* **42** 587.

Zoege von Manteuffel, P. (1955), "Eine Blitzaufnahme," *Orion* **4** 589.

Zou, Y. S. (1989a), "Conditions for producing and maintaining plasma ball lightning in the atmosphere," *Adv. Atmos. Sci.* **6(1)** 62.

Zou, Y. S. (1989b), "The physical explanation to the UFO over Xinjiang, Northern West China," in Y. H. Ohtsuki (ed.) *Science of Ball Lightning (Fireball)*, p. 273, World Scientific, Singapore.

Zou, Y. S. (1995), "Some physical considerations for unusual atmospheric lights observed in Norway," *Physica Scripta* **52(6)** 726.

Zou, Y. S. (1996), "Plasma Rossby wave as a mechanism for atmospheric fireball generation," in *IEEE Conference Record—Abstracts. 1996 IEEE International Conference on Plasma Science*, p. 153, Cat. No. 96CH35939. Institute of Electrical and Electronics Engineers, New York.

Zschokke, W. (1903a), "Ein Kugelblitzphotographie?" *Prometheus* **14** 234.*

Zschokke, W. (1903b), "Ein Kugelblitzphotographie?" *Prometheus* **14** 691.*

Zuccher, M., and Margolle (1876), *Trombes et Cyclones*, p. 101, Hachette et Cie, Paris.

Zurcher, M. (1889a), "Coup de foudre," *La Nature* sep 263.*

Zurcher, M. (1889b), "Coup de foudre," *Météorologie* **37** 261.

Zwart, B. (1972), "Is de bolbliksem gezichtsbedrog?," *Hemel Dampkr.* **70** 20.

Index

GPSR Compliance
The European Union's (EU) General Product Safety Regulation (GPSR) is a set of rules that requires consumer products to be safe and our obligations to ensure this.

If you have any concerns about our products, you can contact us on

ProductSafety@springernature.com

In case Publisher is established outside the EU, the EU authorized representative is:

Springer Nature Customer Service Center GmbH
Europaplatz 3
69115 Heidelberg, Germany

www.ingramcontent.com/pod-product-compliance
Ingram Content Group UK Ltd.
Pitfield, Milton Keynes, MK11 3LW, UK
UKHW021505200726
13877UKWH00027B/491
9781475781809